# 3D Stacked Chips

Ibrahim (Abe) M. Elfadel • Gerhard Fettweis
Editors

# 3D Stacked Chips

From Emerging Processes to Heterogeneous
Systems

 Springer

*Editors*
Ibrahim (Abe) M. Elfadel
Masdar Institute of Science and Technology
Abu Dhabi, United Arab Emirates

Gerhard Fettweis
Vodafone Chair Mobile Communication
Dresden, Germany

ISBN 978-3-319-79305-4      ISBN 978-3-319-20481-9   (eBook)
DOI 10.1007/978-3-319-20481-9

Printed on acid-free paper

This Springer imprint is published by Springer Nature
The registered company is Springer International Publishing AG Switzerland

*To Our Beloved Families*

*YOU, WHO are blessed with shade as well as light, you, who are gifted with two eyes, endowed with a knowledge of perspective, and charmed with the enjoyment of various colours, you, who can actually see an angle, and contemplate the complete circumference of a Circle in the happy region of the Three Dimensions—how shall I make clear to you the extreme difficulty which we in Flatland experience in recognizing one another's configuration?*
*Edwin A. Abott, Flatland*

# Foreword

The research work described in this timely book is the fruit of a deep and successful partnership between Dresden and Abu Dhabi. When this work started in 2012, Dresden, capital of the Free State of Saxony, Germany, was already a global R&D hub of the semiconductor industry. Abu Dhabi, capital of the UAE, was making its first steps on the way to becoming a major R&D hub of the global semiconductor industry. The theme of 3D-integrated circuits was chosen as the subject of collaboration between the two academic partners, Technical University of Dresden and the Masdar Institute of Science and Technology in Abu Dhabi, because of its potential and promise amongst the plethora of more-than-Moore technologies. It is heartening to see this book emerge out of this partnership and collaboration. The publication of this book is a significant event for the UAE, Germany, and GLOBALFOUNDRIES for three reasons:

1. It is a clear illustration of the success of the partnership model that we called Twinlab, which was based on the pairing of leading academics from two different cultural, social, and technological contexts to work closely on advancing subjects of high technological promise and significance.
2. It is a patent proof that the goal we set for ourselves in leveraging such Twinlab partnerships for human capital development has been achieved. The Masdar Institute contributions to the book would not have been possible without such human capital emerging in full possession of the right mix of knowledge and skills needed to execute on research deliverables.
3. It is an unmistakable sign of intellectual leadership in an advanced area of semiconductor technology. Such leadership bodes very well for the continued growth and development of R & D capacity in the Abu Dhabi semiconductor hub.

It is also heartening to see that many of the chips and technology demonstrators described in this volume have used various CMOS technologies provided by GLOBALFOUNDRIES, including its advanced 28 nm technology. This is yet another good omen for the extended partnership that we will strive to establish not only amongst academic institutions in places where GLOBALFOUNDRIES has labs and facilities but also between such institutions and GLOBALFOUNDRIES.

With GLOBALFOUNDRIES expanding and enhancing its technology offerings as a result of the integration of the IBM Microelectronics business, many interesting questions arise as to the positioning of these technologies with respect to the industry road map on 3D chip and package integration. These questions fall squarely within the theme of this book on "heterogeneous 3D integration". Besides its chapters on electrical 3D integration using copper through-silicon vias, the book contains forward-looking chapters on the 3D integration of photonic and optoelectronic components using optical through-silicon vias. Taken all together, the book chapters offer a unique mix of contributions at the cutting edge of semiconductor research in the area of 3D-integrated circuits and 3D packaging.

I very much hope that the path-breaking partnership that has enabled this research will become more bonded and that more contributions of technological relevance will result from it.

Executive Vice President and Chief Strategy Officer                      Ibrahim Ajami
GLOBALFOUNDRIES
Executive Director
Mubadala Technology
Abu Dhabi, United Arab Emirates
July 2015

# Foreword

The Abu Dhabi-Saxony Partnership Committee was established 4 years ago by the two involved governments in order to foster and encourage exchange as well as to build links of joint interest. As one specific result, the Twinlab partnership between the Masdar Institute of Abu Dhabi and the Technische Universität Dresden of Saxony (Twinlab 3D Stacked Chips) was initiated 3 years ago. Over the last years, Abu Dhabi and especially the Masdar Institute of Science and Technology have remarkably and dynamically developed. It is well on the path of being recognised as one of the major international research and development hubs in the UAE. At the same time, Dresden, with its university of excellence, is known and valued throughout Europe and the world as a location of outstanding expertise in the information and communication technology (ICT) sector. This makes two strong partners in a relationship with remarkable potential and excellent future prospects. Within the Twinlab, both partners collaborate on investigating a topic of highest economic interest. As the integration density of semiconductor technology is reaching its limits, the technology of stacking chips within a package allows the continuation of integration density to evolve. However, "3D chip stacks" require the development of many new technologies and underlying conditions, such as highest standards of the communication infrastructure, outstanding design tools, as well as efficient methodologies. These and various other aspects have been the focus of the joint research carried out by experts in Abu Dhabi as well as in Saxony. Another important aspect of this partnership is the generation of highly qualified young people who are considered as one of the key success factors for sustainable development of the semiconductor and ICT industry in both our regions and beyond. Having these motivated young experts on board stimulates further growth of excellent innovative research. Also, it encourages and consolidates the process of building bridges of understanding between the cultures of Abu Dhabi and Saxony which is a vital basis for all our future. The outstanding results illustrated in this book strongly indicate the success of the ongoing partnership. It brings together and also benefits from the joint efforts of researchers with various cultural, social, and technological backgrounds while closely focusing on subjects of significant technological impact. I very much hope that this groundbreaking partnership will become even closer—resulting in further contributions of significant technological relevance in the near and far future.

Saxon State Minister for Higher Education                              Eva-Maria Stange
Research and the Arts, Dresden, Germany
July 2015

# Preface

*So I devoted several months in privacy to the composition of a treatise on the mysteries of Three Dimensions.*

*Edwin A. Abott, Flatland*

The future evolution of the form factors of cell phones, tablets, wearable electronic devices, microsensors, and other similar gadgets of the world's digital fabric requires that we develop integration technologies that enable volume-based assembly of chips, sensors, devices, and interfaces. 3D stacked chips and 3D-integrated circuits belong to such volume-based approaches. These volume-based technologies have already made their way into electronic products such as memory devices and cell phones and are slated to become the dominant trends in microsystem fabrication in the next decade. Many research monographs about this topic have already been published, including a three-volume edited book entitled *Handbook of 3D Integration*, by Wiley-VCH. Springer has also published several research monographs on this topic. One of the most recent ones is the 2013 book by Sung Kyu Lim, entitled *Design for High Performance, Low Power and Reliable 3D Integrated Circuits*.

This book covers recent work on 3D integration conducted by researchers from the Technical University of Dresden, Germany, and the Masdar Institute of Science and Technology, Abu Dhabi, UAE, under the framework of Twinlab 3D Stacked Chip (3DSC), a joint collaborative effort focusing on heterogeneous 3D integration. The book addresses some of the most important challenges in this emerging technology, especially as they pertain to 3D heterogeneous integration. Indeed, one of the promises of 3D chip stacks is to enable the integration, under very small form factors, of chips belonging to different semiconductor technologies (e.g. CMOS vs. SiGe), different design modalities (e.g. digital vs analogue), and different physical domains (e.g. electrical vs. optical). In addition, under the same small form factors, the 3D chip stacks will help extend the integration domain to include not only the processing and communication functions but also the sensing and the power sourcing functions. Recent industrial examples of heterogeneous 3D integration include CMOS imaging sensors and MEMS inertial sensors.

The issues and challenges of heterogeneous 3D integration are addressed at both the process and system levels with particular emphasis on the 3D integration of on-chip high-speed links and optoelectronic systems. In particular, the book contains original material on the use of interposers in 3D-integrated CMOS and Si photonics. Processing, modelling, design, and CAD aspects are all considered and treated in a coherent framework for the first time in the open literature. Of particular interest are interposer process recipes for the manufacturing of high-aspect ratio through-silicon vias (TSVs) that do not require any wafer thinning. Such TSVs may be used for high-speed serial communication in both the electrical and optical domains. Particular attention has been given to the design of transceivers for serial links having TSVs in their path and to the design of digitally assisted clocking circuits for 3D chip stacks. Topical coverage also includes the 3D heterogeneous integration of various photonic devices such as tunable resonators, power sources such as photovoltaic cells, and non-volatile memories based on new materials systems (e.g. ZnO). Finally, the thermal challenges of 3D stacked chips are addressed from the viewpoints of accurate, on-chip, temperature sensing, early physical design planning using thermal TSVs, and the development of athermal photonic components for optoelectronic 3D chip stacks.

The book consists of 17 chapters organised in two parts. The first part, with ten chapters, is devoted to electronic 3D integration using copper TSVs. The second part, with the remaining seven chapters, is devoted to photonic and optoelectronic 3D integration using photonic TSVs. Each part opens with an introductory chapter, I for Part I and II for Part II, positioning the research work in the context of the integrated electronic and Si photonic circuits and systems. Here is a short summary of the content of each part.

In Chap. 2, a process recipe is proposed for the fabrication of high-aspect ratio (up to 20:1) copper TSVs for Si interposer-based 3D integration. The recipe uses atomic layer deposition (ALD) to deposit the copper barrier and the send layer for copper electroplating. Such usage enables the fabrication of high-quality TSVs with uniform cladding deposition along the TSV height.

In Chap. 3, a 3D interposer architecture is used as a platform for the design and implementation of energy-efficient serial communication links across a 3D chip stack. The communication link implementation includes the design of an energy-efficient, low-voltage-swing, multibit capacitive transceiver based on a detailed equivalent circuit model of the TSV channel. Silicon results using GLOBAL-FOUNDRIES 28 nm SLP CMOS technology show that the transceiver design outperforms competing solutions by more than $2\times$ in terms of energy efficiency (Joule per bit per number of TSV channels).

The main motivation for developing serial links for TSV channels is to reduce the number of TSVs needed and thus reduce the area overhead of 3D integration. When multiple signal TSVs are needed in closer proximity to each other, crosstalk will occur and channel equalisation to overcome crosstalk noise on signal TSVs will be needed. This is the main topic of Chap. 4 where equalisation method is implemented in the discrete-time domain and is based on an equivalent circuit model of the capacitive crosstalk between neighbouring TSVs. It also uses realistic assumptions on the IO cells to which the TSVs are connected. Of particular interest in Chap. 4 is the impact of quantisation on the equaliser's performance. It is found that a 3-bit, non-uniform quantiser can outperform a 5-bit uniform quantiser by $2\times$ in terms of crosstalk rejection.

While the quantisation aspects of equalisation pertain to the transmitter, the issue arises as to how to reconstruct the received signal on the TSV channel using the coarsest analogue-to-digital converter (ADC), namely, a 1-bit ADC. In Chap. 5, this issue is addressed in the context of designing energy-efficient receivers for TSV-based communication links.

Heterogeneous integration presents the designer with the challenge of implementing clocking schemes that have to satisfy multiple requirements at the global system level and for each local clock in the system components. Such requirements arise in the multiprocessor system-on-chips.This clock design challenge is addressed in Chap. 6, where an all-digital, phase-locked-loop (ADPLL) architecture is used as the backbone of a clocking solution for a heterogeneous multiprocessor system. The ADPLL is multi-phase and is used to generate fractional frequencies for cores and components with the distinguishing feature of allowing instantaneous changes in the frequency division ratio within a single clock cycle. Such features are important for implementing high-performance dynamic voltage and frequency scaling (DVFS) protocols on a per-core basis. For the ADPLL, silicon results are provided for an implementation in GLOBALFOUNDRIES 28 nm CMOS process that shows a controllable clock from 80 MHz to 2 GHz, having a power consumption of 0.64 mW and occupying an area of 2340 $\mu$m. The frequency synthesiser is also implemented in GLOBALFOUNDRIES 65 nm and 28 nm processes and is shown to achieve competitive figures of merit in area, power consumption, frequency range, and fractional granularity.

DRAM memory cubes were amongst the earliest commercial products using 3D integration technology. These products have been mainly driven by the high-performance computing market and are meant to bridge the performance gap that exists between system memory and CPU/GPU. For non-volatile memory (NVM) such as NAND or NOR flash, no such commercial products exist yet, but R & D work is under way to realise a high-capacity, 3D flash memory. In Chap. 7 of Part I, the technological issues of building a low-power, high-density NVM are addressed. It is shown that Si nanoparticles can enhance charge trapping in NVM and thus can be used to improve retention time and reduce programming and reading voltages. The latter will result not only in lower power consumption but also in mitigating the impact of thermal gradients on stacked NVM layers.

Of course the issue of thermal monitoring and management remains one of the most challenging aspects of 3D ICs. The book devotes three Chaps. 8, 10, and 16, to address this issue at the levels of monitoring, physical design, and device design, respectively. In Chap. 8, the problem of accurate on-chip temperature measurement is investigated, and a novel, compact, temperature sensor is proposed, achieving sub 1 °C accuracy over a temperature range from 0 to 100 °C. A distinguishing feature of the proposed design is the use of the bandgap reference of the temperature sensor as a reference voltage in the 12-bit successive-approximation register ADC. The issue of the number and placement of these on-chip temperature sensors is a research topic of its own and would require full information on the physical design of the 3D IC and its thermal map.

Chapter 10 is devoted to thermal-aware early physical design of 3D IC. The issue of the 3D IC floor planning is considered under the requirement that the resulting floor plan has a temperature map that falls within a predefined set of specifications across the chip stack. This is achieved using thermal TSVs which play, for heat conduction, the role that electrical TSVs play for signal transmission. The proposed floor planning algorithm achieves more than 100 K reduction in temperature for a four-layer stack at a thermal TSV via density of less than 0.5 %.

At the component design level, Chap. 16 in Part II addresses the issue of designing Si photonic components that are insensitive to temperature variations. The challenge here is that temperature impacts not only the index of refraction but also the wavelength at which the Si photonic device operates. For the case of a Mach–Zehnder interferometer (MZI), Chap. 16 proposes an *athermal* design that has a spectral sensitivity of less than 10 pm/K over the 1510–1590 nm wavelength range. The design is based on a mathematical formulation imposing both first-order and second-order sensitivity constraints on the MZI phase condition.

Besides the thermal challenge in 3D chip stacks, the lack of computer-aided design tools that are fully adapted to the 3D design environment has also been a hurdle. Early work on CAD for 3D IC focused on extending IC tools and environments to account for vertical chips stacking using TSVs. Yet, 3D integration is not just a chip technology, it is also a packaging technology, with its "supply chain" including not only IC-centric environments but also packaging and printed-circuit board environments. This viewpoint is adopted in Chap. 9 of Part I, where methodologies for 3D chip-package co-design are described in details using the TSV interposer technology demonstrator of Chaps. 3 and 6 as a case study.

In Part II, after an introductory chapter on the importance of optical communication for interconnect-centric IC design, Chap. 12 offers fabrication recipes of three possible options for manufacturing an optical TSV which are presented. They are an air-filled TSV with Si walls , a polymer-filled TSV with $SiO_2$ cladding, and a hybrid TSV with copper walls. These optical TSVs are fabricated, characterised, and compared in terms of their eye diagrams, bit error rates, and transmitted optical power.

Chapter 13 surveys both the passive and active photonic devices that are the building blocks of photonic communication links. One of the potential benefits of heterogeneous 3D die stacking is the seamless integrations of optical power sources such as III/V semiconductor lasers with Si photonic components using optical TSVs of the type proposed in Chap. 12. In Chap. 13, lasers, photonic modulators, and photodetectors are described along with an overview of the optical, electrical, and optoelectronic measurement techniques for photonic components.

An example of a Si photonic device is given in Chap. 14, where the theory, design, and numerical validation of a tunable silicon microring resonator are presented. Such tunable resonators are essential components for designing filters with controllable resonant frequencies as may be required in waveform-division multiplexing systems. The tuning mechanism adopted in the design is that of a microelectromechanical cantilever. This mechanism has the distinct advantage of being low-power and fully compatible with the Si photonic fabrication platform.

It is well known that photonic transmission is very sensitive to temperature variations. This sensitivity becomes even more problematic in 3D optoelectronic integration where temperature gradients are common due to the blocking of heat conduction paths. As was mentioned previously, Chap. 15 proposes a Mach–Zehnder interferometer (MZI) design that is insensitive to temperature variations. This is achieved by carefully selecting the dimensions of the MZI waveguides so as to satisfy phase invariance conditions with respect to temperature and spectral variations.

The realisation of full on-chip communication links achieving Terabits/s data rates requires that the IC interfaces be of sufficient bandwidth to support the optoelectronic transceivers that are expected to operate in the THz regimes. Such communication systems can be realised only if the promise of 3D heterogenous integration is fulfilled. A case in point is the one discussed in Chap. 16, where the high-bandwidth IC drivers of laser sources are designed. To meet the data and rate and bandwidth specifications, a 130 nm SiGe BiCMOS technology is used. These laser source drivers are validated using the photonic TSV developed in Chap. 12 and are shown to support a data rate of 71 Gbits/s with a power efficiency of 13.4 mW/Gbits/s.

Chapter 17, the last chapter of the book, addresses the issue of integrating power sources and energy harvesters with 3D chip stacks with focus on photovoltaic cells. One promising technology for such integration is the back-contacted hetero-junction solar cell that can achieve up to 26 % conversion efficiency. The chapter is mainly devoted to a parametric study of the performance of such cells as expressed by their fill factors, open-source voltages, and short circuit currents.

To make the most out of the chapters of this book, the reader should have basic understanding of semiconductor processing, IC design, and photonics. Our targeted audience are faculty and graduate students in EECS programmes, engineers and technologists in the semiconductor industry, and R & D managers and leaders interested in keeping apace with the latest in academic research on 3D chip stacking.

We realise there is now a large body of literature devoted to 3D chip stacking and IC integration, and we understand that one more book in this area may pass as a belated expression of "me too-ism" in an already crowded domain. Yet we think that this book offers unique features that set it apart from other distinguished contributions to the 3D integration field. These unique features include the consideration of both electronic and photonic 3D integration, the use of high-speed, on-chip communication as a unifying and motivating theme, and the coverage of topics not typically treated under the 3D integration headline such as thermal sensing and optoelectronic ICs.

The compilation of this book would not have been possible without the dedication, hard work, and commitment of all the contributing authors. To them go our deepest gratitude and warmest thanks!

Abu Dhabi, United Arab Emirates        Ibrahim (Abe) M. Elfadel
Dresden, Germany        Gerhard Fettweis
July 2015

# Acknowledgements

The research work described in this book would not have been possible without the support of the Government of Abu Dhabi, UAE, and the Government of Saxony, Germany, through the Abu Dhabi—Saxony Partnership Committee. The work at the Masdar Institute of Science and Technology was funded by the Mubadala Investment Company, while the work at the Technical University of Dresden was funded by the European Union and the Free State of Saxony through the European Social Fund. All this research has been conducted under the umbrella of Twinlab 3DSC (3D Stacked Chips), a unique concept in international scientific and technological collaboration that has brought together researchers in the UAE and Germany to work on 3D-integrated circuits and packages. Our warmest thanks go to our colleagues, assistants, and students who have made such concepts not only real, concrete, and tangible but also exciting, fun, and rewarding.

We would like to acknowledge Ibrahim Ajami, Sami Issa, Canan Anli, Alden Holden, and Sahar Al-Katheeri for their continuous encouragement and support of Twinlab 3DSC as well as their visionary dedication to the semiconductor R & D ecosystem in the UAE.

Since its inception in April 2012, Twinlab 3DSC has held semi-annual workshops that have alternated between Abu Dhabi in November and Dresden in May or June. These workshops often start with inspiring keynotes delivered by industry, government, or academic leaders. We would like to acknowledge and thank Rafic Makki, Geoffrey Akiki, David McCann, Gerd Teepe, Luigi Capodieci, Rani Ghaida, all from GLOBALFOUNDRIES; Prof. David Pan, University of Texas at Austin; and Ms. Shaima Salem Al Habsi, UAE Embassy, Berlin, for delivering such keynotes.

We also want to thank Mohamed Lakehal from GLOBALFOUNDRIES, Abu Dhabi, for his continuous efforts in providing tighter interlock between the Masdar Institute chip design work and GLOBALFOUNDRIES.

We also thank the senior administrators of our academic institutions, the Masdar Institute and the Technical University of Dresden, for facilitating, enabling, and hosting Twinlab 3DSC over the past 3 years.

Finally, we acknowledge the technical support we have received from Lukas Landau, Friedrich Pauls, and Ronny Henker during the composition of this book as well as the advice and guidance provided to us by Charles Glaser from Springer.

# Contents

# Contributors

**Ayesha A. Al-Shouq** Department of Mechanical and Materials Engineering, Masdar Institute of Science and Technology, Abu Dhabi, United Arab Emirates

**Johann W. Bartha** Technische Universität Dresden, Institute of Semiconductors and Microsystems, Dresden, Germany

**Guido Belfiore** Technische Universität Dresden, Chair for Circuit Design and Network Theory, Dresden, Germany

**Puskar Budhathoki** Department of Electrical Engineering and Computer Science, Institute Center for Smart Infrastructure (iSmart), Masdar Institute of Science and Technology, Abu Dhabi, United Arab Emirates

**Love Cederström** Technische Universität Dresden, Chair of Highly-Parallel VLSI-Systems and Neuro-Microelectronics, Dresden, Germany

**Sujay Charania** Technische Universität Dresden, Institute of Semiconductors and Microsystems, Dresden, Germany

**Marcus S. Dahlem** Department of Electrical Engineering and Computer Science, Institute Center for Microsystems (iMicro), Masdar Institute of Science and Technology, Abu Dhabi, United Arab Emirates

**Nazek El-Atab** Department of Electrical Engineering and Computer Science, Institute Center for Microsystems (iMicro), Masdar Institute of Science and Technology, Abu Dhabi, United Arab Emirates

**Ibrahim (Abe) M. Elfadel** Department of Electrical Engineering and Computer Science, Institute Center for Microsystems (iMicro), Masdar Institute of Science and Technology, Abu Dhabi, United Arab Emirates

**Frank Ellinger** Technische Universität Dresden, Chair for Circuit Design and Network Theory, Dresden, Germany

**Gerhard Fettweis** Technische Universität Dresden, Vodafone Chair Mobile Communication Systems, Dresden, Germany

**Johannes Görner** Technische Universität Dresden, Chair of Highly-Parallel VLSI-Systems and Neuro-Microelectronics, Dresden, Germany

**Adel B. Gougam** Department of Mechanical and Materials Engineering, Masdar Institute of Science and Technology, Abu Dhabi, United Arab Emirates

**Michael Haas** Technische Universität Dresden, Chair for RF Engineering, Dresden, Germany

**Ronny Henker** Technische Universität Dresden, Chair for Circuit Design and Network Theory, Dresden, Germany

**Andreas Henschel** Department of Electrical Engineering and Computer Science, Institute Center for Smart Infrastructure (iSmart), Masdar Institute of Science and Technology, Abu Dhabi, United Arab Emirates

**Sebastian Höppner** Technische Universität Dresden, Chair of Highly-Parallel VLSI-Systems and Neuro-Microelectronics, Dresden, Germany

**Seyedreza Hosseini** Technische Universität Dresden, Integrated Photonic Devices Lab, Dresden, Germany

**Kambiz Jamshidi** Technische Universität Dresden, Junior Professorship Integrated Photonic Devices, Dresden, Germany

**Sebastian Killge** Technische Universität Dresden, Institute of Semiconductors and Microsystems, Dresden, Germany

**Johann Knechtel** Department of Electrical Engineering and Computer Science, Institute Center for Microsystems (iMicro), Masdar Institute of Science and Technology, Abu Dhabi, United Arab Emirates

**Lukas Landau** Technische Universität Dresden, Vodafone Chair Mobile Communication Systems, Dresden, Germany

**Ammar Nayfeh** Department of Electrical Engineering and Computer Science, Institute Center for Microsystems (iMicro), Masdar Institute of Science and Technology, Abu Dhabi, United Arab Emirates

**Niels Neumann** Technische Universität Dresden, Chair for RF Engineering, Dresden, Germany

**Volker Neumann** Technische Universität Dresden, Institute of Semiconductors and Microsystems, Dresden, Germany

**Ali K. Okyay** Department of Electrical Engineering, Bilkent University, Ankara, Turkey

**Friedrich Pauls** Technische Universität Dresden, Vodafone Chair Mobile Communication Systems, Dresden, Germany

**Dirk Plettemeier** Technische Universität Dresden, Chair for RF Engineering, Dresden, Germany

**Sami ur Rehman** Department of Electrical Engineering and Computer Science, Institute Center for Microsystems (iMicro), Masdar Institute of Science and Technology, Abu Dhabi, United Arab Emirates

**René Schüffny** Technische Universität Dresden, Chair of Highly-Parallel VLSI-Systems and Neuro-Microelectronics, Dresden, Germany

**Tobias Seifert** Technische Universität Dresden, Vodafone Chair Mobile Communication Systems, Dresden, Germany

**Ayman Shabra** Department of Electrical Engineering and Computer Science, Institute Center for Microsystems (iMicro), Masdar Institute of Science and Technology, Abu Dhabi, United Arab Emirates

**Hossam Shoman** Department of Electrical Engineering and Computer Science, Institute Center for Microsystems (iMicro), Masdar Institute of Science and Technology, Abu Dhabi, United Arab Emirates

**Laszlo Szilagyi** Technische Universität Dresden, Chair for Circuit Design and Network Theory, Dresden, Germany

**Jaime Viegas** Department of Electrical Engineering and Computer Science, Institute Center for Microsystems (iMicro), Masdar Institute of Science and Technology, Abu Dhabi, United Arab Emirates

**Dennis Walter** Technische Universität Dresden, Chair of Highly-Parallel VLSI-Systems and Neuro-Microelectronics, Dresden, Germany

**Peng Xing** Department of Electrical Engineering and Computer Science, Institute Center for Microsystems (iMicro), Masdar Institute of Science and Technology, Abu Dhabi, United Arab Emirates

# Part I
# Electrical 3D Integration

*At first, indeed, I pretended that I was describing the imaginary experiences of a*
*fictitious person; but my enthusiasm soon forced me to throw of all disguise, and*
*finally, in a fervent peroration, I exhorted all my hearers to divest themselves of*
*prejudice and to become believers in the Third Dimension.*
*Edwin A. Abott, Flatland*

# Chapter 1
# Introduction to Electrical 3D Integration

**Sebastian Killge, Sujay Charania, and Johann W. Bartha**

In 1965, Gordon E. Moore submitted his prediction, now known as *Moore's Law*, on the exponential growth of transistor density in an integrated circuit to the semiconductor industry. It is not clear whether this prediction has become a self-fulfilling prophecy, but it certainly defined a guideline for the entire industrial sector associated with microelectronics, and the industry has kept a steady pace of miniaturization, doubling the device density every 18–24 months.

Yet exponential growth cannot proceed forever. As much was stated by the Semiconductor Industry Association (SIA) in September 2007: "....our ability to shrink down the size of the transistor will be limited by physics sometime within the next 10-15 years." Even back then, it was time to try novel venues to extend this exponential course in new directions. Understanding *Moore's Law* as transistors per chip but considering such chip to consist of stacked and vertically interconnected chips is such a new venue for *Moore's Law*, which can then stay more or less valid even when the transistor size is kept constant. In that sense, 3D integration is the continuation of *Moore's Law* beyond the stopping point of transistor scaling.

It provides the path to further miniaturization through the reduction in footprint, increase in device density, and shortening of interconnect lengths, while enabling higher bandwidth and improving circuit security. The International Technology Roadmap for Semiconductors (ITRS) [2] predicted a steady decrease in chip thickness for three dimensional (3D IC) solutions. Furthermore, it placed a stronger focus on formation of vertical interconnects passing through the silicon substrate, so-called through-silicon vias (TSVs) with high aspect ratios (AR) (10, 15, and

S. Killge (✉) • S. Charania • J.W. Bartha
Technische Universität Dresden, Institute of Semiconductors and Microsystems - IHM,
01062 Dresden, Germany
e-mail: sebastian.killge@tu-dresden.de; Sujay.Charania@tu-dresden.de;
johann.bartha@tu-dresden.de

© Springer International Publishing Switzerland 2016
I.M. Elfadel, G. Fettweis (eds.), *3D Stacked Chips*,
DOI 10.1007/978-3-319-20481-9_1

**Fig. 1.1** Classification of
integration schemes

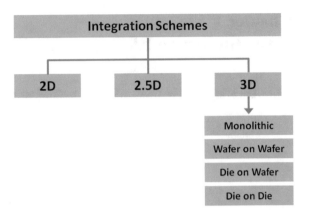

20:1), to enable efficient 3D chip stacking. 3D integrated circuits (ICs) with TSVs offer new levels of advantages in efficiency, power, performance, and form-factor to the semiconductor industry.

In general, integration schemes can be classified into three main categories as shown in Fig. 1.1.

The steady increase in consumer demand for improved performance and smaller sizes has forced a shift from 2D to 2.5D and 3D packaged designs. There are several methods in which one can carry out a real 3D integration. They include Complete-Monolithic, Wafer-on-Wafer, Die-on-Wafer, or Die-on-Die. In all of these methods, the chips are stacked on top of each other.

The terms system-in-package (SiP), silicon-in-package, and multi-chip module (MCM) have all been used to refer to a packaging technology in which multiple dies are mounted on a common substrate that is used to connect them. These technologies started to gain acceptance in the early 1990s. For SiP, the subsystems are individual dies that can be manufactured independently with different nodes, e.g., the CPU could be manufactured in 28 nm, memory in 14 nm, and peripherals in 180 nm node. Later, they are assembled within a single package.

The evolution from conventional packaging to 3D is displayed in Fig. 1.2. A 2.5D IC (also a planar technology) uses specially designed interposer to connect multiple dies before connecting them to the substrate or PCB. The interposer is a specially designed Si or glass substrate with communication structures consisting of dedicated high bandwidth connections and TSV networks which facilitates die-oriented connections. It is the first step towards 3D integration and it means that the ICs are stacked laterally, increasing density per unit area. By using a silicon interposer, the distance between the chips is reduced, the area consumption decrease and the electrical performance gets higher due to reduced thickness. Dies are placed face down on the interposer and connected by micro bumps. The interposer combining multiple small dies is the first approach where multi-functional dies are connected on a single package, i.e., logic and/or memory and/or analog circuit with processor. Interposer technology allows the production of modular and highly manufacturable modular designs that can fully replace large chip designs

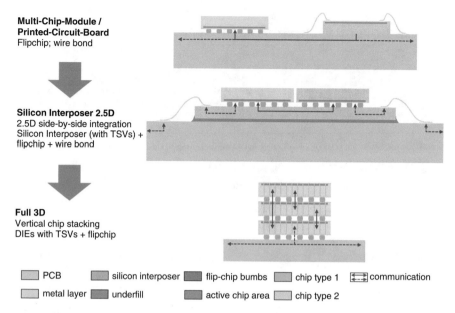

**Fig. 1.2** Advancement towards 3D integration

with performance requirements remaining the same[1]. Such technology makes it possible to manufacture FPGAs that offer bandwidth and capacity exceeding that of the largest possible monolithic FPGA dies but with the manufacturing and time-to-market advantages of smaller dies [3].

2.5D is still a planar technology. From the technological perspective, it is even more advantageous, if the ICs are stacked on top of each other. However there is a difference between 3D system-in-package (SiP) and 3D system on chip (SoC). In the 3D case of SiP, which is also known as a Chip Stack multi-chip module, individual dies are manufactured independently in various technology nodes allowing for integration of an entire system in one package. Each of the dies is a subsystem and is assembled in a single package by stacking them side-by-side or on top of each other. In the 3D SoC, the monolithic design is a continuous die, which offers higher integration and quasi high-speed at the price of loosing the advantage of having a modular design and different technology nodes for subsystems.

Typically, it is manufactured in a single process where all the subsystems, e.g., CPU, memory, peripherals, IOs, are fabricated in the same manufacturing chain. The resulting single die is a 3D SoC that is assembled in its own package for further use. Thus, 3D packaging saves space (very small footprint) by stacking separate dies in a single package. Each individual die is designed with TSVs in order to create vertical interconnects. The length of interconnections is substantially reduced, hence high speeds (low latencies) could be achieved, and therefore power consumption is also reduced.

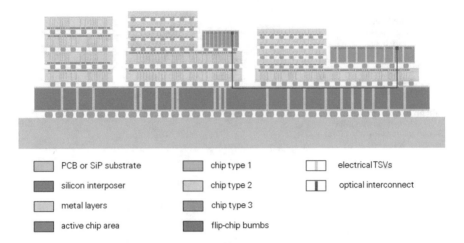

**Fig. 1.3** True 3D stacked IC in combination with an interposer using TSVs and dies of different types of technology

The dies are specifically designed to include TSVs in their interconnections. Thus, irrespective of the type of the die, e.g., memory or logic, heterogeneous or homogeneous, 3D integration is possible (Fig. 1.3). There is an option to use optical connections also with TSVs, which assures very high bandwidths (in the order of ten to hundred GHz). Both of these technologies, 2.5D and 3D, have already representative products in the market such as Xilinx Virtex 7 FPGA for 2.5D integration and the DRAM memory cube from the Hybrid Memory Consortium for 3D integration.

Of course to fully realize all the mentioned advantages of 3D integration, several hurdles have to be overcome. One of the main requirements lies in thermal management. As multiple dies are stacked, the temperature of every layer does not depend only on its own power dissipation but also on that of the surrounding dies. For example, the junction temperature of a memory module (e.g., DRAM) has to be kept below 85 °C for the module to function reliably. On the other hand, the junction temperature of the logic (e.g., CMOS circuits) can easily exceed 100 °C under typical workloads. So in case of heterogeneous integration, where different dies are stacked on top of each other, the temperature variation leads to thermal stress. Furthermore, in contrast with 2D technology where logic and memory are separate, the dies here are stacked densely on top of each other. Since die types and functions may be different in 3D integration, the requirements of power dissipation and performance will differ layer to layer and die to die. These differences produce thermal mismatches that will make thermal emergencies more likely.

Obviously, the design complexity also reaches its peak. The designer has to consider not only one layer of circuit but rather a complete set, including their respective component placements and rules for design. This makes the design extremely challenging, especially the TSV technology creates difficulties. No matter what type of TSVs is used (i.e., optical or electrical), the corresponding optical, electrical, and thermal properties with very small lateral dimensions of a few microns and lengths of a few tens to hundreds of microns, make it extremely challenging to obtain high interconnect densities. In monolithic integration, there is continuous fabrication of various circuits on top of each other by using polysilicon as a base for intermediate circuit slices. Here, a single fault at any stage of the fabrication can lead to complete failure of the chip. Hence the yield remains at stake for this kind of 3D integration. Since the final chip is the combination of various circuits, the testing instruments and algorithms should be designed with utmost specialization and hence testing becomes both complex and costly. Due to the complex supply chain of 3D integration, there is no one market leader for 3D integration, nor is there a clear path to industry standardization among the various actors. As in any merging technology, there is a daunting variability in supply chain that is resulting in lack of compatibility among production methods and products.

Another technical problem is that a copper filled TSV on an active chip causes stress in the silicon substrate. Due to the mismatch, TCE requires a "keep-out zone" (KOZ) around the rim of the TSV where no transistors are allowed [4]. Therefore the loss in yield caused by the integration of TSVs in is much larger than just the area required for the TSV-holes. The KOZ and device variability due to stress–strain patterns in active silicon have to be appropriately accounted for in the 3D process design kit (PDK).

In spite of all the mentioned challenges, 3D integration remains one of the most promising technologies for sustaining *Moore's Law* once device scaling is no longer technologically or economically feasible. The following chapters on the process, design, and characterization of heterogeneous 3D stacked chips are meant to illustrate the well-founded nature of the 3D promise with actual technology demonstrators.

# References

1. P. Dorsey, Xilinx stacked silicon interconnect technology delivers breakthrough FPGA capacity, bandwidth, and power efficiency. Xilinx White Paper: Virtex-7 FPGAs (2010), pp. 1–10
2. International Roadmap Committee et al., International Technology Roadmap for Semiconductors: 2013 edition executive summary. Semiconductor Industry Association, San Francisco, CA (2013). Available at: http://www. itrs. net/Links/2013ITRS/2013Chapters/2013Executive Summary.pdf
3. N. Kim, D. Wu, D. Kim, A. Rahman, P. Wu, Interposer design optimization for high frequency signal transmission in passive and active interposer using through silicon via (tsv), in *2011 IEEE 61st Electronic Components and Technology Conference (ECTC)*, pp. 1160–1167
4. V.F. Pavlidis, E.G. Friedman, 3-d topologies for networks-on-chip. IEEE Trans. Very Large Scale Integr. Syst. **15**(10), 1081–1090 (2007)

# Chapter 2
# Copper-Based TSV: Interposer

**Sebastian Killge, Volker Neumann, and Johann W. Bartha**

## 2.1 Introduction

Through-silicon via (TSV) fabrication consists mainly of the following steps: etching, deposition of insulator, deposition of barrier and seed layers, and electrochemical plating. Depending on the application, the TSV structures differ in size, aspect ratio, density, materials, and technology. Each application has its own requirements which affect the whole processing scheme. The most important parameters for TSV fabrication are aspect ratio and contact density. Their values are specific to each application.

In this chapter, we present a specific process flow for a TSV—interposer realizing through holes down to a diameter of $10\,\mu m$. The fabrication of interconnect is carried out as a through-hole connection. In contrast to the blind hole via integration schemes, the TSVs are etched through to an etch-stop layer. Thus, no grinding, polishing, or etch back processes have to be applied later as it is required in blind hole via integration schemes. Since the fabrication aims at interposer fabrication, no active devices and thus, no restrictions in the thermal budget have to be considered. The complete process flow consists of the following steps and is schematically pictured in Fig. 2.1:

1. The fabrication process is started with a blank silicon wafer. A substrate thickness of $200\,\mu m$ is chosen as a trade-off between an attainable aspect ratio for deep silicon etch and via fill. Also, a removable stop layer is deposited for

S. Killge (✉) • V. Neumann • J.W. Bartha
Technische Universität Dresden, Institute of Semiconductors and Microsystems - IHM,
01062 Dresden, Germany
e-mail: sebastian.killge@tu-dresden.de; volker.neumann@tu-dresden.de;
johann.bartha@tu-dresden.de

© Springer International Publishing Switzerland 2016
I.M. Elfadel, G. Fettweis (eds.), *3D Stacked Chips*,
DOI 10.1007/978-3-319-20481-9_2

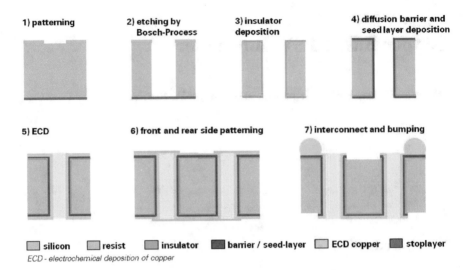

1) patterning   2) etching by Bosch-Process   3) insulator deposition   4) diffusion barrier and seed layer deposition

5) ECD   6) front and rear side patterning   7) interconnect and bumping

☐ silicon   ☐ resist   ☐ insulator   ■ barrier / seed-layer   ☐ ECD copper   ■ stoplayer

*ECD - electrochemical deposition of copper*

**Fig. 2.1** TSV process scheme for thinned wafer with a thermal oxide layer as rear side etch-resist

the subsequent deep reactive ion etch process[1] and a full lithography step is performed to create a photoresist masking layer for the subsequent deep reactive ion etch process.

2. The TSVs are formed by means of deep reactive ion etching. The photoresist and the polymeric sidewall passivation layers created during DRIE processing are stripped wet-chemically. The stop layer is removed by an additional wet-etch process. Furthermore residual polymer is removed and the sidewalls are smoothed by reactive ion etching in an oxygen-nitrogen trifluoride plasma.

3. To prevent leakage between different interconnects, an insulating silicon dioxide layer is grown by thermal processing. Moreover, this process also functions as an annealing process to minimize stress in the crystal lattice caused by the etching process.

4. The diffusion barrier layer and a ruthenium (Ru) seed layer are deposited in situ by atomic layer deposition (ALD).

5. The TSVs are either filled or just enhanced with copper in an electrochemical plating process.

6. Front and back side lithography is used to generate the metallization layer mask for the subsequent patterning processes.

7. The copper layer is structured in a wet etching process and the redistribution lines on the front and back side of the wafer are created by pattern plating. The seed and the barrier layers are removed afterwards in a plasma etch process. In a waferbumping process, SnAg or PbSn solder bumps are fabricated by electrochemical deposition (ECD) and reflow.

---

[1] The properties of different stop layers are investigated in Sect. 2.2 of this chapter.

## 2.2 Deep Reactive Ion Etching

Deep reactive ion etching (DRIE) is widely used to generate MEMS structure, capacitors for deep-trench DRAM, and for fabrication of TSVs. It is an extension of the RIE process, which is a highly anisotropic etch process that is used to generate straight (90°) etch profiles, steep trenches or holes (generally with high aspect ratios) in a substrate. In general, DRIE can be either isotropic or anisotropic. There are two methods available to control isotropy in DRIE: cryogenic and Bosch. In cryogenic DRIE, the wafer is cooled to $-110\,°C$ $(+163\,K)$ to slow down the chemical reaction leading to isotropic etching and generate a sidewall passivization [1]. The second and well-established option is the Bosch DRIE process. It was developed in 1994 at Robert Bosch GmbH [DE 4241045 C1] [2]. This etching process is performed by cycling between a deposition ($C_4F_8$) and an etching ($SF_6$) step and is also known as time multiplexed or pulsed etching. It consists of:

1. Application of a chemically inert passivation layer by plasma induced deposition of a polymer layer using $C_4F_8$ as feed gas.
2. Anisotropic removal followed by isotropic Si chemical etch; $SF_6$ is generally used for etching of Si.

As shown in Fig. 2.2 initially, the polymeric passivation layer covers the resist, the entire structure, and sidewalls and prevents further chemical attack. In the following etching step first the ionic fraction of the plasma assists to remove the passivation on vertical surface (trench bottom) after that the trench bottom is etched isotropically. The isotropic etching is performed for a few seconds, usually with a fluorine based gas ($SF_6$).

This is followed by a deposition step lasting a few seconds, in which the pattern is again covered with a uniform polymeric layer. Then the etching cycle is repeated. Due to the direction of the accelerated ions, the polymeric layer is removed much faster on vertical surfaces than on horizontal surfaces. For the rest of the etching cycle, further etchants start etching the surface vertically, simultaneously the chemically inert polymer layer keeps the sidewalls from further etching and hence the lateral etching or isotropic etching component is substantially reduced. These etching/deposition steps are repeated several times to achieve the required etch depth. The plasma is generated by an inductive plasma source, while the ion bombardment on the substrate is controlled by a capacitively coupled RF power applied to the susceptor plate. Each step contains a significant number of parameters controlling the process properties like gas flows, the power of the inductively coupled plasma or the platen source, time, etc. The optimized etching profiles with slightly positively tapered angles are generated by careful balancing between the two steps.

The advantage of the passivation layer is that the width of a trench can be confined to a certain extent. Otherwise, without the passivation step, the isotropic

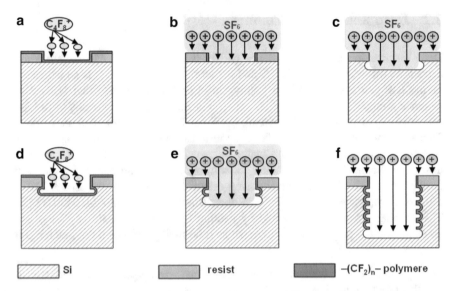

**Fig. 2.2** Bosch process scheme. (**a**) Deposit a conformal $C_4F_8$ passivation layer; (**b**) directed removal of the passivation layer by ions; (**c**) isotropic etching with SF6; (**d**) deposit a conformal $C_4F_8$ passivation layer; (**e**) pass. removal and isotropic etching; (**f**) alternating steps (**b**)–(**e**)

**Fig. 2.3** Bosch scallops created by alternating isotropic passivation and anisotropic opening of the trenches bottom

etching can lead to very high lateral etching, resulting in very broad open structures. Nevertheless this effect can be used for diameter enlargement of the structures in the upper region near the opening.

The cyclic approach using isotropic chemical etching alternating with isotropic passivation and anisotropic opening of the trenches bottom creates a characteristic sidewall shape with a waved profile. These waves are also referred to as *scallops* (Fig. 2.3). The challenge in DRIE is to generate TSVs with a low surface roughness (scallops), a low tilt of the vias, and easily removable CF-polymers residues.

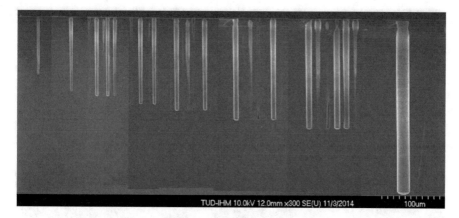

**Fig. 2.4** SEM pictures of the etch rate depending on the dimensions of a feature varying for 5–40 μm TSV fabrication in one etching process

Overall, the sidewall roughness is very critical and should be as low as possible to minimize its impact on the subsequent deposition steps. Indeed, high surface roughness causes a much greater area to be coated, and therefore the coverage of the following layers has to be increased. In addition, residues of CF-polymer may attach to the TSV sidewalls. Another important point is that the etch rate depends on the structure diameter. Thus, it is not possible to etch structures with different diameters or opening surfaces to the same depth.

Dependence of etching on aspect ratio is sometimes called *RIE lag* as it describes the impact of feature dimensions on each rate [3]. This is especially important when features of different dimensions are etched simultaneously. Because of *RIE lag*, smaller structures will be etched slower than larger structures. Figure 2.4 shows the varying etching rate for 5–40 μm TSV fabrication in one etching process.

At the Institute of Semiconductor and Microsystems (TU Dresden), an STS Pegasus machine is used. This is a high-rate-etching machine that is very well suited for fabrication of deep TSVs. The control of the alternating etching and deposition step is the basis to improve the result. As mentioned before, each step consists of a huge range of parameters to control the etching result. To obtain a high uniformity of the etch depth all over the wafer, it is necessary to obtain a highly uniform plasma density which can be controlled with specific fields around the plasma bulk.

Figure 2.5 shows results for optimized process on the STS Pegasus. It is possible to generate TSVs with tapered profiles up to an aspect ratio of 30:1 with low surface roughness. Those TSVs are manufactured with two etching processes. At first, the via is etched to the desired depth. This first process step creates a so-called undercut which is a negative profile in the first 5 μm etching depth. The undercut has to be removed by a second process, because the negative profile makes it difficult to deposit continuous electrical insulator, diffusion barrier and seed layers. The TSVs of the optimized process (Fig. 2.5) have a lower surface roughness at the bottom and in the middle. In the TSVs from the initial test, pillar formation was found.

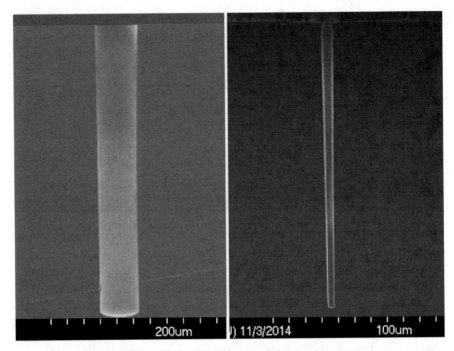

**Fig. 2.5** SEM pictures of the etching results at STS Pegasus DRIE. *Left*: 40 μm diameter; 400 μm deep, AR 10:1. *Right*: 6 μm diameter; 187 μm deep, AR 30:1

The pillar formation is a defect in the TSV sidewalls where the CF-polymer passivation of the TSV sidewalls is burst in the etching cycle and a parasitic etching behind the sidewalls appears.

In the optimized process the pillar formation could be reduced to a minimum. The optimized parameters guarantee a sidewall surface roughness of 20–100 nm at maximum. The integration scheme for the fabrication of the TSV interposer using 200 μm thin wafers enables the creation of through-hole VIAs without additional wafer thinning. Therefore, the etching process must be reliably stopped on the wafer back side without impact on the TSV geometry or the sidewall roughness. Since the temperature control at the wafer relies on He-back side cooling (thin He-gas buffer between wafer susceptor and wafer) the TSV-hole etching must stop on a thin membrane as stop layer.

Applying a photoresist as stop layer is one of the simplest ways to realize this. Spin-coating and bake-out process of the stop layer are carried out prior to the lithography step, which is performed to create the masking layer on the front side of the wafer. However, commonly used photoresists are insulating and the cleaning of the trench bottom after the passivation step requires ion bombardment approaching perpendicular to the surface. An insulating layer will charge up the surface positively and deflect the ions creating a thinning of the sidewall passivation close to the trench bottom and by this a widening of the structures referred to as "notching" (shown in Fig. 2.6).

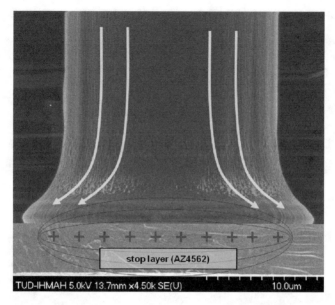

**Fig. 2.6** Notching created by released ion bombardment and back side stop layer charging on trench bottom

So notching is an effect of lateral etching which occurs in high-density plasma etching when approaching a stop layer [4]. The phenomenon is caused by two co-occurring events:

1. The extinction of etching due to the appearance of the non-etchable stop layer causes a gain in radial concentration.
2. The charge accumulation in an insulating stop layer material. The latter is accompanied by an electric field, which then deflects all further impinging ions onto the sidewall of the TSV.

One approach to minimize the notching effect is the application of conductive stop layers [5]. The material commonly used is aluminum [6], which has been the standard material for metallization layers in CMOS fabrication for many years [7]. This greatly simplifies the application since the deposition and etch processes are easily available. In general, film thicknesses of few microns are chosen in order to ensure a sufficient mechanical stability [6, 8].

Besides the prevention of sidewall notching during DRIE, a most important requirement of the etch-stop layer is the ease with which the material is removed subsequent to the TSV formation. In our case an aluminum film thickness of only 50 nm is deposited, and mechanical stability is provided by applying an additional photoresist layer of 5 μm thickness onto the aluminum layer (Fig. 2.7).

As mentioned above, the bad adhesion of polymer layers on the surface has a negative influence on the subsequent films, thus increasing the likelihood of delamination. Furthermore, pillar formations are found that are partly due to

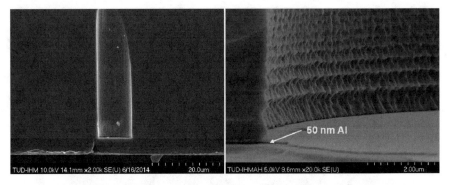

**Fig. 2.7** SEM of etched TSV (20 μm diameter, 200 μm length) at the bottom of the etch-stop layer after DRIE, detailing minimal notching on 50 nm of Al-stop layer

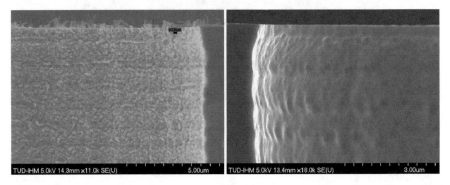

**Fig. 2.8** SEM pictures before and after plasma-enhanced smoothing, using oxygen ($O_2$), argon (Ar), and nitrogen trifluoride ($NF_3$) (max. 20 nm surface roughness at TSV sidewall, 20 μm diameter, 200 μm length)

polymer residues. Their bad adhesion to the via walls often results in delamination of subsequently deposited layers as well. The optimized process reduces the CF-polymer residues. Proper cleaning procedures are necessary and can be employed in two ways: dry plasma or wet chemical cleaning. For the wet chemical step the use of 1-methyl-2-pyrrolidone (NMP) or nonafluoro-4-ethoxy-butane (F7200) is being investigated. In a dry plasma-enhanced cleaning the residual polymer is removed and the sidewalls are smoothed by reactive ion etching in an oxygen-nitrogen trifluoride plasma (Fig. 2.8). The optimized parameters of the wet chemical and dry plasma cleaning steps guarantee a sidewall surface roughness of only 20–100 nm at maximum.

The next step after the generation of the TSV is an enlargement of its diameter in the upper region near the opening. The TSVs are thus manufactured with two subsequent etching processes. At first, the via is etched to the desired depth. This first process step creates a so-called undercut which is like a negative profile in the first 5 μm etching depth. The undercut has to be removed by a second process,

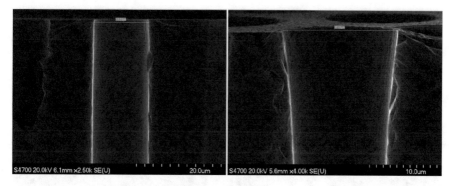

**Fig. 2.9** Removal of the undercut and profile enlargement for tapered TSV by an additional etching step. *Left*: before removal of the undercut. *Right*: after the additional etching for tapered profiles

because the negative profile makes it difficult to deposit a continuous electrical insulator, diffusion barrier and seed layer by conventional deposition techniques. On the one hand, tapered profiles reduce the contact density on the chip. On the other hand, positive profiles make it easier to get good sidewall coverage with subsequent deposition steps (Fig. 2.9). The removal of the undercut is an important prerequisite for good deposition conditions in the PE-CVD, PVD metallization, and ECD steps.

In summary of TSV etching, we can state that by etching to a conductive stop layer, it is possible to fabricate interposer TSV using $200\,\mu$m thin wafers. This allows the creation of TSV structures with diameter sizes ranging from less than $10\,\mu$m up to more than $40\,\mu$m without any additional wafer thinning process (Fig. 2.10).

## 2.3  Insulator, Diffusion Barrier- and Seed-Layer Deposition

The fabrication step following the silicon etching of the TSV is the deposition of insulator, barrier and seed layer. The metallization consists of Cu and is applied by ECD. Since Cu has a high diffusivity in silicon as well as in silicon oxide it requires a defect free conformal barrier layer which simultaneously serves as adhesion or wetting-layer and a conformal conducting layer which serves as starting or seed layer for the ECD process. The fabrication of TSVs with aspect ratios exceeding 10:1 clearly limits the application of conventional deposition processes. Indeed, the production of high aspect ratio TSV requires deposition process enabling the deposition of homogeneous layers on large aspect ratios. In principle, three different deposition processes with a variety of materials are available: physical vapor deposition (PVD), chemical vapor deposition (CVD), and ALD. Depending on the deposition process, the coverage in the high aspect ratio structures can be very inhomogeneous, with increasing loss in film thickness when approaching the trench

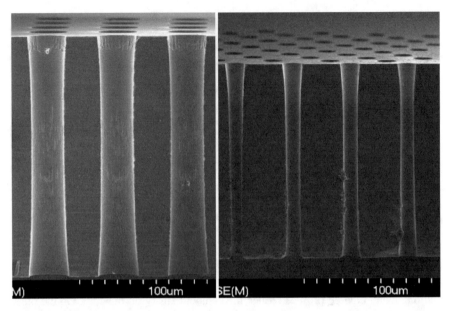

**Fig. 2.10** SEM after silicon via etching onto a stop layer releasable afterwards and additional enlargement of TSV diameter on wafer front- and back-side. *Left*: 20 μm diameter, 200 μm TSV length. *Right*: 10 μm diameter, 200 μm TSV length

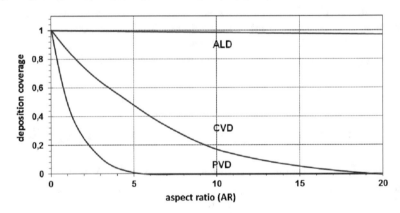

**Fig. 2.11** Schematic graph of deposition coverage in comparison of PVD, CVD, and ALD deposits process

bottom. The film appears discontinuous and does not serve as insulator, barrier or seed layer any more. As shown in Fig. 2.11 the capability to coat trench or a TSV sidewalls by several deposition processes is limited. The step coverage of an optimized process for the TEOS-based plasma-enhanced CVD of silicon dioxide into TSV structures decreases below 30 % for aspect ratios larger than 2.5:1 [9]. Similar values constrain the use of conventional PVD processes for the deposition of barrier and seed layer [10]. The AR limits for the different deposition techniques are as follows: PVD, AR $\leq$ 5:1; CVD, AR $\leq$ 15:1; ALD, AR $\geq$ 10:1.

### 2.3.1  Insulator Deposition

An insulation layer, e.g., $SiO_2$, is necessary to ensure proper electrical functionality of the TSV. After enlargement of the TSV diameter, the rough inner surface of the vias ($\leq$100 nm) is smoothed by a plasma-enhanced chemical vapor deposition (PE-CVD)-TEOS (tetraethylorthosiloxane) process which is required for the electrical insulation of the through contacts. CVD techniques are based on the chemical reaction between the precursor reactants of the gas phase stream and the surface of the substrate or the chemical decomposition of the precursor reactants above the substrate, after which they get adsorbed at the substrate surface. In both cases the adsorbed reactants diffuse along the surface until they find an energy-favorable bonding spot. This is where the film is formed. The by-products of the reactions desorb and are pumped out with the gas stream.

The insulator deposition is carried out as low pressure PE-CVD by means of a microwave assisted ECR-concept (2.45 GHz). It is dependent on the aspect ratio as well. Sufficient step coverage for ARs from 5:1 up to 15:1 was achieved (Fig. 2.12) and the process temperature of around 340 °C resulted in low carbon content of the CVD films. The typical breakdown voltages of the insulator layer range from 9 up to 10 MV/cm.

An alternative to PE-CVD of TEOS is the thermal oxidation of the silicon wafer. A thermal oxide insulator layer offers better film properties, including better electrical and optical properties. In contrast to the PE-CVD of TEOS, the deposition rate in thermal oxidation does not depend on the aspect ratio and the layer thickness of the insulator is constant on the whole wafer. For thermal oxidation an oxidizing agent diffuses to the $SiO_2$/Si interface at high temperature and reacts with the Si substrate. Thermal oxidation of silicon is usually performed at a temperature between 800 and 1200 °C. Thermal oxide is based on a chemical reaction of oxygen and silicon, and requires the diffusion of the oxidant through the already grown $SiO_2$ to the unreacted Si surface. Here, the silicon is converted by the oxygen to silicon oxide. So there is no layer deposited onto the substrate, but the silicon is "consumed." For every unit thickness of silicon consumed, 2.27 unit thicknesses of oxide will appear. If a bare silicon surface is oxidized, 44 % of the oxide thickness will lie below the original surface, and 56 % above it.

To achieve homogeneous layers by thermal oxidation, the TSV must be free of polymer residues and has to have low surface roughness. Thermal oxidation enables the generation of a uniform insulator layer thickness (1 μm or more) on the sidewalls of the TSV irrespective of the aspect ratio. The oxide thickness is proportional to $\sqrt{t}$ where $t$ is the oxidation time. Therefore a thickness of 2 μm appears as a practical limit since even at a maximum furnace temperature of 1100 °C, it takes more than

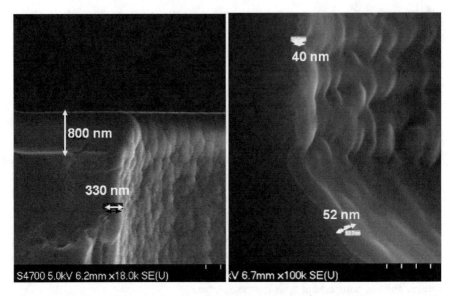

**Fig. 2.12** PECVD TEOS-SiO$_2$ 800 nm. *Left*: TSV-diameter = 10 μm depth 95 μm details: 330 nm on sidewall top. *Right*: TSV-diameter = 10 μm depth 95 μm details: 40 nm on sidewall bottom

8 h to grow 2 μm. It is important to mention that thermal oxidation is applicable to an interposer and trench-first scheme but is not applicable for a trench-middle or trench-last process flow.

## 2.3.2 Atomic Layer Deposition

The TSV geometry generally has high aspect ratio (AR) due to the fact that the TSV required area needs to be minimized. As alluded to before, the challenges associated with high-AR TSV are the DRIE process and the material deposition on the TSV sidewalls. As shown in Fig. 2.11 and discussed before, the capability to coat trench sidewalls by PVD or CVD is limited. An exceptional technology in this respect is ALD.

ALD is a special kind of heterogeneous CVD technology based on a self-limiting monolayer chemisorption of a precursor gas according to a kind of Langmuir adsorption isotherm. This is followed by exposition of the substrate to a second gas reacting with the absorbent to the desired material, enabling again the self-limiting chemisorption of the first precursor [11]. The dosing of the different gases is separated by purging the deposition chamber with inert gas to avoid gas phase (homogeneous) reactions. To grow the desired film, the four-step sequence (Precursor 1-purge-Precursor 2-purge) is applied in a cyclic way (see Fig. 2.13). The growth per cycle is typically less than an Angstrom. Due to limitations in the

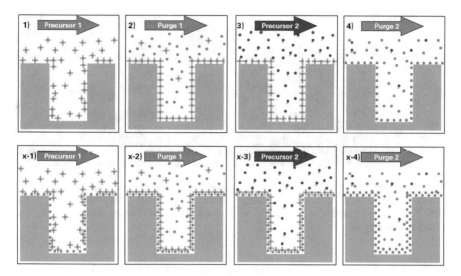

**Fig. 2.13** ALD process phases

adsorption site density or the limited capability to access sites that are blocked by the relatively large organometallic precursor (steric hindrance), the amount of adsorbed effective material is less than a monolayer. Therefore this technique is specifically well suited for the growth of very thin films of a few nanometers.

The described process scheme is generally valid for the so-called homogeneous growth regime considering a layer-by-layer growth of a material on itself. Starting the growth of a material on a different substrate requires a nucleation process that may result in a layer-by-layer mode or in an island formation. The growth can be substrate enhanced or inhibited [12]. It might require up to several hundred ALD cycles to initiate a film growth. Based on the self-limiting adsorption mechanism, the ALD technique shows unique step coverage and a superior uniformity. An improvement of the step coverage is generally possible by extending the exposure time, allowing the precursor more time to find accessible unoccupied adsorption sites.

In relation to the TSV process, ALD could be used to grow the insulator film. However, to avoid parasitic capacitances, the insulator film should be thick, which would take a long time using ALD. Instead, as explained before, thermal oxidation is used of the insulator film. On the other hand, for the copper barrier ALD would be an excellent choice. It is also the adopted choice for the seed layer needed for copper electroplating [13, 14]. Since an ALD process for copper is not available, we have chosen ruthenium (Ru) as seed layer and tantalum nitride (TaN) as barrier material. The properties of these materials are generally known [15, 16]. The major limitation of ALD is its low deposition rate of less than a monolayer per cycle. In particular, the Ru process is hampered by an initial growth delay of about 15 cycles.

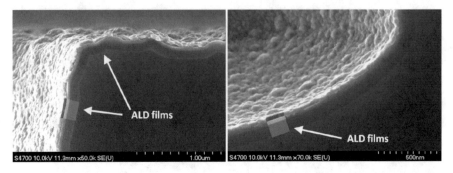

**Fig. 2.14** Examples for an insulator—diffusion barrier- and seed-layer film stack in a TSV with a 20:1 aspect ratio. ALD films *blue*: insulator 100 nm $Al_2O_3$; ALD films *green*: barrier 32 nm TaN; ALD films *red*: seed 33 nm Ru; (TSV-diameter = 5 μm depth 130 μm , details: sidewall top, sidewall bottom) [19]

In this case the nucleation of the seed layer takes place by island formation. Both aspects hamper the formation of closed ultra thin layers.

In the following we present an ALD film stack for a high aspect ratio copper-based TSV. The state of the art copper diffusion barrier in current interconnect technologies is based on a tantalum nitride-based copper diffusion barrier, which is typically produced by PVD [17, 18]. Wojcik et al. showed that ALD films are less conducting than conventional PVD Ta-N films, but they have similar copper diffusion barrier properties [19]. The TaN (or TaC) ALD films are deposited in a showerhead ALD chamber using (*tert*-butylimido)tris(diethylamido)—TBTDET as a precursor. An additional post deposition anneal significantly improves the structural and electrical properties of the TaN-based ALD films [20, 21]. The Ru-ALD films are deposited in a showerhead ALD chamber using (ethylcyclopentadienyl)(pyrrolyl)ruthenium(II)—ECPR as a precursor and molecular oxygen as was presented for the first time in Junige et al. [22] .

For layers down to 33 nm (Ru) and 32 nm (TaN) on 100 nm $Al_2O_3$ insulator ALD films, conformity and continuity have been proven by Knaut et al., using scanning electron microscopy (Fig. 2.14) [19]. The investigations on high aspect ratios (AR>10:1) have shown that a deposition of only 5–10 nm TaN barrier layer and a 5–10 nm Ru seed layer is sufficient for subsequent ECD of copper. In this case, 900 nm $SiO_2$ is used as insulator.

## 2.4 Electrochemical Deposition

Similar to damascene technology, the ECD process in TSV technology uses sulfuric acid–copper sulphate electrolytes with a small amount of chloride anions (usually 30–70 mg/L) and a set of three organic additives: accelerator, suppressor, and leveler. Because of the huge difference in the dimensions of structures to be filled (damascene technology on nm-level, TSV on $\mu$mlevel), the duration of the deposition process is significantly different as well (damascene technology takes seconds to minutes while TSV takes minutes to hours). As a consequence two main parameters in the chemical composition of TSV electrolytes were modified. Firstly, the used copper concentration is increased (often 40–60 g/L) to dampen the effects of copper ion depletion inside the TSV during the deposition process. Secondly, the additive organic set consists of other chemical compounds. Obviously, applied current densities are much lower in TSV technology (1–3 mA/cm$^2$) than in damascene technology (10–30 mA/cm$^2$). The latter is meant to prevent copper ion depletion within the structures when diffusion is the overriding transport process.

The current state of TSV technology is the formation of copper seed layers by plasma-enhanced physical vapor deposition (PE-PVD) followed by Cu-ECD which is realized up to an aspect ratio of 10 reliably. Here, we replaced PE-PVD of Cu by thermal ALD of Ru (see Sect. 2.3.2) aiming at filling TSVs with aspect ratios higher than 10.

In comparison with Cu, Ru as a noble metal shows higher resistivity to acid attack (corrosion) but has a higher specific resistance than CU (Ru: 7.43 $\mu\Omega$ cm [300 K]; Cu: 1.67 $\mu\Omega$ cm) [23]. As expected, the specific resistance of Ru-ALD layers was dependent on film thickness and significantly higher than Ru bulk value. On 10 and 20 nm thick layers (30 $\pm$ 3) $\mu\Omega$ cm, resp., (19 $\pm$ 3) $\mu\Omega$ cm were measured. Ru, in contrast to other noble metals, has high affinity to oxygen. Guo et al. reported a 1 nm Ru oxide film on PVD Ru [24]. In electrolytes, at potentials well below the formation of thick oxide layers, the oxidation of Ru is limited to the top layer of the film. At polycrystalline Ru, it was shown by in situ IR spectra measurements, that even at low potentials in acid solutions OH$^-$ ions of water cover the surface, preventing specific sorption of electrolyte anions over the whole potential range between hydrogen and oxygen evolution [25].

To prevent the negative influence of supposed superficial Ru-O species on Cu-ECD process (especially on Cu nucleation), a cathodic pretreatment in deaerated sulphuric acid followed by a rapid wet transfer to a Cu plating electrolyte was proposed in the literature (e.g., [26] and [27]). We have used a wafer with freshly deposited Ru layers.[2] According to our results under these conditions the Cu deposition performance and adhesion of Cu on Ru proved satisfactory.

Before starting the ECD we prewetted with 10 % Spherolyte$^{TM}$ (Atotech) in 10 % sulphuric acid for 10–20 s, rinsed shortly with deionized water, and then the TSVs were filled with basic electrolyte (i.e., without organic additives) by immersing the wafer under vacuum ($-600$ mbar). The removed wafer was rapidly transferred to a simple bench-top cell (on a magnetic stirrer) and contacted equidistant between two Cu anodes; the transfer took about 1.5–2 min until current flow was guaranteed. The electrolyte was 0.27 M Cu$^{2+}$, 1.73 M H$_2$SO$_4$, 1.45 mM Cl$^-$ with optimized concentrations of commercially available (Intervia$^{TM}$ Viafill Acid Copper, Rohm & Haas) leveler, brightener (i.e., accelerator; SPS), and carrier (i.e., suppressor) [19].

---

[2] Storage of such wafers can be for only a very short time under nitrogen atmosphere or vacuum.

**Fig. 2.15** Focused ion beam
(FIB) cut of Cu-enhanced
TSV: 3.6 μm Cu-ECD after
10 nm thermal Ru-ALD, 5 nm
thermal TaN-ALD seed layer,
and 900 nm $SiO_2$ insulator;
TSV geometry: diameter
20–200 μm depth (AR10:1)

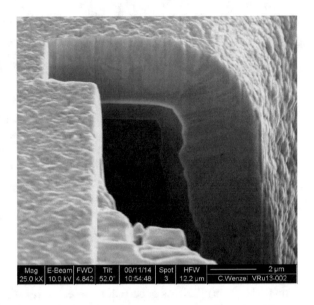

The obtained Cu thickness was about 2 μm (Figs. 2.15 and 2.16) deposited with current densities around 2.5 mA/cm$^2$ for 35–40 min; the ratio between thinnest film thickness in structure in reference to field region on wafer top (step coverage) was 70 %.

## 2.5 Interconnect and Bumping: Waferbumping for Cu and Al Metal Layer on Waferlevel

In this section, we present a process for the creation of interconnect and flip-chip bumps for an Si interposer on wafer scale by using ECD of tin-lead (Sn60Pb40—liquidus temperature 183 °C) or tin-silver (Sn97Ag3—liquidus temperature 221 °C) solder bumps [28]. The process steps are shown in Fig. 2.17. With this technique it is possible to create solder bumps on an interposer with Cu and Al metal pads and a passivation layer. For the ECD a conductive film using 50 nm wolfram titan (WTi) as adhesion layer and 150 nm copper as seed layer is deposited by PVD. A 20–30 μm thick resist layer serves a template for the pattern plating process and enables a uniform deposition of the solder bumps. Afterwards the under-bump metallization is created by ECD of 5 μm copper and 1 μm nickel layer. In the next step, depending on the pad size, an approximately 30 μm thick tin-lead or tin-silver is deposited by ECD. The WTi and Cu overall metallization is removed by a wet chemical etching after the stripping of the resist. For the bump formation the solder bump is reflowed in glycerin using 210 °C/20 s for SnPb and 235 °C/20 s for SnAg (Fig. 2.18).

**Fig. 2.16** Copper enhanced TSV: 2 μm Cu-ECD after barrier and seed layer 5 nm thermal TaN-ALD + 10 nm thermal Ru-ALD TSV geometry: diameter 20–200 μm depth (AR10:1); layer stack: 900 nm SiO$_2$; 5 nm TaN; 10 nm Ru; 2 μm Cu

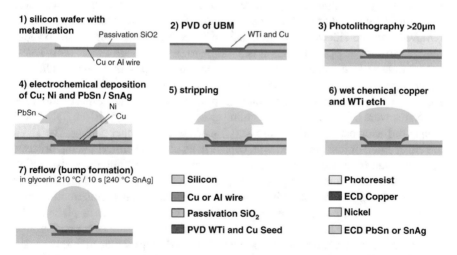

**Fig. 2.17** Wafer-level process for ECD solder bump deposition for interconnect and interposer

**Fig. 2.18** SnAg solder bump deposited by ECD. *Left*: before reflow. *Right*: after reflow. Reflow temperature: 235 °C/20 s in glycerin

## 2.6 Conclusion

This chapter has described a manufacturing process for the creation of an Si interposer using copper-based TSV for 3D integration and chip stacking. A silicon via etching process for aspect ratios from 10:1 to 20:1 using a sacrificial Al-stop layer in 200 μm thick wafers has been demonstrated. The TSVs are formed using deep reactive ion etching (DRIE). The polymeric sidewall passivation layers created during DRIE processing are stripped wet-chemically and the Al-stop layer is removed by an additional wet-etch process. Furthermore, residual polymer is removed and the sidewalls are smoothed by reactive ion etching in an oxygen-nitrogen trifluoride plasma. Thus it is possible to generate high aspect ratio TSVs, having little surface roughness, no notching, and adapted profiles with no additional wafer thinning. To prevent leakage between different interconnects, an insulating silicon dioxide layer is grown by thermal processing. The high aspect ratio TSV structures were coated with an ALD film stack consisting of a 5–10 nm TaN-based copper diffusion barrier and a 10 nm Ru(C) seed layer for copper plating. The TSVs are either filled or just enhanced with copper in an electrochemical plating process. An additional front and back side lithography is used to generate the metallization layer mask for the subsequent patterning processes. The copper layers on both sides of the interposer are structured in a wet etching process and the redistribution lines on the front and the back side of the wafer are created by pattern plating. The seed and the barrier layers are removed afterwards in a plasma etch process. In a waferbumping process, the interconnect and flip-chip bumps for the Si interposer are created on wafer scale using ECD of tin-lead[3] or tin-silver[4] solder. In the subsequent chapters of Part I of the book, it is this interposer TSV process that is used in all

---

[3] Sn60Pb40—liquidus temperature 183 °C.

[4] Sn97Ag3—liquidus temperature 221 °C.

the circuits, systems, and CAD technology demonstrators (Chaps. 3–6, and 9). The electrical characterization of the TSV, including pads and bumps, is presented in the next chapter.

**Acknowledgements** The authors would like to acknowledge A. Hiess, W. Haas, M. Junige, U. Merkel, K. Richter, A. Jahn, S. Waurenschk, F. Winkler, C. Wenzel from the Institute of Semiconductors and Microsystems, TU Dresden, for their assistance and hard work. Without their research, this project would have been impossible.

# References

1. J.W. Bartha, J. Greschner, M. Puech, Ph. Maquin, Low temperature etching of Si in high density plasma using $SF_6/O_2$. Microelectron. Eng. **27**(1), 453–456 (1995)
2. F. Lärmer, A. Schilp, German patent no. DE-4241045, 26.5.1994
3. S.L. Lai, D. Johnson, R. Westerman, Aspect ratio dependent etching lag reduction in deep silicon etch processes. J. Vac. Sci. Technol. A **24**(4), 1283–1288 (2006)
4. G.S. Hwang, K.P. Giapis, On the origin of the notching effect during etching in uniform high density plasmas. J. Vac. Sci. Technol. B **15**(1), 70–87 (1997)
5. C.-H. Kim, Y.-K. Kim, Prevention method of a notching caused by surface charging in silicon reactive ion etching. J. Micromech. Microeng. **15**(2), 358 (2005)
6. N. Lietaer, P. Storås, L. Breivik, S. Moe, Development of cost-effective high-density through-wafer interconnects for 3D microsystems. J. Micromech. Microeng. **16**(6), S29 (2006)
7. K. Buchanan, The evolution of interconnect technology for silicon integrated circuitry. Proc. of International Conference on Compound Semiconductor Manufacturing - GaAsMAN-TECH **44**, 1–3 (2002)
8. F.E. Rasmussen, J. Frech, M. Heschel, O. Hansen, Fabrication of high aspect ratio through-wafer vias in CMOS wafers for 3-D packaging applications, in *Proceedings of the 12th International Conference on Solid-State Sensors, Actuators and Microsystems (Transducers)*, vol. 2 (IEEE, Piscataway, New Jersey, 2003), pp. 1659–1662
9. D. Archard, K. Giles, A. Price, S. Burgess, K. Buchanan, Low temperature PECVD of dielectric films for TSV applications, in *Electronic Components and Technology Conference (ECTC)* (2010), pp. 764–768. ISBN 3-932434-75-7
10. K. Powell, S. Burgess, T. Wilby, R. Hyndman, J. Callahan, 3D IC process integration challenges and solutions, in *International Interconnect Technology Conference (IITC)* (IEEE, Piscataway, New Jersey, 2008), pp. 40–42
11. M. Ritala, M. Leskelä, H.S. Nalwa, *Handbook of Thin Film Materials*, vol. 1, Chap. 2 (Academic, San Diego, 2001), p. 103
12. S.K. Kim, J.H. Han, G.H. Kim, C.S. Hwang, Investigation on the growth initiation of Ru thin films by atomic layer deposition. Chem. Mater. **22**(9), 2850–2856 (2010)
13. M. Leskelä, M. Ritala, Atomic layer deposition (ALD): from precursors to thin film structures. Thin Solid Films **409**(1), 138–146 (2002)
14. S.M. George, Atomic layer deposition: an overview. Chem. Rev. **110**(1), 111–131 (2009)
15. D.K. Schroder, *Semiconductor Material and Device Characterization* (Wiley, New York, 2006)
16. H. Wojcik, R. Kaltofen, U. Merkel, C. Krien, S. Strehle, J. Gluch, M. Knaut, C. Wenzel, A. Preusse, J.W. Bartha et al., Electrical evaluation of Ru–W (-N), Ru–Ta (-N) and Ru–Mn films as Cu diffusion barriers. Microelectron. Eng. **92**, 71–75 (2012)
17. D. Edelstein, C. Uzoh, C. Cabral Jr., P. DeHaven, P. Buchwalter, A. Simon, E. Cooney, S. Malhotra, D. Klaus, H. Rathore et al., A high performance liner for copper damascene interconnects, in *Proceedings of the IEEE 2001 International Interconnect Technology Conference* (IEEE, Piscataway, New Jersey, 2001), pp. 9–11

18. S.M. Aouadi, Y. Zhang, P. Basnyat, S. Stadler, P. Filip, M. Williams, J.N. Hilfiker, N. Singh, J.A. Woollam, Physical and chemical properties of sputter-deposited $TaC_xN_y$ films. J. Phys. Condens. Matter **18**(6), 1977 (2006)
19. M. Knaut, M. Junige, V. Neumann, H. Wojcik, T. Henke, C. Hossbach, A. Hiess, M. Albert, J.W Bartha, Atomic layer deposition for high aspect ratio through silicon vias. Microelectron. Eng. **107**, 80–83 (2013)
20. D. Schmidt, M. Knaut, C. Hossbach, M. Albert, C. Dussarrat, B. Hintze, J.W. Bartha, Atomic layer deposition of Ta–N-based thin films using a tantalum source. J. Electrochem. Soc. **157**(6), H638–H642 (2010)
21. C. Hossbach, S. Teichert, J. Thomas, L. Wilde, H. Wojcik, D. Schmidt, B. Adolphi, M. Bertram, U. Mühle, M. Albert et al., Properties of plasma-enhanced atomic layer deposition-grown tantalum carbonitride thin films. J. Electrochem. Soc. **156**(11), H852–H859 (2009)
22. M. Knaut, M. Junige, M. Albert, J.W. Bartha, In-situ real-time ellipsometric investigations during the atomic layer deposition of ruthenium: a process development from [(ethylcyclopentadienyl)(pyrrolyl) ruthenium] and molecular oxygen. J. Vac. Sci. Technol. A **30**(1), 01A151 (2012)
23. M. Beblo, A. Berktold, U. Bleil, H. Gebrande, B. Grauert, U. Haack, V. Haack, H. Kern, H. Miller, N. Petersen et al., Landolt-Börnstein: numerical data and functional relationships in science and technology-new series, in *Landolt-Bornstein: Group 6: Astronomy*, vol. 1 (Springer, Berlin, 1982)
24. L. Guo, A. Radisic, P.C. Searson, Electrodeposition of copper on oxidized ruthenium. J. Electrochem. Soc. **153**(12), C840–C847 (2006)
25. N.S. Marinkovic, M.B. Vukmirovic, R.R. Adzic, Some recent studies in ruthenium electrochemistry and electrocatalysis, in *Modern Aspects of Electrochemistry* (Springer, New York, 2008), pp. 1–52
26. T.P. Moffat, M. Walker, P.J. Chen, J.E. Bonevich, W.F. Egelhoff, L. Richter, C. Witt, T. Aaltonen, M. Ritala, M. Leskelä et al., Electrodeposition of Cu on Ru barrier layers for damascene processing. J. Electrochem. Soc. **153**(1), C37–C50 (2006)
27. P.P. Sharma, I.I. Suni, Impedance studies of Ru oxide reduction in sulfuric acid. J. Electrochem. Soc. **158**(2), H111–H114 (2011)
28. Ch. Wenzel, K. Drescher, Applikation flip-chip-bumping, in *Interdisziplinäre Methoden in der Aufbau- und Verbindungstechnik*, ed. by K.-J. Wolter, S. Wiese (Ddp Goldenbogen, Dresden, Germany, 2003), pp. 157–170. ISBN 3-932434-75-7

# Chapter 3
# Energy Efficient Electrical Intra-Chip-Stack Communication

Johannes Görner, Dennis Walter, Michael Haas, and Sebastian Höppner

## 3.1 Introduction

Two developments in microelectronic systems can be observed over the last decades. The first development is that the complexity for minimum component cost in the digital domain increases at an exponential rate (Moore's Law), resulting in highly integrated system-on-chip (SoC). The second development is the increase in diversity in non-digital content (RF, MEMS, mixed signal) which can be integrated with CMOS-based speciality technologies (More than Moore), resulting in highly integrated heterogeneous systems-in-package (SiP).

The construction of chip stacks is a 3D integration technology, which enables more compact and broadband intraconnect networks, leads to more miniaturization and even more heterogeneous integration capabilities. The International Technology Roadmap for Semiconductors (ITRS) 2011 [18] defines a "3D-integration tech-

---

J. Görner (✉)
Technische Universität Dresden, Chair of Highly-Parallel VLSI-Systems and
Neuro-Microelectronics, Dresden, Germany
e-mail: Johannes.Goerner@tu-dresden.de

D. Walter
Technische Universität Dresden, Chair of Highly-Parallel VLSI-Systems and
Neuro-Microelectronics, Dresden, Germany
e-mail: Dennis.Walter@tu-dresden.de

M. Haas
Technische Universität Dresden, Chair for RF Engineering, Dresden, Germany
e-mail: Michael.Haas@tu-dresden.de

S. Höppner
Technische Universität Dresden, Chair of Highly-Parallel VLSI-Systems and
Neuro-Microelectronics, Dresden, Germany
e-mail: Sebastian.Hoeppner@tu-dresden.de

© Springer International Publishing Switzerland 2016
I.M. Elfadel, G. Fettweis (eds.), *3D Stacked Chips*,
DOI 10.1007/978-3-319-20481-9_3

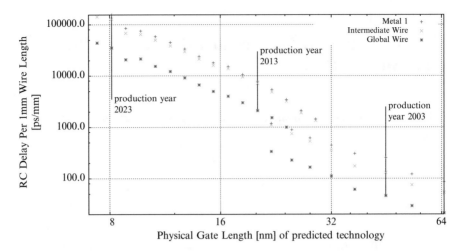

**Fig. 3.1** RC delay in (ps) for a 1mm wire on different interconnect-levels for different predicted Interconnect Technology Requirements based on data from the International Technology Roadmap for Semiconductors [14–19] for MPU/ASIC processes

nology" for interconnects as a "via" technology that allows the stacking of basic circuit components in the third dimension, not only interconnect planes. The ITRS Roadmap categorizes 3D integration technologies based on the interconnect-level as follows:

**3D-Packaging**   Interconnect-Level: Package
Traditional packaging technology is used for interconnects, e.g., wire-bonded die stacks, package-on-package stacks, no through-silicon vias (TSVs)

**3D-Wafer-Level-Package**   Interconnect-Level: Bond-pad
TSVs are manufactured after the IC fabrication (via last process), interconnects on bond-pad level

**3D-System-on-Chip(SoC)**   Interconnect-Level: Global
Large circuit blocks (components:IP-blocks) are stacked over different physical layers, unbuffered I/O drivers, little or small ESD protection, TSV pitch requirement 4–16 μm

**3D-Stacked-Integrated-Circuits (SIC)**   Interconnect-Level: Intermediate
Stacking of smaller circuit blocks, IP blocks are partially distributed over different layers, wafer-to-wafer stacking, TSV pitch: 1–4 μm

**3D-IC**   Interconnect-Level: Local
Stacking of multiple front-end device layers with a common back end metal interconnect layer stack, TSV density on local interconnect-level

The integration of heterogeneous chips from different technologies into a chip stack would enable the partitioning of analog and digital components on different chip layers and different technologies. Therefore, digital circuits can be implemented in leading edge CMOS technologies, because they benefit the most from technology scaling in terms of integration density. On the other side, analog circuits do not scale as well as digital circuits and it is hard and time-consuming to implement analog functionality into leading edge CMOS technologies with a low power supply voltage. They could be realized in a different technology optimized for these kinds of circuits.

With the miniaturization of CMOS technologies, the metal cross section and the wire pitch gets smaller. This leads to higher serial resistance, longer global interconnect delays, and higher power consumption due to more parasitic capacitances. With technology scaling, a trade-off between serial resistance and coupling capacitance has to be done. In Fig. 3.1 the predicted RC-delay for a 1 mm wire is shown for different predicted technologies and for local, intermediate, and global metal wires.

Through 3D chip stack integration, global interconnects get shorter with additional vertical interconnects, promising smaller global interconnect delays and less parasitic capacitances. In multi-processor SoC (MPSoC) designs several cores and components have to communicate with each other. The state-of-the-art approach for an optimized use of interconnect resources is to use a Network-on-chip (NoC), which can be extended to 3D chip stacks.

Many traditional 2D planar designs are pin limited and are either wire bond or flip chip designs. They have to maintain a minimum pad size and pitch and therefore have a limited I/O capacity, which is usually a bottleneck. Smaller dimensions, smaller pitches, and an increased pin count will result in a boost of I/O bandwidth between different dies in a chip stack.

The stacking of chips leads to a higher count of active elements per area. An increased number of active elements will result in increased power consumption, which leads to more heat generation. By splitting one active layer of a 2D chip design into several stacked active layers of a 3D chip stack with only a fraction of the original footprint, the heat dissipation capability is reduced dramatically. Therefore, energy efficient design and power management become even more important compared to 2D SoC designs.

A TSV is mainly a low ohmic interconnect having a capacitance against its surrounding substrate. Driving a TSV leads to charging and discharging of this parasitic capacitance, which results in power dissipation. Due to the possible high number of TSVs used for signaling and the requirement of power efficiency to limit heat dissipation, the design of power-efficient transceivers for TSV signal transmission is very much worth considering. Such power-efficient transceivers are described in Sect. 3.4, while the physical effects, electromagnetic simulations, and behavioral simulations are explained in Sect. 3.3.

The basic architecture of such a point-to-point TSV link is similar to the one shown in Sect. 3.2, which describes a NoC link solution for long distance interconnects, which may be part of a planar chip or a 3D chip stack. By employing a common interface, the network does not need to be aware of the physical implementation details.

## 3.2    Network-on-Chip Links

### 3.2.1    *Long Distance Interconnects for Network-on-Chip Designs*

The increasing complexity of modern multi-processor systems-on-chip (MPSoC) designs leads to new challenges for on-chip communication. It has been shown that following the NoC approach is a promising solution for ever increasing bandwidth and energy-efficiency requirements [1]. In heterogeneous systems, the interconnect length between different cores and routers can be in the range from a few μm to a few mm and thereby show significant differences in delay and power. In addition, huge MPSoC systems are often designed as globally asynchronous, locally synchronous (GALS) [27, 31] which eases dynamic voltage and frequency scaling (DVFS) [9] but poses further challenges for interconnect design as no high-speed synchronous clock signal is available.

From a NoC perspective, all these interconnects are just point-to-point connections. If these links are properly black-boxed, the surrounding logic does not need to be aware of their physical implementation and just sees a packet-in, packet-out interface. While continued technology scaling enables higher integration densities, on-chip wires do not scale as fast, and this becomes a bottleneck for system performance and energy efficiency. These scaling trends justify putting increased effort in specialized driver and receiver circuits for these long range interconnects. High-speed serial point-to-point interconnects offer several advantages over conventional full-swing buffered CMOS signaling:

- High bandwidth point-to-point connection without the need for buffering and thereby no dependency on active circuitry, different voltage islands, or power gated regions in between.
- Computing cores and memory macros can be completely bridged on the top metal layers, and there is no need for feedthrough-channels or detours.
- High energy efficiency by using low-swing signaling.
- High flexibility by packet-in, packet-out approach and black-boxing physical implementation. Furthermore, changes can be made late in the implementation phase by replacing high-speed serial links by parallel ones or vice versa.

### 3.2.2    *System Level Integration*

For easy system level integration it is mandatory to be able to switch between different physical realization schemes, i.e., full-swing buffered CMOS vs high-speed low-swing serial, without changing the NoC topology. Therefore, a packet-in, packet-out interface is defined which is identical for all physical implementation schemes. Each interface is clocked by the local clock signal and separated from each other by internal FIFOs, which are used for data synchronization between the

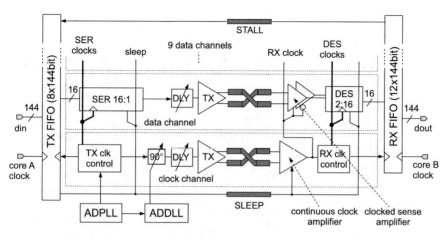

**Fig. 3.2** NoC link architecture showing packet-in, packet-out approach with integrated serializer (SER) and deserializer (DES) circuits. Besides an on-chip All-Digital Phase-Locked-Loop (ADPLL) for clock generation, an All-Digital Delay-Locked-Loop (ADDLL) is needed for 90° clock phase shifting

different clock domains. An overview of the serial link architecture and its clocking is given in Fig. 3.2. It is shown that the clock domain of the high-speed link is completely independent from the surrounding logic and all clocks can be scaled without interdependencies [30]. The link consists of one clock lane and several parallel data lanes, the number of these data lanes can be adjusted according to the NoC packet size. The packets are pushed into the FIFO memory at transmitter side from where they are fetched by the link logic, transmitted and finally written to a FIFO again at receiver side. For correct double-data-rate (DDR) operation of the high-speed link an on-chip delay-locked-loop (DLL) is needed which enables shifting the clock signal by 90° relative to the data signals. The link is fully source-synchronous, for each data bit a corresponding clock edge is transmitted in parallel. This leads to high jitter tolerance and enables a per-packet clock gating so that no unnecessary line driver and receiver power is wasted for clock toggles during idle times. Both, data and clock, are transmitted as low-swing signals. These low-swing signals are amplified again to full-swing at receiver, the clock signal with a time continuous amplifier and the data signals by clocked sense amplifiers. The link includes per data channel 16-to-1 SERDES circuits and corresponding clock control logic.

For improved energy efficiency, a full-swing low-speed SLEEP signal is transmitted to the receiver for power-down mode activation as from a system level perspective, link utilization is often quite low and idle times dominate total link usage. In the opposite direction, a STALL signal is used to stop data transmission if the receiver FIFO is almost full and thereby prevents data loss.

One advantage of uninterrupted point-to-point links is the bridging of predefined macro blocks like computing cores or memories. This active circuitry can be bridged

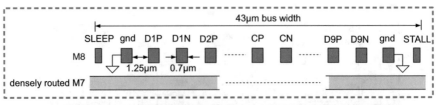

**Fig. 3.3** Physical dimensions of the transmission lines for a 6 mm uninterrupted connection with 1 clock and 9 data channels, ground shields used only between the transmission lines and the full-swing SLEEP and STALL signals. The 3D plot shows the NoC link physical routing on upper metal layers within the power mesh

on the upper metal layers as illustrated in Fig. 3.3, where the physical dimensions of the transmission lines are given. The NoC link with its clock and data lanes is embedded in the power mesh. Due to the perpendicular routing, crosstalk from neighboring layers is not an issue for the differential signal line levels as it just translates into common mode noise. The metal lines are sized for highest throughput per area while still maintaining good RC delay (bandwidth) of the transmission lines.

### 3.2.3   Transceiver Architecture

It has been shown that low-swing differential interconnects provide high bandwidth per area and uninterrupted transfer of lengths up to 10 mm [13, 21, 26, 35]. Capacitively driven links are especially promising because of their built-in pre-emphasis, thereby mitigating the low-pass behavior of RC limited on-chip interconnect [7, 26, 28].

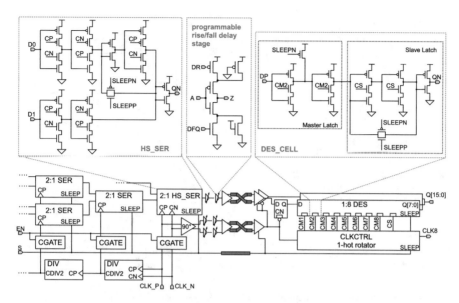

**Fig. 3.4** NoC transceiver SERDES clocking architecture and calibration circuits

In Fig. 3.4 the detailed clock and SERDES architecture is shown. At the transmitter side, 16-to-1 serialization is performed by a 2-to-1 multiplexer tree, where at each stage the clock frequency is doubled. The clock signal is shifted by 90° for DDR sampling at transmitter side and all parallel transmission lines can be individually adjusted in their relative delay by programmable rise/fall delay cells to compensate for mismatch effects. At the receiver, the DDR data signals are converted to SDR signals by the two parallel clocked sense amplifiers, active on opposite clock edges. Data is further deserialized by two 8 bit parallel master/slave latches where the slave latch is enabled after 16 bits of data per channel have been received. For best energy efficiency, the SERDES circuits are implemented as dynamic latches which can be made static during sleep mode to prevent unnecessary switching activity or high short circuit currents due to floating nodes.

The proposed driver circuit [30] employs large coupling capacitances which define the signal swing on the differential lines. It is shown in Fig. 3.5. By choosing the ratio between the mainly parasitic line capacitance and the driver capacitance the voltage swing is defined, which is typically in the range of 80–200 mV. As the capacitive coupling only transfers AC signals on the transmission lines, additional circuitry is needed for well-defined absolute voltages and steady state behavior, i.e., time intervals with no toggling activity. Therefore, a resistive driver built from PMOS transistors is employed in parallel to the coupling capacitances. This weak high resistive driver circuit pulls the transmission lines to either $V_{DD}$ or $V_{SSnoc}$ and thereby defines the differential DC level. The voltage difference $V_{DD} - V_{SSnoc}$ should match the signal swing defined by the capacitance ratio of line capacitance and coupling capacitance. The additional 4 PMOS transistors between the two

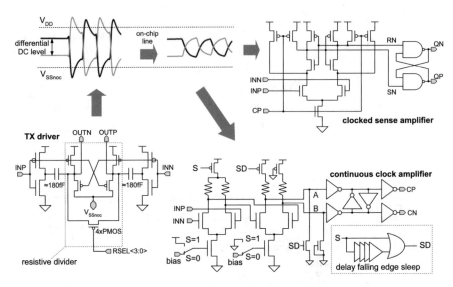

**Fig. 3.5** NoC transceiver with capacitive line driver, low-swing differential signaling on transmission lines, time continuous clock amplifier, and clocked data sense amplifier

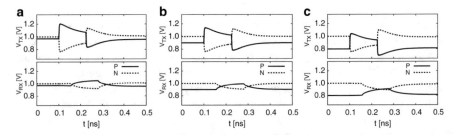

**Fig. 3.6** Voltage response of transmission line at transmitter (*upper*) and receiver (*lower*) side for different DC swing voltages, demonstrating importance of proper DC level calibration. (**a**) DC swing too low. (**b**) DC swing optimum. (**c**) DC swing too high

differential transmission lines can be used to form a resistive divider circuit with the weak high resistive driver. If this divider is switched on, the DC level can be adjusted by selectively switching the 4 PMOS transistors on or off via RSEL individually per link, which allows for a single $V_{SSnoc}$ to be used for multiple different links with different AC swings.

The importance of proper DC level definition is demonstrated in Fig. 3.6. If the DC swing is either too low or too high, the data eye closes at receiver side and cannot be properly detected. Only an optimum calibration ensures equal eye width and height for both polarities. The voltage overshoot at transmitter side due to the capacitive coupling helps mitigate the low-pass behavior of the transmission lines and acts as a built-in pre-emphasis.

At the receiver side the clock signal is amplified by the time continuous amplifier shown in Fig. 3.5, which can be completely shut-off during sleep mode. To prevent functional failures during sleep activation or returning to active mode the integrated sleep scheme ensures that no false toggles are propagated to the full-swing clock outputs. Data signals are sampled by the shown clocked sense amplifier at the receiver side.

### 3.2.4  Calibration and Test

As shown in Fig. 3.4 the serial link includes programmable delay cells for each channel which allow for individually adjusted rise and fall delays to compensate for process mismatches. While these cells can be adjusted in the picosecond range, choosing the best delay configuration for most reliable data transmission is difficult because of the challenge to accurately measure these small delay differences.

While asynchronous sub-sampling is a feasible way for delay measurements with picosecond resolution and minimal required on-chip hardware overhead, it requires additional external components and test time [8, 10]. A complete self-contained on-chip implementation is difficult because of the need for an additional asynchronous sample clock and the huge amounts of sampling data that needs to be stored for statistical analysis.

For a realistic implementation scenario in a huge MPSoC system, there needs to be an automatic built-in self-test (BIST) and built-in self-calibration (BISC). The main purpose of this logic is to ease hardware test by offering a simple digital test interface which just delivers the final results of automatic test and calibration. A possible realization of such a BIST/BISC controller is presented in detail in [12], the basic calibration algorithm is shown in Fig. 3.7. The calibration is based on measuring the bit error rate (BER) at different data and clock rise/fall delay settings $(d_{\text{rise}}, d_{\text{fall}}, d_{clk,rise}, d_{clk,fall})$ and determine the settings with lowest BER. Besides the

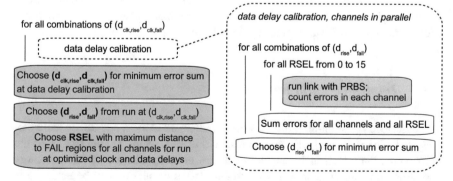

**Fig. 3.7** Simplified test and calibration algorithm for automatic on-chip BIST/BISC of clock and data channels

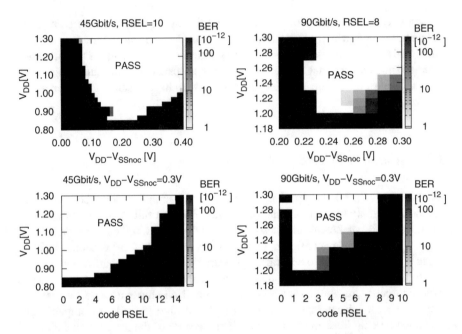

**Fig. 3.8** Measured BER plots from a 65 nm implementation at 45 and 90 Gb/s link speeds by sweeping $V_{SSnoc}$ or RSEL

per channel delay settings, also the signal swing on the transmission lines needs to be calibrated, which is achieved by sweeping the digital control value RSEL for the resistive divider within these measurements to determine the best overall combination of delay and swing adjustment. Hardware requirements are quite low as only some control logic including a bit error counter, a small per data channel memory and a pseudo-random bit stream generator (PRBS) are needed.

By employing such a calibration scheme, not only can the basic functionality be tested but also the robustness against temperature and voltage variations can be improved. Because of the optimal calibration values, data transfer includes more safety margin which can be used for better robustness, higher speed, or improved energy efficiency by lowering the supply voltage.

### 3.2.5 Implementation Results

The described serial on-chip link has been implemented in 65 and 28 nm technologies [12]. The measured BER plots presented in Fig. 3.8 are results from a 65 nm implementation which includes additional test circuitry like a loop-back memory and BER counters. It is shown that reliable data transmission can be achieved at different speeds by selecting a proper DC level. This can be configured by either

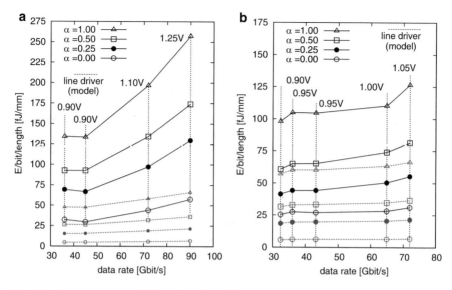

**Fig. 3.9** Measured energy efficiency for 65 and 28nm implementations at different link speeds, supply voltages, and activity factors $\alpha$, *black lines* indicating total link efficiency, *blue lines* indicating the energy ratio for the line driver only. (**a**) 65 nm 6 mm link. (**b**) 28 nm 6 mm link

selecting a matching $V_{DD} - V_{SSnoc}$ voltage directly by adjusting $V_{SSnoc}$ or by using the resistive divider and selecting a proper RSEL value. The BER plots also indirectly show that the AC swing depends on the supply voltage $V_{DD}$ as this defines the amount of charge transfered by the capacitive driver.

Measured energy efficiency for 6 mm links in 65 and 28 nm is plotted in Fig. 3.9. It can be seen that by reducing link speed and supply voltage energy efficiency is improved. While the 65 nm implementation shows huge overhead for FIFOs, SERDES, and clocking, which can be estimated by the difference between the corresponding black (total energy efficiency) and blue (line driver only) data points, for 28 nm this overhead is significantly reduced. This is the main contributor for the improved energy efficiency of the 28 nm implementation and shows that technology scaling helps in reducing the overhead for on-chip SERDES links. The data also supports the claim that on-chip wires do not scale well with technology. For comparable bandwidth, the wire dimensions of the 28 nm version are roughly identical to the 65 nm one and show equivalent resistance and capacitance. Therefore, link speed and driver power do not show any improvements.

Overall, serial high-speed on-chip links are a viable solution for high bandwidth interconnect requirements in modern MPSoC systems. With continued technology scaling, the overhead for SERDES circuits compared to total link power is reduced while bandwidth and power of transmitting data over on-chip wires do not scale. By forwarding the clock source-synchronous in parallel to the data signals, the link is compatible with GALS systems and offers a drop-in replacement for long parallel buffered on-chip busses.

## 3.3 TSV Channel

### 3.3.1 TSV Behavior Description

For an accurate modeling of TSV interconnects in 3D chip stacks as well as for high frequencies of 100 GHz several modes which are supported by the interconnect and TSV effects must be considered. In this section, an overview of the supported modes should be given and several TSV effects like the depletion effect, the proximity effect, the skin effect, and the influence of a substrate contact on the transmission behavior shall be explained and classified.

- Modes
  TSVs are basically metal-insulator-semiconductor (MIS) structures which support slow-wave, dielectric quasi-TEM, and skin effect modes. The transition frequency, $f_e$, from slow-wave to dielectric quasi-TEM mode depends only on the material parameters $\sigma_{Si}$ and $\epsilon_{Si}$ of the silicon substrate [4] and is given by

$$f_e = \frac{\sigma_{Si}}{2\pi\epsilon_{Si}}. \tag{3.1}$$

  In contrast the transition frequency, $f_\sigma$, from the dielectric quasi-TEM mode to the skin effect mode depends on the material parameters $\sigma_{Si}$ and $\mu$ of the silicon substrate and the TSV geometry (the thickness of the substrate layer between the TSVs) [4]

$$f_\sigma = \frac{1}{\left(p - 2(r_{via} + t_{ox} + t_{dep})\right)^2 \pi\mu\sigma_{Si}}. \tag{3.2}$$

  where $p$ is the pitch of the TSV array. The silicon considered in this work has a conductance $\sigma_{Si}$ of approximately 10 S/m. Therefore, the transition frequency $f_e$ for a $\epsilon_{Si}$ of 11.9 is about 15.1 GHz and must be considered while $f_\sigma$ for a reasonable distance between neighboring TSVs below 200 μm is higher than 700 GHz and can therefore be neglected.
- Depletion effect
  The TSV forms a cylindrical metal-oxide-semiconductor (MOS) capacitor whose capacitance depends on the bias voltage of the TSV. The potential difference of the TSV and the silicon substrate leads to a depletion region around the TSV oxide. This region can be modeled as an additional insulator layer with the relative permittivity of silicon due to the lag of mobile charge carriers. Therefore, the sidewall capacitance of the TSV is reduced (series circuit of two capacitances). The relative change of the sidewall capacitance depends on the thickness of the oxide layer. A thinner oxide layer causes a higher oxide layer capacitance which is in series with a much wider depletion

layer (i.e., smaller depletion layer capacitance) leading to a strongly reduced side wall capacitance. For the maximum depletion width, the planar MOS capacitor formula can be used [4]:

$$t_{\text{dep}} = \sqrt{\frac{4\epsilon_0\epsilon_{r,\text{Si}}kT}{q^2 N_a} \ln\left(\frac{N_a}{n_i}\right)}.$$  (3.3)

With a Boltzmann constant $k = 1.38 \cdot 10^{-23}$ J/K, a room temperature $T = 300$ K, an electron charge $q = 1.602 \cdot 10^{-19}$ C, an intrinsic semiconductor concentration $n_i = 1.08 \cdot 10^{10}$ cm$^{-3}$, and a doped acceptor concentration $N_a = 1.25 \cdot 10^{15}$ cm$^{-3}$, this leads to a maximal depletion width of approximately $0.8\,\mu$m.

The influence of the voltage-dependent capacitance can be strongly reduced by choosing oxide thicknesses higher than the depletion width which is especially advisable for the design of capacitive driver circuits.

- Skin effect

  The skin effect describes the distribution of AC currents near the surface of the conductor due to eddy currents induced by changing magnetic fields. The reduced area (the skin depth) in which the current flows increases the resistance of the conductor. The skin depth can be calculated using

$$\delta = \frac{1}{\sqrt{\pi f \mu_0 \mu_r \sigma}}$$  (3.4)

  where $\sigma$ is the electrical conductivity of the conductor. In copper which is often used as a metal for TSVs the skin depth is $2\,\mu$m at 1 GHz, $0.66\,\mu$m at 10 GHz, and $0.2\,\mu$m at 100 GHz with a conductivity of $58 \cdot 10^6$ S/m. Due to the very low DC conductance of a TSV, the skin effect will influence the transmission characteristics of the TSV at frequencies higher than 20 GHz.

- Proximity effect

  If the TSV-to-TSV distance is small in relation to the TSV radius the AC current is crowded to a smaller region of the conductor so that the resistance increases. Dependent on the direction of the current flowing through the conductor the current is crowded to the outer (away from the other conductor) or in the inner half (facing to the other conductor) due to the interplay of the magnetic flux of conductors which are close to each other. Like the skin effect the increase in resistance scales with the frequency. The proximity effect can be neglected for TSV configurations where the TSV pitch is equal or higher than six times the TSV radius. This means that for a TSV radius of $10\,\mu$m the pitch should be equal or higher than $60\,\mu$m. In this work this is always satisfied, so the proximity effect can be neglected.

- Substrate contacts

  In a chip, the substrate is normally electrically contacted to ground to have a fixed substrate potential. This can be modeled by an additional admittance from the TSVs sidewall capacitance to the ground contact which results in an increased transmission line capacitance in comparison, for example, to a ground signal TSV configuration without ground contact, where the sidewall capacitance is halved due to the two TSV sidewall capacitances in series. With the ground contact, only the sidewall capacitance of the signal TSV defines the capacitance of the transmission line. Additionally, the ground contact reduces the influence of coupling between TSVs.

### 3.3.2   TSV Modeling

For RF probing, ground signal (GS) and ground signal ground (GSG) RF probes are normally available up to several 100 GHz. GSG probes generally have a higher bandwidth than GS probes. Therefore, GSG probes were chosen for all measurements. Accurate TSV models, based on the equivalent circuit shown in Fig. 3.10, are needed to compare the model predictions with measured data. In the following a model for a GSG configuration from the literature is introduced. Then the modeling of the depletion effect to cover the nonlinear behavior of the TSVs, the modeling of the substrate contact, and a matrix method to determine the coupling admittance between TSVs.

- GSG model
  From Lu [25] a signal TSV surrounded by multiple ground TSVs model was developed based on a GS configuration. The silicon conductance and silicon capacitance of the GS model is scaled by a scaling factor $M$ and the inductance by a factor $1/M$. $M$ depends on the ratio of pitch to TSV diameter and the number of ground TSVs. The factor $M$ was determined using Ansys Q3D simulations. In a later publication from Lu [24] an analytical expression was derived for the GSG configuration. The GS and GSG models were verified with a double sided probing station performing measurements up to 40 GHz. The values from the TSV model can be determined using the following equations:

$$L = \frac{\mu h_{\text{via}}}{\pi} \cosh^{-1}\left(\frac{p}{2r_{\text{via}}}\right) \frac{1}{M} \tag{3.5}$$

$$C_{\text{Si}} = \frac{\pi \epsilon_0 \epsilon_{r,\text{Si}}}{\cosh^{-1}\left(\frac{p}{2r_{\text{via}}}\right)} M \tag{3.6}$$

**Fig. 3.10**  Equivalent circuit of a TSV

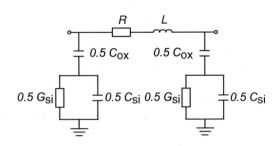

$$G_{Si} = \left( \frac{\sigma_{Si}}{\epsilon_0 \epsilon_{r,Si}} + \omega \tan \delta_{Si} \right) C_{Si} \qquad (3.7)$$

$$C_{ox} = \frac{\pi \epsilon_0 \epsilon_{r,ox} h_{via}}{\ln \left( 1 + \frac{r_{via} + t_{ox}}{r_{via}} \right)} \qquad (3.8)$$

$$R = \sqrt{R_{dc}^2 + R_{ac}^2} \qquad (3.9)$$

where $p$ is the TSV pitch, $R_{dc} = h/\left(\sigma \pi r_{via}^2\right)$ is the DC resistance, and $R_{ac} = h/\left(\sigma \pi \delta(d - \delta)\right)$ the AC resistance. The analytical expression for $M$ using a GSG configuration is

$$
\begin{aligned}
M_{GSG} &= C_{Si,GSG}/C_{Si,GS} \\
&= \frac{4\pi \epsilon_0 \epsilon_{r,Si} h_{via}/(3 \log(p/r_{via}) - \log(2 \sin(\theta/2)))}{\pi \epsilon_0 \epsilon_{r,Si} h_{via}/(\cosh^{-1}(p/(2r_{via})))}
\end{aligned}
\qquad (3.10)
$$

where $\theta$ equals 180° for a GSG configuration with all TSVs in a row.
• Depletion effect model
The modeling of the depletion effect is detailed in [20, 32]. In both, the cylindrical Poisson equation

$$\frac{1}{r} \frac{d}{dr} \left( \frac{d\psi}{dr} \right) = \frac{qN_a}{\epsilon_0 \epsilon_{Si}} \qquad (3.11)$$

is solved.
Three regions can be distinguished in the solution domain: the accumulation region ($V_{TSV} < V_{FB}$), the depletion region ($V_{FB} \leq V_{TSV} < V_{Th}$), and the minimum depletion region ($V_{TSV} > V_{Th}$). The flatband voltage $V_{FB}$ is defined as:

$$V_{FB} = \phi_{ms} - \frac{(r_{via} + t_{ox})qQ_{tot}}{\epsilon_0 \epsilon_{r,ox}} \ln \left( \frac{r_{via} + t_{ox}}{r_{via}} \right) \qquad (3.12)$$

where $Q_{tot}$ is the total oxide charge and $\phi_{ms}$ is the difference between two work functions. The first is from the TSV metal to the barrier metal (4.25 eV for Tantal), and the second is from the TSV metal to the doped substrate (approximately 4.89 eV). The amount of total oxide charge depends on the manufacturing process of the silicon dioxide. It is approximately 0 for thermally grown silicon dioxide [32]. The threshold voltage $V_{Th}$ can be calculated using:

$$V_{Th} = V_{FB} + 2V_t \ln \left( \frac{N_a}{n_i} \right) + \frac{qN_a \left( r_{max}^2 - (r_{via} + t_{ox})^2 \right)}{2\epsilon_0 \epsilon_{r,ox}} \ln \left( \frac{r_{via} + t_{ox}}{r_{via}} \right)$$

$$(3.13)$$

where $V_t = kT/q$ and $r_{max} = r_{via} + t_{ox} + t_{dep,max}$.

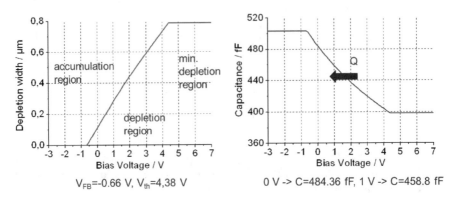

**Fig. 3.11** Calculated depletion width (*left*) and resulting TSV capacitance (*right*) for $r_{via}$ = 10 μm, $t_{ox}$ = 0.9 μm, Tantal as barrier layer and an oxide charge of 0 (thermally grown silicon dioxide)

The capacitance in the accumulation region is the oxide capacitance of the TSV. In the depletion region, it is the series combination of the oxide and the depletion capacitances, and so it depends on the actual depletion width. In the minimum depletion region, the capacitance is the series combination of the oxide capacitance and the minimum depletion capacitance (maximum depletion width). The depletion width in the depletion region can be calculated using the formulas in [20, 32] or by linear interpolation between the minimum and the maximum depletion width. The interpolation introduces small errors but eases the implementation effort drastically. Figure 3.11 shows the results using the formulas in [20, 32]. By increasing the oxide charges the curves will be shifted to the left (negative bias voltages). For 1 V, the sidewall capacitance is reduced from approximately 503 to 458.8 fF.

- Substrate contact model

  The substrate contact can be modeled as an additional admittance. An analytical expression was introduced in [33]

$$Y_{sub} = \left(\sigma_{Si} + j\omega\epsilon_0\epsilon_{r,Si}\right) \frac{4\pi h_{via}}{2\ln\left(h_{via}/r_0\right) - 2 + 3r_0/h_{via}} \tag{3.14}$$

where $r_0 = r_{via} + t_{ox} + t_{dep}$ and $h_{via} \gg r_0$. For a TSV radius $r_{via} = 10$ μm, an oxide thickness $t_{ox} = 0.9$ μm, a depletion width $t_{dep} = 0$ μm, a TSV height $h_{via} = 200$ μm, and a silicon substrate with a conductivity $\sigma_{Si} = 10$ S/m, a capacitance of approximately 66.5 fF and a conductance of approximately 6.3 mS are calculated.

- Coupling effects with substrate contact

  The method was introduced by Engin [3] in 2013 for multiple TSVs with at least one return path TSV. The method can be modified for arrangements with

**Fig. 3.12**  3 TSV
configuration with substrate
contact

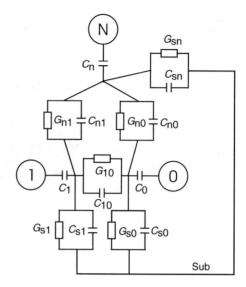

a substrate contact which works as a reference node instead of the return path
TSV. A 3 TSV configuration with substrate contacts is shown in Fig. 3.12.
The imaginary part of the low frequency impedance matrix looks as follows

$$Z_{\text{ox},l} = \begin{pmatrix} Z_0 & 0 & \dots & 0 \\ 0 & Z_1 & \dots & 0 \\ \vdots & \vdots & \vdots & \vdots \\ 0 & 0 & \dots & Z_N \end{pmatrix} \tag{3.15}$$

where $Z_i = \frac{1}{j\omega C_i}$ is the impedance of the side wall capacitance. Then $Z_{\text{ox},l}$ can be
calculated for high frequencies $Z_{\text{ox},h}$ and deembedded form the high frequency
matrix $Z_h$

$$Z_{\text{Si}} = Z_h - Z_{\text{ox},h} \tag{3.16}$$

where the resulting matrix $Z_{\text{Si}}$ is the coupling impedance matrix which can be
transformed to an admittance matrix $Y_{\text{Si}} = Z_{\text{Si}}^{-1}$. The admittance matrix looks
as follows

$$Y_{\text{Si}} = \begin{pmatrix} \sum_{i=1}^{N} Y_{0i} + Y_{S,0} & -Y_{01} & \dots & -Y_{0N} \\ -Y_{10} & \sum_{i=0,i\neq1}^{N} Y_{1i} + Y_{S,1} & \dots & -Y_{1N} \\ \vdots & \vdots & \vdots & \vdots \\ -Y_{N0} & -Y_{N1} & \dots & \sum_{i=0}^{N-1} Y_{Ni} + Y_{S,N} \end{pmatrix} \tag{3.17}$$

Fig. 3.13 HFSS simulation
setup to obtain the impedance
matrix

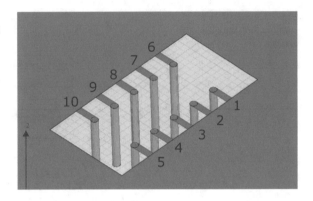

Keeping in mind that

$$Y = G + j\omega C \qquad (3.18)$$

the coupling and substrate elements can be calculated from the coupling admittance matrix. The non-diagonal element represents the GC coupling elements between the corresponding TSVs. By subtracting the other row elements from the diagonal elements the GC substrate elements can be distinguished. Note that the impedance matrix used to calculate the coupling and substrate contact admittances can be obtained from HFSS by simulating the TSV array plus substrate contact with ports on one side and opens at the other side as in Fig. 3.13.

### 3.3.3 TSV Measurements

In this subsection test structures have been fabricated to validate the manufacturing process and to characterize the manufactured TSVs. In this subsection, the RLCG parameter extraction from the measured data is described and measurement results for the fabricated TSVs are shown.

- RLCG parameter extraction

  For RF structures, S-parameters are normally measured. They can be related to the RLCG parameters by using the ABCD matrix transformation for a transmission line:

$$R = \mathrm{Re}\left( Z_0 \frac{(1 + S_{11})(1 + S_{22}) - S_{12}S_{21}}{2S_{21}} \right) \qquad (3.19)$$

$$L = \frac{\mathrm{Im}}{\omega}\left( Z_0 \frac{(1 + S_{11})(1 + S_{22}) - S_{12}S_{21}}{2S_{21}} \right) \qquad (3.20)$$

$$G = \text{Re}\left(\frac{1}{Z_0} \frac{(1 - S_{11})(1 - S_{22}) - S_{12}S_{21}}{2S_{21}}\right) \tag{3.21}$$

$$C = \frac{\text{Im}}{\omega}\left(\frac{1}{Z_0} \frac{(1 - S_{11})(1 - S_{22}) - S_{12}S_{21}}{2S_{21}}\right). \tag{3.22}$$

For noisy data, it is beneficial to optimize the RLGC parameters with a nonlinear optimization algorithm (e.g., Matlabs lsqnonlin) in such a way that they fit the measured S parameters. Therefore, the transmission line parameters $\gamma$ and $Z$ are used with

$$\gamma = \sqrt{(R + j\omega L)(G + j\omega C)} \tag{3.23}$$

$$Z = \sqrt{\left(\frac{R + j\omega L}{G + j\omega C}\right)}. \tag{3.24}$$

The S-parameter matrix can be calculated as follows [2]:

$$[S] = \frac{1}{D_S}\begin{pmatrix} (Z^2 - Z_0^2)\sinh(\gamma) & 2ZZ_0 \\ 2ZZ_0 & (Z^2 - Z_0^2)\sinh(\gamma) \end{pmatrix} \tag{3.25}$$

where $D_S = 2ZZ_0\cosh(\gamma) + (Z^2 + Z_0^2)\sinh(\gamma)$. The computed RLCG parameters based on the measured data are shown in the following section.

• GSG TSVs

A test structure to characterize TSVs can be seen in Fig. 3.14. It was manufactured on a 200 μm thick wafer with an oxide thickness of approximately 0.9 μm.

It consists of two coplanar waveguides (CPW) feeding lines, a two-time TSV transition through the substrate and a CPW connection between the TSVs on the backside. The CPW reference line has approximately the same length as all CPWs in the test structure together. The CPW signal line width $w$ is 50 μm, the gap size $s$ is 30 μm, and the width of the ground line $g$ is 90 μm. The measured S-parameters in comparison to HFSS simulations for different silicon conductivities are shown in Fig. 3.15.

For $S_{21}$ the HFSS simulation results for a silicon conductivity of 7 S/m are fitting very well to the measurement results while for $S_{11}$ a silicon conductivity of 6 S/m leads to better results. For the RLCG parameter comparison a silicon conductivity of 7 S/m was chosen.

The RLCG parameters can be extracted by measuring the whole test structure, calculating the RLCG parameters, measuring the reference CPW, calculating the RLCG parameters, subtracting the RLCG parameters of the reference line from the whole test structure, and dividing the values by 2. For a test structure with CPW parameters $w = 30$ μm, $s = 20$ μm, and $g = 90$ μm the extracted parameters are shown in Fig. 3.16 and compared with the model of Lu [25] using $M_{\text{GSG}} \approx \sqrt{2}$.

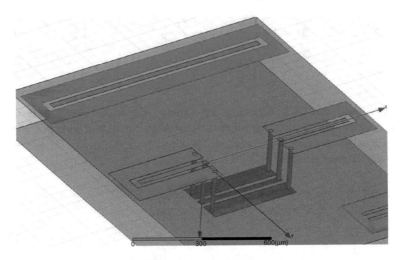

**Fig. 3.14** Test structure with two TSV transitions and CPW reference line

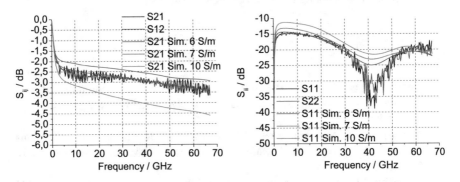

**Fig. 3.15** Comparison of measured and simulated $S_{21}$ (*left*) and $S_{11}$ (*right*) for the test structure in Fig. 3.14 for different silicon conductivities ($r_{via} = 10\,\mu m$, $t_{ox} = 0.9\,\mu m$ and *pitch* $= 100\,\mu m$)

For the admittance parameters, the extracted measured values and the values of the model are fitting well. For the inductance, a good fit for the lower frequency region can be seen while the resistance of the model and the measured data differ significantly. The large difference is probably due to the very small contribution of the resistance to the S-parameters in this frequency region. The resistance is relatively low over the whole frequency band for both the model and the hardware.

Another method to verify the TSV model is to fit the CPW reference line to a transmission line, combine the TSV model with that of the transmission line, and compare the results with the measured data of the combined structure. For the case when we neglect the skin effect, i.e., only the DC resistance is used, and ignore the coupling between the horizontal and vertical transmission lines, the results are given in Fig. 3.17. The amplitudes of the measured and simulated $S_{21}$

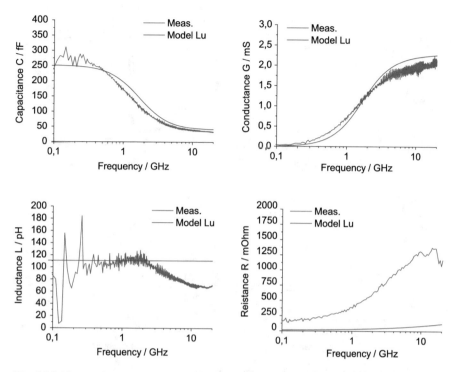

**Fig. 3.16** Measured capacitance (*upper left*), conductance (*upper right*), inductance (*bottom left*), and resistance (*bottom right*) of a GSG TSV ($r_{via} = 10\,\mu m$, $t_{ox} = 0.9\,\mu m$ and *pitch* = $80\,\mu m$)

and $S_{11}$ are matching very well in the low-frequency region and differ slightly for higher frequencies by less than 0.25 dB for $S_{21}$ and less than 1.8 dB for $S_{21}$. The measured and simulated phases of $S_{21}$ are also matching in the low-frequency region and differ for higher frequencies by less than $10°$, while $S_{11}$ is shifted in the frequency by approximately 2 GHz. However, the overall trends of the model and the measurements are matched quite well.

## 3.3.4  Equivalent Circuit Extraction

The starting point for building the equivalent circuit is the HFSS model in Fig. 3.18a which will be used to fit the equivalent circuit using Keysight's ADS to the HFSS S-parameters. The HFSS model consists of 4 GSG connections: Rx and Clock and Tx and Clock. The brown contact line is the substrate contact to ground. The geometric parameters are as follows. The small pitch between the TSVs is $75\,\mu m$, the large pitch is $150\,\mu m$, the TSV radius is $10\,\mu m$, the oxide thickness is $0.9\,\mu m$, the line width of the copper lines is 20 and $25\,\mu m$, the diameter of the small pads is $40\,\mu m$, and the diameter of the large pads is $80\,\mu m$.

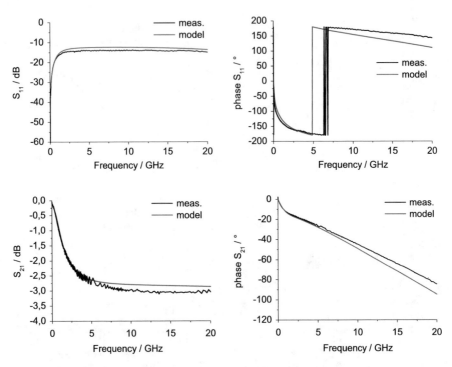

**Fig. 3.17** Comparison of the TSV Model and measured data

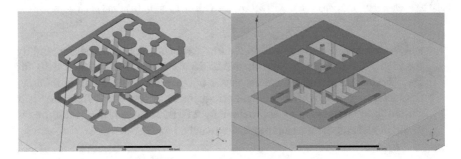

**Fig. 3.18** HFSS model of Titan3D (**a**), reduced model without pads (**b**)

The HFSS model in Fig. 3.18a was first reduced to the model shown in Fig. 3.18b without pads to obtain the values for the TSVs with substrate contact. The equivalent circuit of the demonstrator consists of a pad model using a normal transmission line model and a GSG TSV model as a T circuit with included substrate contact admittances (Fig. 3.19).

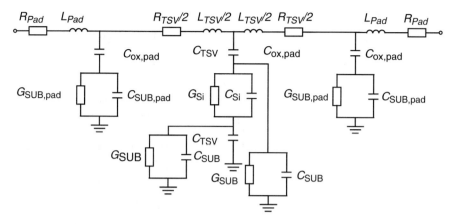

**Fig. 3.19** Equivalent circuit of one GSG link

The methodology for extracting the circuit parameters is as follows:

1. Compute the S-parameters of the structure shown in Fig. 3.18b
2. Use the RLGC values from the Lu GSG model
3. Compute circuit parameters, using the ADS nonlinear optimization algorithm, and fit the ADS model to the HFSS S-parameters. At this stage, do not include pads, but include the capacitive coupling and the substrate contact admittance,
4. Compute the S-parameter of the structure shown in Fig. 3.18a
5. Add pad parameters to the model built in 3, using open/short simulations of the pads
6. Re-optimize the circuit parameters, including pad admittance, by applying the ADS nonlinear optimization algorithm to fit the full model to the HFSS S-parameters of 4.

The obtained parameters are summarized in Table 3.1 and the S-parameter results are shown in Fig. 3.20.

The amplitudes and phases of $S_{11}$ and $S_{21}$ are matching quite well. The capacitance of the link is especially important for the transmitter and receiver design (see Sect. 3.4). The capacitance results at low frequencies are summarized in Table 3.2.

Analytical expressions and simulations fit very well. Only a small deviation of approximately 5.5 % for the pad capacitance occurs. The L2L deembedding was used to deembed the transmission line from the TSV transition. The mathematical expressions can be found in [29]. Finally, the optimized circuit, including the simplified depletion model, has been implemented in Verilog-A and used in time-domain Spice simulations.

**Table 3.1** Obtained values from the model and the optimization for the equivalent circuit using a silicon conductivity of 10 S/m

| Parameter | Value | Determined by |
|---|---|---|
| $R_{TSV}$ | 11 mΩ | Model Lu (DC value) |
| $L_{TSV}$ | 112 pH | Model Lu |
| $C_{ox} = 2 \cdot C_{ox,Lu}$ | 503.5 fF | Model Lu |
| $C_{Si} = C_C$ | 28 fF | Optimization |
| $G_{Si} = G_C$ | 2.62 mS | Optimization |
| $C_{SUB}$ | 35 fF | Optimization |
| $G_{SUB}$ | 3.34 mS | Optimization |
| $t_{dep}$ | 0.28 μm | Depletion model |
| $C_{TSV}$ | 460 fF | Depletion model |
| $R_{Pad}$ | 72 Ω | Short pad simulation |
| $L_{Pad}$ | 65 pH | Short pad simulation |
| $C_{ox,Pad}$ | 271 fF | Open pad simulation |
| $C_{SUB,Pad}$ | 14 fF | Optimization |
| $G_{SUB,Pad}$ | 1.33 mS | Optimization |

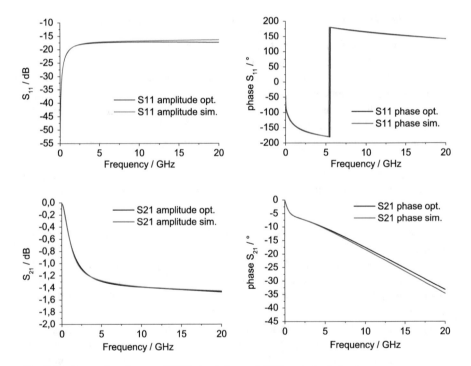

**Fig. 3.20** Comparison between HFSS simulation and ADS optimized circuit

**Table 3.2** Capacitance distribution of the whole link

| Element | Capacitance fF (analytical) | Verification |
|---|---|---|
| Pads bottom side | 253 fF (plate capacitor approximation) | 268 fF (HFSS open pad simulation) |
| Pads upper side | 253 fF (plate capacitor approximation) | 268 fF (HFSS open pad simulation) |
| TSV oxide capacitance | 503 fF (Xu [GS], Lu [GSG] model) | Approx. 500 fF (HFSS open via and L2L deembedding for GS configuration) |
| Overall result | 1009 fF | 1045 fF (HFSS), 1045fF (ADS equ. circ.) |

## 3.4  Capacitive Multi-Bit TSV Transceiver

TSVs provide electrical interconnects between dies in a 3D chip stack with properties as described in Sect. 3.3. 3D chip stacks have a limited heat dissipation capability. TSVs on global interconnect-level (3D-SOC) or on bond-pad level have a significant parasitic TSV capacitance. Driving this capacitance contributes to power dissipation and generates heat. Therefore, energy efficient transceivers are required. A low voltage swing signal can be generated by building a capacitive voltage divider by inserting coupling capacitors in series with capacitive TSVs. Therefore, no additional power mesh or power domain resources are required. By using a binary weighted capacitive coupling and a double data rate scheme, 4 bits can be transmitted per clock cycle and data lane. The transceiver can be integrated into the link architecture described in Sect. 3.2. System integration is simplified by the fact that the NoC links and NiCS link have the same interfaces.

### 3.4.1  Architecture

TSVs have a certain pitch, diameter, and keep out zone (KOZ) width, therefore they occupy active die area. If we are considering electrical TSV integration on global interconnect-level (3D-SIC/3D-SOC), the TSVs are large compared to transistors of modern CMOS technologies. In order to reduce the overall footprint, a serialized data transmission is preferable. The use of the same point-to-point link architecture as NoC links for network in chip stack (NiCS) links simplifies the system integration. An electrical TSV chip stack setup is shown in Fig. 3.21. Transmitter and receiver circuits are placed on different dies and are interconnected by electrical TSVs. A TSV provides a low ohmic interconnect from one die to an adjacent die with a high capacitance between TSV and surrounding substrate. This parasitic capacitance has a significant influence on the power dissipation of the TSV transceiver. By using a low voltage swing signaling scheme with capacitive

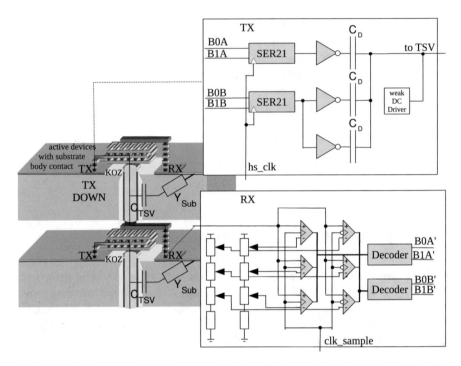

**Fig. 3.21** Possible setup of a dual input capacitive coupling transceiver in a data lane within a 3D chip stack for point-to-point links for 3D-NoC communication

coupling the dynamic power dissipation can be reduced. The series combination of coupling capacitance and TSV capacitance reduces the effective load capacitance for the CMOS driver. By using two binary weighted capacitive drivers, a four level amplitude modulated signal can be generated. An eye-diagram of this signal is shown in Fig. 3.22.

The capacitive 4-PAM transceiver shown in Fig. 3.23 uses a source-synchronous clocking scheme. It is composed of several data paths and one shared clock path. The clock path is used to transmit data via the TSV-channel and to sample data in phase. The transmitter in the clock path and the data paths are the same. The receiver in the clock path has to recover the edges of the low-swing signal without additional timing uncertainties. Therefore a continuous clock amplifier has to be used which has a static current during active mode. Due to the source-synchronous clocking scheme, clocked comparators (sense amplifiers) can be used to sample the data channels. Clocked sense amplifiers can be implemented as CMOS devices and thereby consuming power mainly due to switching events and only low static leakage power otherwise.

Per data channel 4 bits of data are serialized by 4-to-2 (DDR), converted into a 2 bit amplitude modulated voltage level by a capacitive divider DAC and transmitted through a low ohmic TSV. The swing of the clock signal can be selected through a

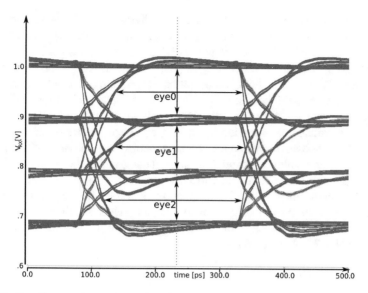

**Fig. 3.22** Eye-diagram of a 4-pulse-amplitude-modulated (PAM) signal from a dual input capacitive coupling transmitter; simulation with a parasitic extracted view, with $k \cdot V_{DD} = 300\,mV$, frequency $f = 2\,GHz$, and a data rate of 8 Gb/s

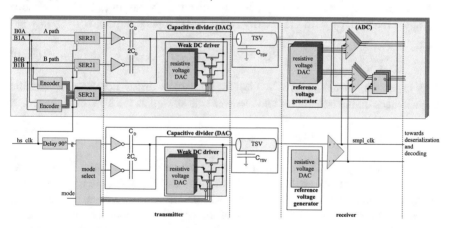

**Fig. 3.23** Integration of transmitter (TX) and receiver (RX) into a packet-based NiCS link

mode select circuit. The weak DC driver can have an RC time constant that is orders of magnitude higher than the capacitive driver, but it should be strong enough to prevent leakage currents altering the voltage on the TSV node.

On the receiver side, 6 clocked comparators convert the 4 voltage level symbol into 3 bit thermometer values by comparing the input level against 3 distinct intermediate voltage thresholds for the rising and falling edge, respectively. A thermometer decoder is used to convert the data back to binary format. The clock can be generated by an on-chip clock generator [11] and forwarded from the transmitter in a source-synchronous clocking scheme [30].

The resulting voltage step of the capacitive divider depends on the capacitance value $C_{TSV}$ of the transmission channel composed of ESD-structures, redistribution layers, bumps, and TSVs. The resistive divider has to be designed to match the voltage levels accordingly to enable the output node to stay at the same voltage level, even if the same symbol is transmitted over a long period. Mismatch effects and design uncertainties between $C_{TSV}$ and $C_D$ lead to a deviation from the nominal capacitive voltage divider value. Mismatch between the voltage levels provided by the resistive voltage divider and the capacitive voltage divider leads to scattering of the symbol voltage levels, which results in narrow eye-diagrams. An adjustable resistive divider enables an adaption of the voltage levels of the weak DC driver to the voltage levels generated by the capacitive divider.

In addition, mismatch effects lead to an offset voltage in every sense amplifier, which must be compensated in order to compare values within a small decision range. The individual offset of the sense amplifiers can be compensated through a digitally configurable potentiometer.

### 3.4.2   Power Dissipation Analysis

The dynamic power dissipation $P_{dyn}$ for charging and discharging a load capacitance $C_L$ at the frequency $f$ with the activity rate $\alpha$ and complete settling is:

$$P_{dyn} = \alpha \cdot C_L \cdot V_{DD}^2 \cdot f \tag{3.26}$$

The capacitive load $C_{LFS}$ for a CMOS full-swing driver consists of parasitic capacitances on the transmitter and receiver side as well as the TSV capacitance.

$$C_{LFS} = C_{TXPAR} + C_{TSV} + C_{RXPAR} \tag{3.27}$$

The dielectric layer and the depletion region between the TSV and the substrate cause a capacitance $C_{TSV}$ between TSV and substrate [32]. TSVs are relatively large compared to the size of CMOS devices in sub-$\mu$m technology nodes, therefore, it is assumed that charging and discharging of the capacitance between the transmitter and receiver is the main contributor to the power dissipation of the transceiver circuit.

In this section, we compare the dynamic power dissipation of the driver circuit between a capacitive voltage divider circuit, as shown in Fig. 3.21, and a full-swing CMOS driver.

In the capacitive load model of the transmitter shown in Fig. 3.24, $\gamma$ is the input side and $\beta$ the output side parasitic capacitance factor for the driver capacitance $C_D$. Due to proper sizing of the driver capacitance $C_D$ a voltage range divider ratio $k$ can be realized. $k$ describes the ratio between the total voltage swing at the receiver node and the input node of the capacitive divider. The voltage of the input node of

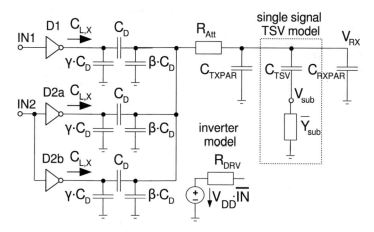

**Fig. 3.24** Capacitive load model of the capacitive driver circuit, values used within this design: $\gamma \approx 0.08$, $\beta \approx 0.505$, $C_{\text{TXPAR}} = 150\,\text{fF}$, $C_{\text{RXPAR}} = 170\,\text{fF}$

the capacitive divider is between $V_{\text{SS}} = 0\,\text{V}$ and $V_{\text{DD}}$. $X$ is the number of input bits and $S \cdot V_{\text{DD}}$ is the voltage spacing between adjacent symbols.

$$k = S \cdot \left(2^X - 1\right) = \frac{\Delta V_{\text{RX}}}{\Delta V_{\text{TX}}} = \frac{2^X - 1}{\left(\beta + 1\right) \cdot \left(2^X - 1\right) + \frac{C_{\text{LFS}}}{C_{\text{D}}}} \tag{3.28}$$

The coupling capacitor $C_{\text{D}}$ can be chosen to realize a certain voltage swing ratio $k$. The capacitive load seen by one capacitive CMOS driver within a transmitter with $X$ inputs is

$$C_{\text{L},X} = C_{\text{D}} \cdot \left(1 + \gamma - \frac{1}{(2^X - 1) \cdot (\beta + 1) + \frac{C_{\text{LFS}}}{C_{\text{D}}}}\right) \tag{3.29}$$

$$C_{\text{L},X} = \frac{C_{\text{LFS}}}{1 - k \cdot (1 + \beta)} \cdot \left(1 + \gamma - \frac{k}{2^X - 1}\right) \cdot \frac{k}{2^X - 1} \tag{3.30}$$

In Fig. 3.25 the capacitive load of one CMOS driver normalized to the capacitive load of one full-swing driver is plotted vs. the relative symbol level distance $S$. The capacitive load of one CMOS driver in a dual input capacitance transmitter is always higher than in the single input capacitance transmitter, because an increased number of drivers increases the total capacitive load on the TSV node. For each CMOS driver the dynamic power dissipation $P_{\text{dyn}}$ [Eq. (3.26)] is proportional to its capacitive load $C_{\text{L},X}$. The CMOS inverter can be sized smaller compared with a full-swing transmitter, due to the reduced capacitive load. For smaller relative symbol spacing $S$, the relative overhead for a dual input driver compared to a single input driver gets smaller. The total dynamic power dissipation is:

**Fig. 3.25** Normalized
capacitive load $C_{L,X}$ of one
CMOS driver for a single and
dual input capacitive
transmitter

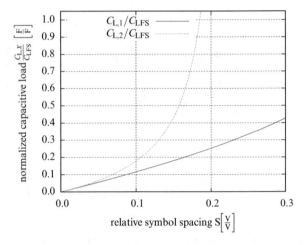

**Fig. 3.26** Normalized energy
per bit for different
configurations of the
capacitive transmitter

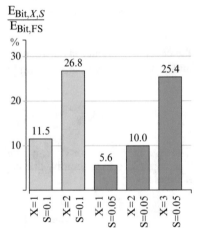

$$P_{\text{dyn},X} = \sum_{i=1}^{X} 2^{i-1} C_{L,X} \alpha V_{\text{DD}}^2 f = (2^X - 1) C_{L,X} \cdot \alpha \cdot V_{\text{DD}}^2 \cdot f \qquad (3.31)$$

The data rate at a multi input capacitance driver increases with the number of input
ports $X$. The number of capacitive drivers increases with $2^X - 1$. The capacitive load
seen by the capacitive driver depends on the number of inputs $X$ and the voltage
divider ratio $k$. Figure 3.26 compares the energy per bit for different configurations.
A single input capacitive coupling driver with a voltage swing of $0.1\,V_{\text{DD}}$ consumes
11.5 % of the energy per bit consumption compared to a full-swing driver. A dual
input capacitive coupling driver with the same symbol voltage distance consumes
26.8 % of the energy per bit of a full-swing driver. From this the following design
guidelines can be concluded:

- If the design is limited by the energy per bit, then the single-input capacitive transmitter is preferable.
- If the design is limited by the number of available TSVs and additionally minimum symbol spacing $S \cdot V_{DD}$ is required (e.g., due to RX offset), then the dual-input capacitive transmitter is preferable.

## 3.4.3 Design

### 3.4.3.1 Reference Voltage Generation

Reference voltages are required on the receiver and the transmitter sides. A resistive voltage digital to analog converter is used to provide these reference voltages. The resistive voltage DAC consists of a resistive voltage divider ladder followed by a transmission gate select tree network to be able to adjust the reference voltage. On the receiver side the single ended low voltage swing signal has to be compared with reference voltages to restore a full-swing signal. On the transmitter side a weak DC level driver is required in order to define the DC voltage level of the otherwise floating TSV node. The resistance of the resistive ladder should be high to reduce the static current but still low enough to prevent leakage currents from altering the node level. One of four tap positions of the resistive divider is connected to the TSV node via PMOS transistor switches dependent on the transmitted symbol. Therefore, an uncoded transmitted data stream can be paused at any moment.

### 3.4.3.2 Transmitter

In order to design the capacitively coupled transmitter, the model of the channel has to be known. The method described in Sect. 3.3.4 can be used to get an equivalent circuit description of the channel between the receiver die and the transmitter die. Simplified equivalent circuit representations can be constructed from layout views of transmitter and receiver components with the aid of parasitic extraction tools and testbenches.

The transfer function of a simplified equivalent circuit representation of a capacitive TSV driver shown in Fig. 3.24 with a non-low-ohmic substrate contact can be calculated. The transfer function has two poles $p_3 < p_2 < 0$ and one zero at $a$ and can be used for design parameter considerations.

$$G(s) = k \frac{\frac{\omega_n^2}{a} \cdot (s + a)}{s^2 + 2 \cdot \zeta \cdot \omega_n \cdot s + \omega_n^2} \tag{3.32}$$

$$a = \frac{G_{\text{SUB}}}{C_{\text{TSV}} + C_{\text{SUB}}} \tag{3.33}$$

The driver design target is to achieve a fast settling signal on the receiver side. Therefore, oscillation and large overshooting should be prevented. This can be prevented if $\zeta > 1$ and the zero $a$ is smaller, than the dominant pole $p_2$ [22]. If $a + p_2 = 0$, then the system is a first order system with a time constant of $1/p_3$. The dominant pole $p_2$ can be altered through $R_{\text{DRV}}$. Overshoots in a multi-bit capacitive driver can result in narrowing adjacent eyes in the eye-diagram.

An attenuation resistance $R_{\text{Att}}$ (see Fig. 3.24) prevents the LC circuit consisting of TSV inductance and capacitances on the receiver and the transmitter side from weakly damped oscillations. If a TSV connection has an ESD protection, than $R_{\text{Att}}$ is redundant with the ESD protection serial resistance.

In the proposed design in a 28 nm CMOS technology, the coupling capacitance is constructed from metal finger capacitors within the metal stack and overlapping capacitance of minimal length MOS transistors. The metal stack above a KOZ can be used effectively for placing metal finger capacitors.

In Fig. 3.27 a layout in 28 nm CMOS technology of a TSV transmitter including clock path, data path, serializer, and ESD protection is shown.

Coupling effects with surrounding active circuits or with neighbor signaling TSVs have an influence on the TSV signal behavior. This coupling effects can be reduced by adding shielding elements like grounded TSVs, guard rings, or substrate contacts around signal TSVs.

**Fig. 3.27** Layout of a TSV transmitter including data and clock path, serializer, and ESD protection. The macro has a size of $64\,\mu\text{m} \times 160\,\mu\text{m}$

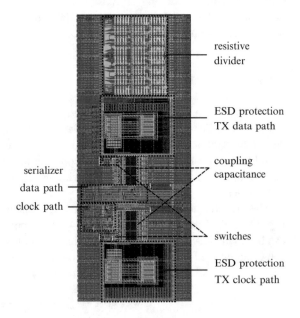

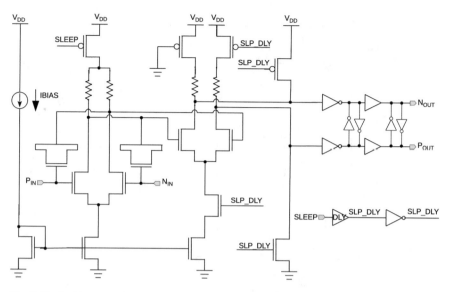

**Fig. 3.28** Architecture of the continuous amplifier for clock signal full-swing recovery on the receiver side

### 3.4.3.3   Receiver

A time continuous amplifier is required for the clock path of the receiver in order to restore the clock edge. We have used a differential continuous amplifier as shown in Fig. 3.28 with several amplifier stages. It has a static current consumption. During sleep mode the static current consumption is reduced and the clock signal level is held. One input node is connected to a TSV clock signal and the other input node is connected to a high ohmic voltage reference node. Additional MOS capacitances have been added between the input nodes and output nodes of the first amplifier stage, in order to reduce the overall effects of capacitive coupling to the high resistive reference node.

A clocked sense amplifier is used to compare the data signal with a reference voltage. The architecture is shown in Fig. 3.29. It has no static currents apart from intrinsic MOS leakage currents. One input node is connected to a TSV data signal and the other input node is connected to a high ohmic voltage reference node. In order to reduce the clock feed-through effects to the input nodes a complementary dummy input stage has been added, shown on the right side.

## 3.4.4   Influence of Design and Technology Parameters

In 3D chip stacks, TSVs are embedded into the substrate of a die. In a bulk CMOS substrate there is coupling through the TSV capacitance and the substrate to the

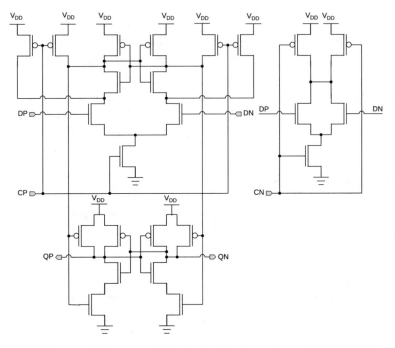

**Fig. 3.29** Architecture of the sense amplifier for data signal sampling on the receiver side

active layer of the die (see Fig. 3.21). In the case of a p-substrate the substrate is usually connected to ground via body contacts on the surface. Therefore, placement and geometrical properties of TSVs and properties of substrate affect the properties of the TSV channel.

We have investigated the influence of substrate conductivity and the equivalent driver resistivity on the eye pattern of the TSV signal [6]. For model parameter extraction a configuration with a signal TSV surrounded by two ground connected TSVs with a pitch of 75 μm has been used. The TSV height $H$ is 200 μm. The TSV radius $r_0$ is 10 μm and the oxide layer $t_{ox}$ is 0.9 μm thick. A depletion zone thickness of 0.28 μm is estimated at the operational point, resulting in an overall TSV sidewall capacitance $C_{TSV}$ of 460 fF. The substrate connection to ground has been modeled with planar metal on the surface. The model parameters are extracted from 3D field solver (HFSS) simulations (see Sect. 3.3.4), the resulting TSV model is shown in Fig. 3.30. The conductance $Y_{SUB}$ from the surrounding substrate of the signal TSV to the surface bulk ground connection is represented by $G_{SUB}$ and $C_{SUB}$ within this model. $G_C$ and $C_C$ represent the substrate coupling from the signal TSV to the ground connected TSVs, and $R_{TSV}$ and $L_{TSV}$ represent the serial resistance and inductance of the TSV, respectively. The substrate conductances $G_{SUB}$ and $G_C$ are dependent on the size ($r_0$, $H$) of the TSV and are proportional to the substrate material constant $\sigma_{SUB}$.

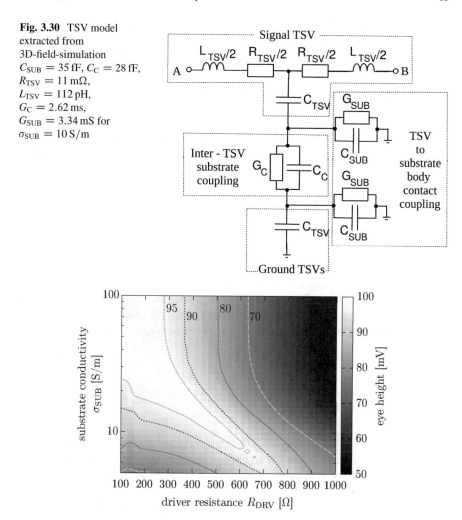

**Fig. 3.30** TSV model extracted from 3D-field-simulation $C_{SUB} = 35\,fF$, $C_C = 28\,fF$, $R_{TSV} = 11\,m\Omega$, $L_{TSV} = 112\,pH$, $G_C = 2.62\,ms$, $G_{SUB} = 3.34\,mS$ for $\sigma_{SUB} = 10\,S/m$

**Fig. 3.31** The minimum height of eye 0, 1, and 2 as function of the driver resistance $R_{DRV}$ and the substrate conductivity $\sigma_{SUB}$ at a frequency of 2 GHz, the TSV model is shown in Fig. 3.30 and the equivalent circuit diagram is shown in Fig. 3.24

A common horizontal eye center is used for all three eyes, their width and height are defined at the eye center as shown in Fig. 3.22.

The circuit model shown in Fig. 3.24 has been simulated with the simplified inverter model and the extracted TSV model (Fig. 3.30). Figure 3.31 shows a map of the eye height at the receiver side as function of the resistance $R_{DRV}$ of the inverter model and the substrate conductivity $\sigma_{SUB}$. For a given substrate conductivity $\sigma_{SUB}$ an optimal inverter exists which maximizes the eye opening for a given transmission frequency $f$. The substrate conductivity $\sigma_{SUB}$ defines the conductance $G_{SUB}$, which influences the settling behavior of the TSV sidewall capacitance $C_{TSV}$. The equivalent resistance $R_{DRV}$ influences the slope of the immediate signal transition. Both $R_{DRV}$ and $G_{SUB}$ contribute to the settling of the TSV node in the circuit model shown in Fig. 3.24. If $R_{DRV}$ is large, it dominates the settling time of the TSV and thereby minimizes the eye opening. If $R_{DRV}$ is too small compared to the substrate resistance, then the signal on the TSV exhibits an overshoot due to the fact that charge coupled in by the driver cannot be redistributed in the substrate immediately, this narrows adjacent eyes. For a maximum eye opening an optimized $R_{DRV}$ has to be chosen, this is especially the case for substrates with smaller conductance or higher data rates. The equivalent resistance $R_{DRV}$ of the driving inverter can be modified with the width of the driving CMOS devices.

### 3.4.5 Comparison

In Table 3.3 the power efficiency of the capacitive multi-bit transmitter for one data path at a substrate conductivity $\sigma_{SUB}$ of 100 S/m is compared with a full-swing (FS) transmitter and published TSV transceivers. Note that the values of our work do not include a receiver and the TSV channel consists of a TSV model only without ESD or pad structures. The power consumption has been obtained from post layout Spice simulations.

## 3.5 Conclusions

A point-to-point link architecture has been presented which provides a unified packet-in, packet-out interface for horizontal on-chip and vertical intra-chip communication in MPSoCs. It is well suited for GALS systems as input and output interfaces can be controlled by different clock domains. A source-synchronous data transmission leads to high jitter tolerance and prevents unnecessary power dissipation during idle times. As part of the link architecture, the transceiver must be adapted to the transmission channel characteristics. For long distance on-chip connections a low-swing differential signaling with capacitive pre-emphasis and a weak high resistive driver is presented. The NoC link was implemented in 65 and 28 nm. Measurement results have shown that serial high-speed on-chip links are a viable solution for high bandwidth interconnect requirements in modern MPSoC systems. An overview over several TSV effects like depletion effect, proximity effect, skin effect, and effect of substrate contacts has been given. Models have been developed to describe the behavior of TSVs within a contacted substrate and a good fit between S-parameters of measured and simulated TSV structures has been shown. A method to extract an equivalent circuit of a TSV arrangement has been described. The architecture of an energy efficient capacitively coupled multi-bit transceiver has also been presented with an analysis of the interplay between number of inputs, signal swing, and energy efficiency. Design details about

**Table 3.3** TSV transmitter comparison

| Reference | Our work [6] | | [23] | [23] | [5] | [34] |
|---|---|---|---|---|---|---|
| Type | LVSE-CAP-PAM4 | FS | FS | LVSE | LVSE | LVSE-CAP |
| TSV $H$, $r_o$, $t_{ox}$ (μm) | TX only (simulation) 200, 10, 0.9 | | TX+RX – | | 50, 5, 0.5 | Emulated |
| $C_{TSV}$ (fF) | 460 | | 200 | | 90 | – |
| Energy efficiency $\left(\frac{fJ}{\text{Bit}\cdot\#\text{TSVs}}\right)$ | 50 | 197 | 220 | 110 | 170 | – |
| Energy efficiency $\left(\frac{fJ}{\text{Bit}\cdot\#\text{TSVs}\cdot 1\,\text{pF}}\right)$ | 109 | 428 | 1100 | 550 | 1890 | 149 |
| @Data rate (Gb/s) | 8 | 4 | 6 | 6 | 1 | 8 |
| Technology | 28 nm | | 45 nm SOI | | 65 nm | 65 nm |
| Voltage swing | 0.3 V | 1.0 V | $V_{DD}$ | 0.3 V | 0.4 $V_{DD}$ | 0.04 V |

the implementation in 28 nm technology have been given. It was shown that the driver strength and the substrate conductivity influence the eye pattern opening. A comparison with other TSV transceivers concludes this chapter.

# References

1. W. Dally, B. Towles, Route packets, not wires: on-chip interconnection networks, in *Proceedings Design Automation Conference*, 2001
2. W. Eisenstadt, Y. Eo, S-parameter-based IC interconnect transmission line characterization. IEEE Trans. Compon. Hybrids Manuf. Technol. **15**(4), 483–490 (1992)
3. A. Engin, Equivalent circuit model extraction for interconnects in 3D ICs, in *18th Asia and South Pacific Design Automation Conference (ASP-DAC)* (2013), pp. 1–6
4. A. Engin, S. Narasimhan, Modeling of crosstalk in through silicon vias. IEEE Trans. Electromagn. Compat. **55**(1), 149–158 (2013)
5. F. Furuta, K. Osada, 6 Tbps/W, 1 Tbps/mm2, 3D interconnect using adaptive timing control and low capacitance TSV, in *IEEE International 3D Systems Integration Conference (3DIC), 2011* (2012), pp. 1–4
6. J. Görner, S. Höppner, D. Walter, M. Haas, D. Plettemeier, R. Schüffny, An energy efficient multi-bit TSV transmitter using capacitive coupling, in *21st IEEE International Conference on Electronics Circuits and Systems 2014 (ICECS 2014)* (2014), pp. 650–653
7. R. Ho, T. Ono, R. Hopkins, A. Chow, J. Schauer, F. Liu, R. Drost, High speed and low energy capacitively driven on-chip wires. IEEE J. Solid State Circuits **43**(1), 52–60 (2008)
8. S. Höppner, D. Walter, G. Ellguth, R. Schüffny, Mismatch characterization of high-speed NoC links using asynchronous sub-sampling, in *International Symposium on System on Chip 2011 (SoC)*, 2011
9. S. Höppner, C. Shao, H. Eisenreich, G. Ellguth, M. Ander, R. Schüffny, A power management architecture for fast per-core DVFS in heterogeneous MPSoCs, in *IEEE International Symposium on Circuits and Systems (ISCAS)* (2012), pp. 261–264
10. S. Höppner, D. Walter, G. Ellguth, R. Schüffny, On-chip measurement and compensation of timing imbalances in high-speed serial NoC links. Int. J. Embed. Real-Time Commun. Syst. (IJERTCS) Volume 3, Issue 4, pp. 42–56 (2012)
11. S. Höppner, S. Haenzsche, G. Ellguth, D. Walter, H. Eisenreich, R. Schüffny, A fast-locking ADPLL with instantaneous restart capability in 28-nm CMOS technology. IEEE Trans. Circuits Syst. Express Briefs **60**(11), 741–745 (2013)
12. S. Höppner, D. Walter, T. Hocker, S. Henker, S. Hänzsche, D. Sausner, G. Ellguth, J.-U. Schlussler, H. Eisenreich, R. Schüffny, An energy efficient multi-Gbit/s NoC transceiver architecture with combined AC/DC drivers and stoppable clocking in 65 nm and 28 nm CMOS. IEEE J. Solid State Circuits **50**, 749–762 (2015)
13. H. Ito, J. Seita, T. Ishii, H. Sugita, K. Okada, K. Masu, A low-latency and high-power-efficient on-chip LVDS transmission line interconnect for an RC interconnect alternative, in *IEEE International Interconnect Technology Conference* (2007), pp. 193–195
14. ITRS, International technology roadmap for semiconductor interconnect (2002)
15. ITRS, International technology roadmap for semiconductor interconnect (2004)
16. ITRS, International technology roadmap for semiconductor interconnect (2006)
17. ITRS, International technology roadmap for semiconductor interconnect (2008)
18. ITRS, International technology roadmap for semiconductor interconnect (2011)
19. ITRS, International technology roadmap for semiconductor interconnect (2013)
20. G. Katti, M. Stucchi, K. de Meyer, W. Dehaene, Electrical modeling and characterization of through silicon via for three-dimensional ICs. IEEE Trans. Electron Devices **57**(1), 256–262 (2010)

21. B. Kim, V. Stojanovic, A 4Gb/s/ch 356fJ/b 10mm equalized on-chip interconnect with nonlinear charge-injecting transmit filter and transimpedance receiver in 90nm CMOS, in *IEEE International Solid-State Circuits Conference Digest of Technical Papers (ISSCC)* (2009), pp. 66–67,67a

22. B.-M. Kwon, M.-E. Lee, O.-K. Kwon, On nonovershooting or monotone nondecreasing step response of second-order systems. Trans. Control Autom. Syst. Eng. **4**, 283–288 (2002)

23. Y. Liu, W. Luk, D. Friedman, A compact low-power 3D I/O in 45nm CMOS, in *IEEE International Solid-State Circuits Conference Digest of Technical Papers (ISSCC)* (2012), pp. 142–144

24. K.-C. Lu, T.-S. Horng, Comparative modeling of single-ended through-silicon vias in GS and GSG configurations up to v-band frequencies. Prog. Electromagn. Res. **143**, 559–574 (2013)

25. K.-C. Lu, T.-S. Horng, H.-H. Li, K.-C. Fan, T.-Y. Huang, C.-H. Lin, Scalable modeling and wideband measurement techniques for a signal TSV surrounded by multiple ground TSVs for RF/high-speed applications, in *IEEE 62nd Electronic Components and Technology Conference (ECTC)* (May 2012), pp. 1023–1026

26. E. Mensink, D. Schinkel, E. Klumperink, E. van Tuijl, B. Nauta, Power efficient gigabit communication over capacitively driven RC-limited on-chip interconnects. IEEE J. Solid State Circuits **45**(2), 447–457 (2010)

27. B. Noethen, O. Arnold, E. Perez Adeva, T. Seifert, E. Fischer, S. Kunze, E. Matus, G. Fettweis, H. Eisenreich, G. Ellguth, S. Hartmann, S. Höppner, S. Schiefer, J.-U. Schlusler, S. Scholze, D. Walter, R. Schüffny, A 105GOPS 36mm2 heterogeneous SDR MPSoC with energy-aware dynamic scheduling and iterative detection-decoding for 4G in 65nm CMOS, in *IEEE International Solid-State Circuits Conference Digest of Technical Papers (ISSCC)* (2014), pp. 188–189

28. J.-s. Seo, R. Ho, J. Lexau, M. Dayringer, D. Sylvester, D. Blaauw, High-bandwidth and low-energy on-chip signaling with adaptive pre-emphasis in 90nm CMOS, in *IEEE International Solid-State Circuits Conference Digest of Technical Papers (ISSCC)* (2010), pp. 182–183

29. X. Shi, K. Seng Yeo, M. Anh Do, C. Chye Boon, Characterization and modeling of on-wafer single and multiple vias for CMOS RFICs. Microw. Opt. Technol. Lett. **50**(3), 713–715 (2008)

30. D. Walter, S. Höppner, H. Eisenreich, G. Ellguth, S. Henker, S. Hänzsche, R. Schüffny, M. Winter, G. Fettweis. A source-synchronous 90Gb/s capacitively driven serial on-chip link over 6mm in 65nm CMOS, in *IEEE International Solid-State Circuits Conference Digest of Technical Papers (ISSCC)* (2012), pp. 180–182

31. M. Winter, S. Kunze, E. Adeva, B. Mennenga, E. Matus, G. Fettweis, H. Eisenreich, G. Ellguth, S. Höppner, S. Scholze, R. Schüffny, T. Kobori, A 335Mb/s 3.9mm2 65nm CMOS flexible MIMO detection-decoding engine achieving 4G wireless data rates, in *IEEE International Solid-State Circuits Conference Digest of Technical Papers (ISSCC)* (2012), pp. 216–218

32. C. Xu, H. Li, R. Suaya, K. Banerjee, Compact AC modeling and performance analysis of through-silicon vias in 3-D ICs. IEEE Trans. Electron Devices **57**(12), 3405–3417 (2010)

33. C. Xu, R. Suaya, K. Banerjee, Compact modeling and analysis of through-Si-via-induced electrical noise coupling in three-dimensional ICs. IEEE Trans. Electron Devices **58**(11), 4024–4034 (2011)

34. I.-M. Yi, S.-M. Lee, S.-J. Bae, Y.-S. Sohn, J.-H. Choi, B. Kim, J.-Y. Sim, H.-J. Park, A 40-mv-swing single-ended transceiver for tsv with a switched-diode rx termination. IEEE Trans. Circuits Syst. Express Briefs **61**(12), 987–991 (2014)

35. L. Zhang, J.M. Wilson, R. Bashirullah, L. Luo, J. Xu, P.D. Franzon, A 32-Gb/s on-chip bus with driver pre-emphasis signaling. IEEE Trans. Very Large Scale Integr. Syst. **17**(9), 1267–1274 (2009)

# Chapter 4
# Multi-TSV Crosstalk Channel Equalization with Non-uniform Quantization

**Tobias Seifert, Friedrich Pauls, and Gerhard Fettweis**

## 4.1 Introduction

3D chip-stacks have added a new dimension to circuit design. Those designs can benefit from reduced length of global interconnects, resulting in decreased power consumption and increased area efficiency [1]. The signal transfer between two integrated circuit (IC) layers is usually realized by through-silicon via (TSV) links. Since the TSVs are directly placed in the substrate and surrounded by a thin insulator, undesired electrical coupling interference (crosstalk) can occur amongst neighboring TSVs.

In order to preserve most chip area for active circuits and to provide high vertical bandwidth, it is desirable to arrange multiple signal TSVs in a highly dense TSV cluster, exacerbating the problem of TSV-to-TSV coupling. Additionally, many applications demand for high energy efficiency, which requires capacitively driven transceivers with low voltage swings to limit the dynamic power consumptions [2]. Thus, even low coupling interference between signal TSVs can cause critical signal distortion that affects the functional correctness of the whole 3D IC system.

Crosstalk mitigation techniques have therefore been studied intensively, mostly concentrating on conventional signal shielding by adding ground TSVs [3] or using grounded guard rings and coaxial TSVs [4, 5]. In [6, 7], lossy passive contacts are used to compensate the capacitive effect of TSV interconnects and perform a partial equalization of the TSV frequency response. However, these approaches usually

T. Seifert (✉) • F. Pauls • G. Fettweis
Technische Universität Dresden, Vodafone Chair Mobile Communication Systems,
Georg-Schumann-Straße 11, 01187 Dresden, Germany
e-mail: tobias.seifert2@tu-dresden.de; friedrich.pauls@tu-dresden.de;
gerhard.fettweis@ifn.et.tu-dresden.de; Gerhard.Fettweis@tu-dresden.de

© Springer International Publishing Switzerland 2016
I.M. Elfadel, G. Fettweis (eds.), *3D Stacked Chips*,
DOI 10.1007/978-3-319-20481-9_4

lack acceptable area efficiency or do not adapt well to arbitrary TSV coupling channels. Furthermore, the introduced shielding capacitances reduce the achievable TSV bandwidth.

This chapter introduces a novel crosstalk compensation approach, originally presented in [8], that uses digital signal processing to equalize the TSV voltage signals. This approach is directly inspired by the field of mobile communications where signals are spatially multiplexed, transmitted over a time-dispersive multiple-input multiple-output (MIMO) radio channel, and finally equalized at the receiver [9]. In analogy to wireless communication, we regard the multi-TSV channel as a MIMO channel to which equalization as well as channel estimation techniques can be directly applied. We also show the robustness of this approach against uniform analog-to-digital converter (ADC) impairments, namely ADC quantization noise and ADC clock jitter. Furthermore, the scope is extended to non-uniform quantization, demonstrating that significant crosstalk reduction can be achieved, even if only few comparators are available for analog-to-digital conversion. Hence, the aim of this chapter is to prove the concept of digital equalization and to show a new method to deal with crosstalk in multi-TSVs. Aspects concerning specific implementation issues as well as optimized equalizer designs are outside the scope of this chapter.

The remainder of this chapter is organized as follows: Sect. 4.2 describes the multi-TSV system as well as the general channel equalization concept. Section 4.3 deals with a basic TSV-to-TSV lumped circuit model and its crosstalk channel frequency response, while Sect. 4.4 shows simulation results of the equalization performance. In Sect. 4.5, the equalization performance is investigated for non-uniform quantization. Section 4.6 concludes the book chapter.

## 4.2 Concept of Crosstalk Compensation Based on MIMO Channel Equalization

We consider two ICs which are vertically connected by $N$ TSV signal links, resulting in a 3D chip-stack. Note that the TSV array is not restricted to be arranged on a regular grid structure.

From a communications perspective it is possible to regard the upper IC layer as a transmitter, the $N$-TSV array as a MIMO channel, and the lower IC layer as a receiver. Thus, for each TSV the upper chip *transmits* a voltage signal $x_i(t)$ and the lower chip *receives* a corresponding voltage signal $y_j(t)$ with $i, j = 1, \ldots, N$. Their input–output relation is given by the channel impulse response $h_{ij}(t)$. Since the signal $y_j(t)$ might be interfered by transmit signals of other TSV signal links $i \neq j$, it has to be described as the superposition of all input signals, respectively, convolved by individual channel impulse responses, leading to

$$y_j(t) = \sum_{i=1}^{N} (h_{ij} * x_i)(t). \qquad (4.1)$$

Stacking up all $N$ output signals and transforming them to frequency-domain results in the compact matrix description

$$\underbrace{\begin{bmatrix} Y_1(f) \\ \vdots \\ Y_N(f) \end{bmatrix}}_{\mathbf{Y}} = \underbrace{\begin{bmatrix} H_{11}(f) & \cdots & H_{1N}(f) \\ \vdots & \ddots & \vdots \\ H_{N1}(f) & \cdots & H_{NN}(f) \end{bmatrix}}_{\mathbf{H}} \underbrace{\begin{bmatrix} X_1(f) \\ \vdots \\ X_N(f) \end{bmatrix}}_{\mathbf{X}}. \qquad (4.2)$$

Note that the channel matrix $\mathbf{H}(f)$ is assumed to be symmetric due to channel reciprocity. Each frequency response $H_{ii}(f)$ on the main diagonal represents the direct channel of TSV $i$, while each off-diagonal function $H_{ij}(f)$ corresponds to the coupling channel between input port of TSV $i$ and output port TSV $j$, causing crosstalk voltage at the output port of TSV $j$. In this chapter we show how the crosstalk interference can be significantly reduced by jointly equalizing the received signal vector $\mathbf{y}(t)$.

Channel equalization is a well-known technique in digital communications [9, 10], deployed to tackle intersymbol interference (ISI) and to improve the reliability of the estimated transmit symbols. For this purpose, $\mathbf{Y}(f)$ is filtered by the inverse channel matrix, approximated by a filter

$$\mathbf{G} = \hat{\mathbf{H}}^{-1}, \qquad (4.3)$$

the so-called zero-forcing (ZF) equalizer [10]. $\hat{\mathbf{H}}$ denotes the approximation or estimate of $\mathbf{H}$. Since the TSV channel does not change over time, the estimation of the channel as well as ZF-filter determination can be carried out as an initial calibration procedure.

### 4.2.1 Channel Equalization: Continuous-Time View

Assuming that $\mathbf{G}$ is a matrix of analog filter functions $G_{ij}(f)$, the resulting received and equalized signal vector in frequency-domain is

$$\hat{\mathbf{X}}(f) = \mathbf{G}(f)\mathbf{Y}(f). \qquad (4.4)$$

When $\hat{\mathbf{H}} = \mathbf{H}$, the equalized received signals are identical to the original transmit signals ($\hat{\mathbf{X}} = \mathbf{X}$) and the crosstalk is completely compensated.

### 4.2.2 Channel Equalization: Discrete-Time View

In communications, the equalization is usually carried out by a digital filter, assuming that a discrete-time signal $y_j[k]$ is received at the output port of TSV $j$. Similarly, the TSV channel is given by $M + 1$ channel coefficients $h_{ij}[m]$ with $m = 0, \ldots, M$. Note that $M = 0$ refers to a memoryless channel. Taking these changes into account, (4.1) turns into a discrete convolution, explicitly expressed by

$$y_j[k] = \sum_{i=1}^{N} \left( h_{ij}[0] x_i[k] \right) + \underbrace{\sum_{m=1}^{M} \sum_{i=1}^{N} \left( h_{ij}[m] x_i[k - m] \right)}_{y_j^{\mathrm{IF}}[k]}, \tag{4.5}$$

where $y_j^{\mathrm{IF}}[k]$ denotes the crosstalk interference caused by previously transmitted signal values $x_i[k - m]$, with $m \geq 1$. This interference can be subtracted in advance from the received signal by approximating the transmit signal $x_i[k - m] = \hat{x}_i[k-m]$ and the channel coefficients $h_{ij}[m] = \hat{h}_{ij}[m]$. Subsequently, the interference-reduced received signals $y_j'[k]$ are jointly equalized by means of the linear ZF-filter $\mathbf{G}_0 = \hat{\mathbf{H}}_0^{-1}$, where $\hat{\mathbf{H}}_0 \in \mathbb{R}^{N \times N}$ is composed of the estimated fundamental channel coefficients $\hat{h}_{ij}[0]$. The matrix description for the equalized signal vector $\hat{\mathbf{x}}[k] = [\hat{x}_1[k] \hat{x}_2[k] \ldots \hat{x}_N[k]]^{\mathrm{T}}$ in time-domain is

$$\hat{\mathbf{x}}[k] = \mathbf{G}_0 \, \mathbf{y}'[k], \qquad \text{with } \mathbf{y}'[k] = \mathbf{y}[k] - \sum_{m=1}^{M_{\mathrm{L}}} \hat{\mathbf{H}}[m] \, \hat{\mathbf{x}}[k - m], \tag{4.6}$$

where $M_{\mathrm{L}} \leq M$ is the channel memory length considered for equalization.

## 4.3 TSV-to-TSV Crosstalk Coupling

In the following we consider two neighboring TSVs which corresponds to $N = 2$. The focus is on TSV-to-TSV coupling, determined by the crosstalk channel frequency response $H_{\mathrm{c}}(f)$. The direct channel frequency response is given by $H_{\mathrm{D}}(f)$, leading to the frequency-domain TSV channel matrix

$$\mathbf{H} = \begin{bmatrix} H_{\mathrm{D}}(f) & H_{\mathrm{c}}(f) \\ H_{\mathrm{c}}(f) & H_{\mathrm{D}}(f) \end{bmatrix}. \tag{4.7}$$

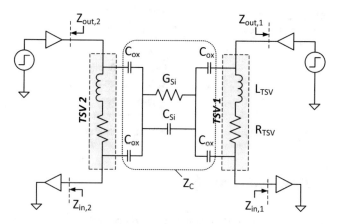

**Fig. 4.1** Lumped circuit model of two parallel TSVs, coupled by the impedance $Z_C$ and terminated by input and output impedances $Z_{in}$ and $Z_{out}$ of an I/O cell

## 4.3.1  Lumped Circuit Model

Two adjacent TSVs can be modeled by lumped elements, as shown in Fig. 4.1. The capacitance $C_{ox}$ of the silicon dioxide insulator, the substrate capacitance $C_{si}$, and substrate conductance $G_{si}$ are derived from material properties and TSV dimensions. Since the TSV impedance is very low in comparison with the coupling impedance, the components $R_{TSV}$ and $L_{TSV}$ are neglected [11]. Further the latency is neglected, leading to flat amplitude and constant phase. For the sake of simplicity, we therefore assume that the direct channel frequency response is ideally constant, i.e., $H_D(f) = 1$.

Based on the geometry and material properties of the TSVs, the remaining circuit components are calculated by

$$C_{ox} = \frac{\pi \epsilon_0 \epsilon_{ox} l}{\ln \left( \frac{r+t}{r} \right)} \tag{4.8}$$

$$C_{si} = \frac{\pi \epsilon_0 \epsilon_{si} l}{\ln \left( \frac{d}{2(r+t)} + \sqrt{\frac{d^2}{4(r+t)^2} - 1} \right)} \tag{4.9}$$

$$G_{si} = \frac{\sigma}{\epsilon_0 \epsilon_{si}} C_{si}. \tag{4.10}$$

Throughout this chapter, the values for the TSV length $l$, the TSV radius $r$, insulator thickness $t$, and TSV center-to-center distance $d$ are set as stated in Table 4.1.

**Table 4.1** TSV dimensions and I/O cell characteristic used in this chapter

| Parameter | | Value |
|---|---|---|
| TSV height | $l$ | $75\,\mu m$ |
| TSV radius | $r$ | $2.5\,\mu m$ |
| Insulator SiO$_2$ thickness | $t$ | $0.5\,\mu m$ |
| TSV center-to-center distance | $d$ | $10\,\mu m$ |
| Supply voltage | $V_{DD}$ | $1.0\,V$ |
| I/O input capacitance | $C_{in}$ | $48\,fF$ |
| I/O output resistance | $R_{out}$ | $1\,k\Omega$ |

### 4.3.2 I/O Termination Analysis

As shown in Fig. 4.1, TSVs are connected to I/O cells whose input impedance $Z_{in}$ and output impedance $Z_{out}$ have to be taken into account when the total transmission behavior is analyzed. If we assume that the ports of each TSV are terminated by identical I/O cells, we obtain the crosstalk channel frequency response

$$H_c(f) = \frac{Z_{out}Z_{in}^2}{Z_{out}^2 Z_c + Z_{out}Z_{in}(3 + 2Z_c) + Z_{in}^2(2Z_{out} + Z_c)}, \qquad (4.11)$$

where the coupling impedance is given by

$$Z_c = \frac{G_{Si}}{G_{Si}^2 + (\omega C_{Si})^2} - j\left(\frac{\omega C_{Si}}{G_{Si}^2 + (\omega C_{Si})^2} + \frac{1}{\omega C_{ox}}\right). \qquad (4.12)$$

Often S-parameters are used to describe the transfer behavior in the frequency-domain. For this purpose, the input and output ports of the TSVs are terminated by $50\,\Omega$. However, in a real 3D IC, TSVs are connected to I/O cells which typically represent a much higher impedance. Hence, the termination condition of the model is changed to input impedance $Z_{in} = -j(1/C_{in})$ and output impedance $Z_{out} = R_{out}$ (see Table 4.1) of a possible 65 nm I/O cell.

Figure 4.2 depicts the magnitude of $H_c(f)$ for both cases. The curves show that the high impedance of a realistic I/O termination exposes a significant larger coupling level (more than 30 dB for $f \leq 2\,\text{GHz}$) compared to the case of $50\,\Omega$ termination.

Throughout the rest of the chapter, I/O cell termination is used as described.

## 4.4 TSV-to-TSV Crosstalk Channel Equalization

The simplified lumped circuit leads to the transmission system model shown in Fig. 4.3. The received signals $y_1(t)$ and $y_2(t)$ are sampled and uniformly quantized by an ADC and subsequently equalized in the digital domain. Note that $Z_{in}$ represents now the input impedance of the ADC.

**Fig. 4.2** Termination analysis for the TSV-to-TSV coupling channel. In the 50 Ω case, the input and output impedances are $Z_{in} = Z_{out} = 50\,\Omega$. For the *I/O cell* case $C_{in} = 48\,\text{fF}$ and $R_{out} = 1\,\text{k}\Omega$

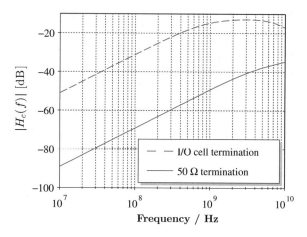

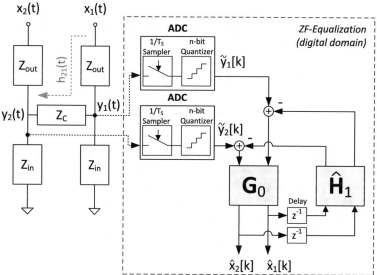

**Fig. 4.3** Transmission system model of the two parallel TSV links, including ADCs and the equalization unit according to (4.6). The considered memory length of the depicted equalizer is $M_L = 1$

### 4.4.1   Continuous-Time View

Applying a 1 GHz rectangular signal of magnitude 1.0 V to the input $x_1(t)$ leads to a crosstalk voltage signal at the output $y_2(t)$, as depicted in Fig. 4.4. A peak voltage of 180 mV is observed.

In order to illustrate the equalization principle as described in Sect. 4.2.1, the ADC sampling rate is assumed to be ideally high enough to resolve all frequencies of an arbitrary input signal, hence we are regarding initially continuous-time signals.

**Fig. 4.4** Illustration of the crosstalk signal and the equalized signal observed at TSV 2 when an 1 GHz signal is applied to TSV 1. The equalization is carried out for the ADC output signal $\tilde{y}_2(t)$, uniformly quantized within the range $[-0.2\,\text{V}, 1.2\,\text{V}]$

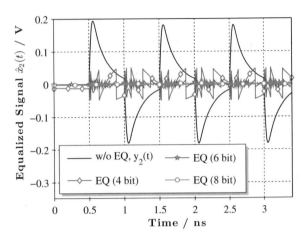

However, a limited number of ADC quantization stages is considered, resulting in quantized output signals $\tilde{y}_{1,2}(t)$. It is assumed that the crosstalk channel $H_c(f)$ is perfectly known to the equalizer.

By using (4.3) and (4.4), equalization results in

$$\hat{X}_2(f) = \frac{1}{\det(\mathbf{H})} \left( H_D(f)\,\tilde{Y}_2(f) - H_c(f)\,\tilde{Y}_1(f) \right) \tag{4.13}$$

in frequency-domain. Figure 4.4 depicts the equalized time signal $\hat{x}_2(t)$, depending on various ADC quantization. It can be seen that the larger the error due to bit width limitation, the more flickering is exhibited in the equalized signal. However, even for 4 bit resolution the observed quantization noise is always much lower than the peak voltage in the non-equalized case.

### 4.4.2  Discrete-Time View

When TSVs are used as network-on-chip links, $y_j(t)$ represents a digital signal. One or multiple comparators (each adjusted to the respective threshold level) are usually deployed at the output ports of the TSVs to decide for the correct binary sequence.

However, the induced crosstalk interference may lead to wrong decisions, depending on the magnitude and decay time of the interference signal and the threshold level distances. In this regard, capacitively driven transceivers are especially susceptible to interference and noise [2].

In order to show the potential of equalization for the system depicted in Fig. 4.3, we assume no 1-bit comparator but a configurable ADC. The ADC samples the TSV signal $y_j(t)$ at half the symbol period $T_S$ but its clock is impaired by jitter, modeled as a gaussian random variable $n$ with distribution $\mathcal{N}\left(0, \sigma^2\right)$. Hence, we obtain the discrete output sequence

$$\tilde{y}_j[k] = \lfloor y_j(t = t_k) \rceil, \tag{4.14}$$

with $t_k = \left(k + \frac{1}{2}\right) T_S + n, \ k \in \mathbb{N}$.

$\lfloor \cdot \rceil$ denotes the quantization operation and $n$ the clock jitter. The ADC output is equalized as shown in (4.6). To quantify the equalization performance, the root-mean-square deviation (RMSD) for a single TSV, defined by

$$\text{RMSD}\left(\hat{x}_i[k]\right) = \sqrt{\frac{\text{E}\|\hat{\mathbf{x}}_k - \mathbf{x}_k\|^2}{N}} \tag{4.15}$$

is used, where $\mathbf{x}_k$ is the reference transmit vector with $x_i[k] \in \{0\,\text{V}, 1\,\text{V}\}$. Note that the channel coefficients $h_{ij}[k]$ are perfectly known to the equalizer.

The performance analysis has been carried out for the TSV-to-TSV scenario with $N = 2$ and a symbol period of $T_S = 0.5\,\text{ns}$. The mean square deviation was determined by averaging over $3 \cdot 10^5$ symbol transmissions and jitter realizations. The plots in Fig. 4.5 depict the equalization performance depending on the ADC clock jitter. We see that at least two channel coefficients ($M_L \geq 1$) have to be considered to benefit from equalization. A small performance increase for $M_L = 2$ can only be observed if the clock jitter and quantization noise are low enough. Otherwise, the ADC impairments dominate the overall performance. For 4-bit resolution and jitter $\sigma < 0.08\,\text{ns}$, the crosstalk interference is reduced by more than 50 %. By increasing the signal resolution up to 8-bit, the quantization noise decreases and the jitter noise becomes dominant, as shown in Fig. 4.5c. If no jitter noise is present, the remaining deviation is less than $2\,\text{mV}$ and the crosstalk interference is almost perfectly compensated.

## 4.5 Impact of Non-uniform Quantization on Equalization Performance

The distortion due to signal quantization can significantly be reduced by a proper quantization level distribution, which depends on the probability density of the analog signal [12]. To enhance the performance of the proposed equalization approach, different non-uniform ADCs with adjustable quantization level positions are considered and compared.

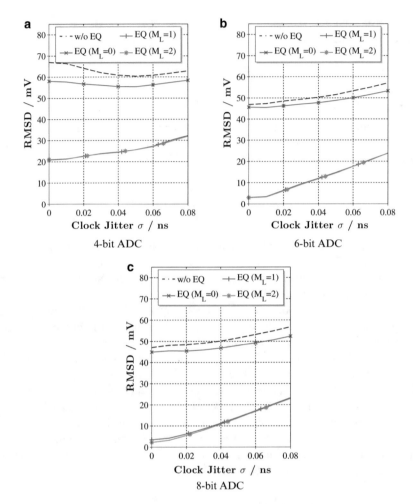

**Fig. 4.5** Illustration of the RMSD for the equalized (EQ) $\hat{x}_{1/2}[k]$ and non-equalized (w/o EQ) $\tilde{y}_{1/2}[k]$ received signal, plotted over the ADC clock jitter standard deviation $\sigma$. A (**a**) 4-bit, (**b**) 6-bit, and (**c**) 8-bit uniform quantizer with the range $[-0.2\,\text{V}, 1.2\,\text{V}]$ is used. Additionally, the impact of different channel memory lengths $M_L$ is also investigated

## 4.5.1 Quantizer Types

Besides a standard uniform quantizer, two non-uniform quantizers are considered. Since both non-uniform quantizers have uniform quantization level distribution around the corresponding signal levels, they can be seen as partially uniform.

- *Uniform Quantizer*:
  This quantizer is a *n*-bit quantizer whose quantization levels are uniformly distributed within the range $[-x_{q,min}, x_{q,max}]$, with $x_{q,max} = 1\,\text{V} + x_{q,min}$. This quantizer has been used for all simulations in Sect. 4.4 (Fig. 4.6).

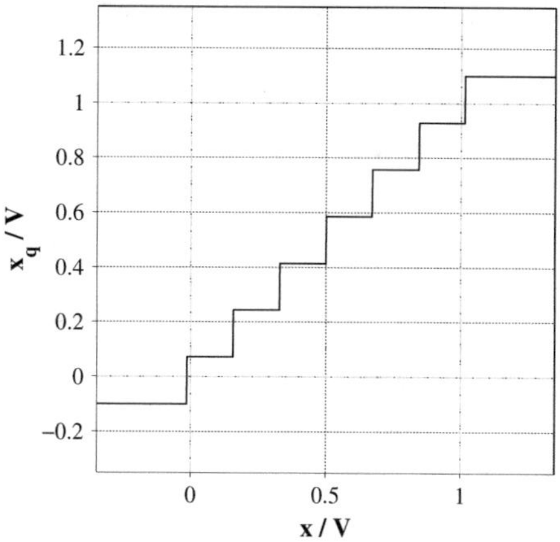

**Fig. 4.6** Illustration of the step curve of the uniform quantizer for 3 bit and a minlevel of $x_{q,min} = 100\,mV$

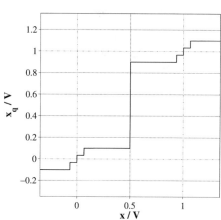

**Fig. 4.7** Illustration of the step curve of the partial mid-rise quantizer for 3 bit and a minlevel of $x_{q,min} = 100\,mV$

- *Non-uniform Quantizer (Mid-rise):*
  Here, the quantization levels are uniformly distributed at the respective signal levels 0 and 1 V. Therefore, two uniform subranges $[-x_{q,min}, x_{q,min}]$ and $[1\,V - x_{q,min}, 1\,V + x_{q,min}]$ exist. Since the quantization stages are grouped symmetrically for each signal level, this non-uniform quantizer is referred to *mid-rise* (Fig. 4.7).

- *Non-uniform Quantizer (Mid-tread):*
  Here, the quantization levels are uniformly distributed at the respective signal levels 0 and 1 V as well. In contrast to the non-uniform mid-rise quantizer, each signal level is identical with one quantization level. The remaining quantization levels are positioned towards the threshold $V_{DD}/2$. Hence, this non-uniform quantizer is referred to as *mid-tread*. The ranges are given by $[-x_{q,min}, x_{q,min} + \Delta q]$ and $[1\,V - x_{q,min} - \Delta q, 1\,V + x_{q,min}]$, where $\Delta q$ denotes the distance between two adjacent quantization stages (Fig. 4.8).

**Fig. 4.8** Illustration of the step curve of the partial mid-tread quantizer for 3 bit and a minlevel of $x_{q,min} = 100\,mV$

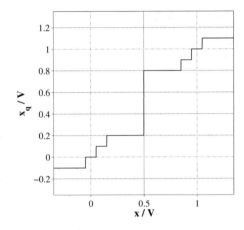

**Fig. 4.9** Uniform quantizer: equalization performance in terms of RMSD, depending on the number and positions of the quantization stages

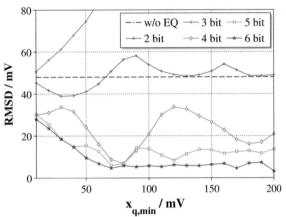

## 4.5.2  Performance Analysis

In the following, the equalization performance is investigated for the two different non-uniform ADCs with adjustable quantization levels. The uniform ADC serves as a reference. It is assumed that the ADC clock jitter can be neglected, especially when a low number of bits is used for signal quantization.

The performance of all three quantizers has been investigated depending on the number of bits and the quantization level positions. The corresponding performance results are shown in Figs. 4.9, 4.10, and 4.11. The dashed line indicates the resulting RMSD when no equalization is applied. This deviation due to crosstalk is about 48 mV and confirms the result in Fig. 4.5c.

As we can see from Fig. 4.9, an increase in bitwidth results typically in lower RMSD values. However, to reach a significant crosstalk reduction of RMSD < 10 mV, at least 4 bits would be needed and $x_{q,min}$ should be set to 80 mV. The more bits are used for quantization, the less important is the exact quantization

**Fig. 4.10** Non-uniform quantizer (mid-rise): illustration of the equalization performance in terms of RMSD, depending on the number and positions of the quantization stages

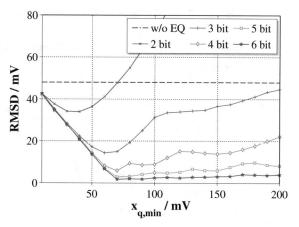

**Fig. 4.11** Non-uniform quantizer (mid-tread): illustration of the equalization performance in terms of RMSD, depending on the number and positions of the quantization stages

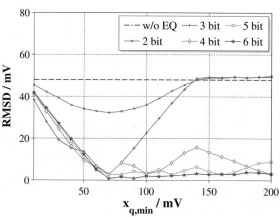

level positioning, expressed by $x_{q,min}$. However, the RMSD value is unlikely to get reduced any further. In Fig. 4.10, the equalization performance for the non-uniform mid-rise quantizer is shown. Already 2 bit can reduce the RMSD, even if not significantly. When 3 bit are spent for quantization, the RMSD value can be reduced down to 15 mV. Using the non-uniform mid-tread quantizer, the situation is even better. If $x_{q,min}$ is properly set to 70 mV, a RMSD value of less than 5 mV can be observed. Further increase in bitwidth hardly allows any further crosstalk reduction.

These results are very promising as they prove that even a low number of quantization levels results in significant crosstalk compensation performance provided the quantization levels are properly positioned.

Finally, Table 4.2 gives an overview of the minimum RMSD values that can be reached for the various $n$-bit quantizers.

**Table 4.2** Crosstalk compensation performance depending on $n$-bit quantizers

| Quantizer | $n$ (bit) | RMSD (mV) | RMSD-reduction factor |
|---|---|---|---|
| Uniform | 2 | 51 | 1.0 |
| | 3 | 39 | 1.2 |
| | 4 | 7 | 6.8 |
| | 5 | 6 | 8.0 |
| Non-uniform mid-rise | 2 | 34 | 1.4 |
| | 3 | 14 | 3.4 |
| | 4 | 6 | 8.0 |
| | 5 | 3 | 16.0 |
| Non-uniform mid-tread | 2 | 32 | 1.5 |
| | 3 | 2.9 | 16.0 |
| | 4 | 2.8 | 17.1 |
| | 5 | 2.7 | 17.7 |

The RMSD reduction is determined in comparison with the unequalized case where RMSD = 48 mV

## 4.6  Conclusion

This chapter has presented a novel approach to perfectly compensate for crosstalk effects in TSV arrays by means of digital equalization of the multi-TSV channel. The main advantage of this approach is its ability to adapt to arbitrary coupling channels while taking mismatch into account. Moreover, since crosstalk is not avoided in advance, TSVs can be arranged very compactly, leading to high area efficiency.

To prove the described concept, the equalization performance was evaluated for a TSV-to-TSV scenario based on a lumped circuit model. Although both ADC clock jitter and ADC quantization noise limits the performance, remarkable crosstalk reduction could be observed even for as few as 4 bits of quantization. Moreover, the performance can be further optimized using a non-uniform quantizer, making this approach feasible with low additional circuit complexity.

## References

1. J.W. Joyner, et al., Impact of three-dimensional architectures on interconnects in gigascale integration. IEEE Trans. Very Large Scale Integr. Syst. **9**(6), 922–928 (2001)
2. D. Walter, et al., A source-synchronous 90Gb/s capacitively driven serial on-chip link over 6mm in 65nm CMOS, in *IEEE International Solid-State Circuits Conference* (2012), pp. 180–182
3. Y.-J. Chang, et al., Novel crosstalk modeling for multiple through-silicon-vias (TSV) on 3-D IC: experimental validation and application to Faraday cage design, in *IEEE Conference on Electrical Performance of Electronic Packaging and Systems* (2012), pp. 232–235

4. N.H. Khan, S.M. Alam, S. Hassoun, Through-silicon via (TSV)-induced noise characterization and noise mitigation using coaxial TSVs, in *IEEE International Conference on 3D System Integration* (2009), pp. 1–7
5. J. Cho, et al., Modeling and analysis of through-silicon via (TSV) noise coupling and suppression using a guard ring, in *IEEE Transactions on Components, Packaging and Manufacturing Technology* (2011), pp. 220–233
6. J. Kim, et al., Through silicon via (TSV) equalizer, in *IEEE Conference on Electrical Performance of Electronic Packaging and Systems* (2009), pp. 13–16
7. R.-B. Sun, C.-Y. Wen, R.-B. Wu, Passive equalizer design for through silicon vias with perfect compensation, in *IEEE Transactions on Components, Packaging and Manufacturing Technology* (2011), pp. 1815–1822
8. T. Seifert, G. Fettweis, Multi-TSV crosstalk compensation based on digital MIMO channel equalization, in *IEEE Conference on Electrical Performance of Electronic Packaging and Systems* (2014), pp. 49–52
9. J. Proakis, D. Manolakis, *Digital Signal Processing*, 4th edn. (Prentice Hall, Englewood Cliffs, 2006)
10. G. Kaleh, Channel equalization for block transmission systems. IEEE J. Sel. Areas Commun. **13**, 110–121 (1995)
11. T. Song, Analysis of TSV-to-TSV coupling with high-impedance termination in 3D ICs, in *IEEE International Symposium on Quality Electronic Design* (2011), pp. 1–7
12. P.F. Panter, W. Dite, Quantization distortion in pulse-count modulation with nonuniform spacing of levels, in *Proceedings of the IRE* (1951), pp. 44–48

# Chapter 5
# Energy Efficient TSV Based Communication Employing 1-Bit Quantization at the Receiver

Lukas Landau and Gerhard Fettweis

## 5.1 Energy Efficient Multigigabit-Per-Second Transceiver

When considering a short range multigigabit-per-second communication link such as given in a through silicon via (TSV) scenario, the overall energy consumption depends strongly on the type of analog-to-digital conversion. Among various combinations of sampling rate and quantization resolution the special case of energy efficient 1-bit quantization is the most promising alternative to high resolution quantization that is commonly used in communications. This is due to several reasons: On the one hand the switching delay of mainstream CMOS technology has become extremely short which allows for increasing sampling rates. On the other hand the supply voltage is decreasing more and more aiming for lower power consumption. As a consequence a supply voltage of nowadays roughly 1.0–1.4 V provides only a little headroom for sophisticated analog processing for the amplitude resolution. Hence, the resolution of a signal edge transition in time domain can be superior to voltage resolution of analog signals [13]. In what follows, communication concepts are investigated which are based on a discrete time channel whose output is quantized with only 1-bit of resolution. Indeed, the coarse quantization corresponds to a significant loss in terms of achievable rate. However, this loss can be compensated partially by an increase of sampling rate. The advantage of oversampling the sign of the received signal is already investigated in [10] in a theoretical fashion. However, practical concepts are still unknown and are thus addressed in this work (Fig. 5.1).

L. Landau (✉) • G. Fettweis
Technische Universität Dresden, Vodafone Chair Mobile Communication Systems,
Georg-Schumann-Straße 11, 01187 Dresden, Germany
e-mail: lukas.landau@tu-dresden.de; Gerhard.Fettweis@tu-dresden.de

© Springer International Publishing Switzerland 2016
I.M. Elfadel, G. Fettweis (eds.), *3D Stacked Chips*,
DOI 10.1007/978-3-319-20481-9_5

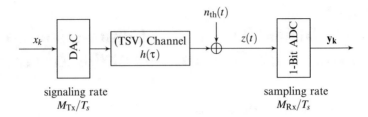

**Fig. 5.1** System model for communications between stacked chips, e.g., using a through silicon via; $M_{Rx} = MM_{Tx}$; $n_{th}(t)$ thermal noise process according to noise power density spectrum

## 5.2 System Model

Due to charging and discharging of the inherent capacitance of the considered communication channel the signal at the channel output will not follow immediately the signal transitions induced by the signal source. This behavior is taken into account by considering a channel with memory where its impulse response in time domain is denoted as $h(\tau)$. Further, the output signal is disturbed by an additive thermal noise process $n_{th}(t)$ characterized by its noise power spectrum. While the transmitter generates transmit symbols $x_k$ with the signaling rate $\frac{M_{Tx}}{T_s}$, the sampling frequency at the receiver is considered as $\frac{M_{Rx}}{T_s}$, where $M_{Rx} = MM_{Tx}$ holds. In this regard, $M$ is a design parameter being an integer value that relates signaling and sampling rate. $T_s$ is the reference time interval. The digital-to-analog converter (DAC) is modeled by the filter impulse response denoted as

$$g_{DAC}\left(\tau, \frac{M_{Tx}}{T_s}\right) = \begin{cases} \sqrt{\frac{M_{Tx}}{T_s}}, & 0 \leq \tau < \frac{T_s}{M_{Tx}} \\ 0, & \text{else.} \end{cases} \tag{5.1}$$

Combining channel and DAC yields the effective waveform filter $v(\tau) = g_{DAC}(\tau) * h(\tau)$. Indeed the channel can be characterized by the TSV structure hence it will be termed TSV channel in what follows. The analog received signal is described by

$$z(t) = \sum_{k=-\infty}^{\infty} x_k v\left(t - k\frac{T_s}{M_{Tx}}\right) + n_{th}(t), \tag{5.2}$$

which is then sampled and quantized with a symmetric 1-bit ADC

$$y_{k,m} = Q_1\left(z\left(k\frac{T_s}{M_{Tx}} + \frac{mT_s}{M_{Rx}}\right)\right), \quad m = 0, \ldots, M-1, \tag{5.3}$$

where $Q_1(z \geq 0) = 1$, respectively, $Q_1(z < 0) = -1$ denotes the quantization. The $M$ samples of the received signal with respect to a $k$th symbol duration are denoted in vector notation $\mathbf{y}_k = [y_{k,0}, \ldots, y_{k,M-1}]^T$. In what follows, the notation $x^n = [x_1, \ldots, x_n]^T$ is used for stacked scalars especially sequences of input symbols. Stacked output vectors are denoted using bold letters $\mathbf{y}^n = [\mathbf{y}_1, \ldots, \mathbf{y}_n]^T$. Furthermore, subsequences are denoted as $x_{k-L}^k = [x_{k-L}, \ldots, x_k]^T$. The introduced communication link characterized by a total channel memory of length $L$ can be fully described by an equivalent discrete time model given by

$$\mathbf{z}_k = \mathbf{V}\mathbf{U}x_{k-L}^k + \mathbf{G}\mathbf{n}_{k-\xi}^k, \tag{5.4}$$

where $\mathbf{V}$ is the joint filter matrix of DAC and TSV channel and $\mathbf{G}$ is the filter matrix which colorizes the white noise samples $\mathbf{n}_{k-\xi}^k$. In this regard, $\xi + 1$ describes the number noise samples being correlated. $\mathbf{U}$ represents the $M$-fold upsampling matrix with the dimension $(L+2)M - 1 \times L + 1$ and its elements are given by

$$U_{i,j} = \begin{cases} 1 & \text{for } i = j \cdot M \\ 0 & \text{else.} \end{cases} \tag{5.5}$$

The filter matrices containing the discrete time representation of the reversed filter impulse response function vector $\mathbf{v}_r = \text{reverse}(\mathbf{v})$ and $\mathbf{g}_r = \text{reverse}(\mathbf{g}_{DAC})$ have Toeplitz structure as follows

$$\mathbf{V} = \begin{pmatrix} [\,\mathbf{v}_r^T\,] & 0 \cdots & 0 \\ 0 & [\,\mathbf{v}_r^T\,] & 0 \cdots 0 \\ & \ddots & \ddots & \ddots \\ 0 \cdots & & 0 & [\,\mathbf{v}_r^T\,] \end{pmatrix}, \quad \mathbf{G} = \begin{pmatrix} [\,\mathbf{g}_r^T\,] & 0 \cdots & 0 \; 0 \\ 0 & [\,\mathbf{g}_r^T\,] & 0 \cdots 0 \; 0 \\ & \ddots & \ddots & \ddots \\ 0 \cdots & & 0 & [\,\mathbf{g}_r^T\,] \; 0 \end{pmatrix}, \tag{5.6}$$

where $\mathbf{V}$ has the dimension $M \times M(L+2) - 1$ and $\mathbf{G}$ has the dimension $M \times M(\xi+1)$. The vector $\mathbf{n}_k = [n_{k,0}, \ldots, n_{k,M-1}]$ describes independent and identically distributed (i.i.d.) noise realizations with variance $\sigma_n^2$. It is assumed that the noise filter has a unit energy property such that the filtered noise samples also have variance $\sigma_n^2$. Finally the received signal is converted into a vector of binary symbols with value 1 or $-1$ by

$$\mathbf{y}_k = Q_1\{\mathbf{z}_k\}. \tag{5.7}$$

From literature it is known, that Markov processes can asymptotically achieve the capacity of intersymbol interference channels [5]. Indeed the considered channel is characterized by memory such that it is considered that an input signal $x_k$ is the output of a Markov source, such that each transmit symbol depends on a finite number $N$ of previous input symbols $P(x_k|x_{k-N}^{k-1})$. The achievable rate for different signaling methods, respectively, modulation strategies will be studied. For simplicity the following study will consider that the comparator at the receiver is symmetric around 0 and $E\{x_k\} = 0$.

## 5.3  Transmission Schemes

The fully connected $n$-multiprocessor has to cope with a large number of connection links given by $\sum_{i=1}^{n-1}(n-i)$. For larger $n$ a communication bottleneck appears since connections cannot be realized physically when satisfying a requirement on a maximum ratio between area required for connectivity and processing area. Furthermore a high connection density induces impairments such as crosstalk which need to be compensated like in Chap. 4. A direct consequence of a limited number of physical links is the need for multigigabit-per-second transmission rate at an affordable energy consumption. In what follows, two advanced communication concepts will be reviewed, which are suitable for energy efficient chip-to-chip communication utilizing the 1-bit quantizer at the receiver. As there is no bandpass transmission considered for the chip-to-chip communication the investigation is based on a real valued baseband. In this regard, real valued modulation schemes have been investigated which are based on amplitude modulation (4-ASK) or faster-than-Nyquist signaling. Sophisticated sequence designs, matched to the 1-bit quantizer, will be discussed for both cases.

### 5.3.1  ASK Sequences

In this section it is aimed to increase the transmission rate by employing multiple amplitude levels [7]. Especially 4-ASK sequences are considered, which are in the conventional fashion hardly detectable with a 1-bit receiver. This issue can be tackled by oversampling and considering the waveform and introducing a sophisticated sequence design as follows. One example of a sequence design is carried out for a parameter selection which has been found out to be well fitting for communications. While a signaling rate of $\frac{1}{T_s}$ is considered at the transmitter, the receiver employs a sampling rate of $\frac{M}{T_s}$ with $M = 3$. For a proof of concept, in the following it is assumed that the TSV channel impulse response is given by

$$h(\tau) = \begin{cases} \sqrt{\frac{1}{T_s}}, & 0 \leq \tau < T_s \\ 0, & \text{else}, \end{cases} \tag{5.8}$$

which is a moving average filter having low-pass characteristic. As a consequence the entire discrete channel which corresponds to (5.1) and (5.8) is denoted as $\mathbf{v} = \left[\frac{1}{3}, \frac{2}{3}, 1, \frac{2}{3}, \frac{1}{3}, 0\right]^T$. The sequence design principle does not take into account noise. Later for numerical evaluations a truncated cosine function will be considered which results in a more realistic channel model and smoother transitions. Also additive colored noise will be considered.

### 5.3.1.1   Independent and Uniformly Distributed Input

For a channel which is not restricted by quantization, a 4-ASK input would be chosen as independent and uniformly distributed (i.u.d.) symbols. This corresponds to 4 input symbols, resp., an input entropy rate 2 bits per symbol. The possible signal transitions with the given waveforms are illustrated in Fig. 5.2, where two vertical bars in the middle of two sequential symbols indicate the additional sampling points in time domain. With respect to the constrained receiver design, the signal reception suffers from the coarse quantization and a number of sequences cannot be reconstructed unambiguously. In order to gain understanding of the reconstruction problem it is useful to distinguish between classes of symbol transitions. In the present example 3, respectively, 4 classes have been identified and they are called transition states. Each individual sequence can be represented as a realization of a state machine drawn in Fig. 5.3. According to the binary observation at the receiver with threefold oversampling the transition states named "A", "B", "C" and "D" have special properties in terms of sequence reconstruction:

- A: $x_{k-1}$ and $x_k$ can be directly reconstructed based on the current channel output ("Decision")

**Fig. 5.2**  i.u.d. symbol transitions, 4-ASK symbols, triangular waveform

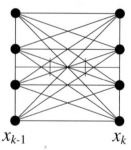

$$x_{k-1} \qquad\qquad x_k$$

**Fig. 5.3**  State machine describing i.u.d. sequences

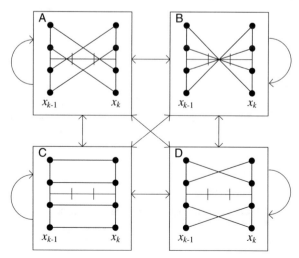

- B: $x_{k-1}$ and $x_k$ can be reconstructed when $x_{k-1}$ or $x_k$ is known at the receiver ("Forward")
- C: possible ambiguity with state D ("Ambiguity1")
- D: possible ambiguity with state C ("Ambiguity2")

Due to the fact that not all sequences can be distinguished at the receiver the achievable rate will be below the source entropy rate of 2 bits per symbol. In the next sections the transition states are utilized for the design of structured sequences that can be reconstructed.

### 5.3.1.2 Reconstructable Sequences

The sequence reconstruction problem occurs when the ambiguity states are passed in a random way. One approach to cope with the problem is to apply a special rule for one of the ambiguity states "C" or "D". In the proposed approach the state "D" has been selected. The special rule implies that after passing the "D" transition state and until passing the next "A" transition state, only "B" transition states are allowed to occur. "B" transition states which occur subsequently to a "D" are named "B*". The "A" state ("Decision") is a terminal state that disambiguates a received segment of a sequence. The corresponding modified state machine is drawn in Fig. 5.4. The adjacency matrix, which indicates a connection in the modified state machine with a 1 entry, is given by

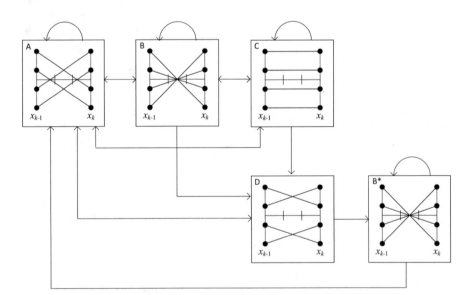

**Fig. 5.4** State machine describing reconstructable sequences with $P(x_k|x^{k-1})$, some connections are unidirectional

$$\mathbf{D} = \begin{bmatrix} 1\ 1\ 1\ 1\ 0 \\ 1\ 1\ 1\ 1\ 0 \\ 1\ 1\ 1\ 1\ 0 \\ 1\ 0\ 0\ 0\ 1 \\ 1\ 0\ 0\ 0\ 1 \end{bmatrix}, \tag{5.9}$$

where the row corresponds to the current state and the column corresponds to the subsequent state. Shannon showed in [11] that the maximum entropy rate of the proposed state machine is given by,

$$H_{\max} = \lim_{n \to \infty} \frac{1}{n} \log_2 \sum_{i,j} [\mathbf{D}^n]_{i,j}$$

$$= \log_2(\lambda)$$

$$= 1.7716 \text{ bit per symbol}, \tag{5.10}$$

where $\lambda$ is the largest real valued eigenvalue of $\mathbf{D}$. In order to obtain the reconstruction property some redundancy is added because the maximum input entropy rate for 4-ASK sequences is 2 bit per symbol. Furthermore following [11] the corresponding transition probabilities are given by

$$P_{i,j} = \frac{b_j}{b_i} \cdot \frac{D_{i,j}}{\lambda}, \tag{5.11}$$

where $b_i$ and $b_j$ are entries of the right-sided eigenvector corresponding to the eigenvalue $\lambda$.

The symbol probability of the $k$th symbol depends on previous symbol realizations $P(x_k|x^{k-1})$. Nevertheless, the number of previous symbols is not limited. As a result this source model cannot be described by a state model where each state equals to a number of previous symbols, which might be favorable in terms of analysis of the achievable rate. An alternative model providing this property is presented in the next section.

### 5.3.1.3 Reconstructable Finite State Markov Source

For simplicity a source model where each output depends on a finite number $N$ of previous symbols is beneficial. In this case the transition probabilities can be written as

$$P(x_k|x^{k-1}) = P(x_k|x_{k-N}^{k-1}). \tag{5.12}$$

In order to obtain this further modifications can be included. For specific $N$ the following requirements need to be fulfilled:

- $N = 1$ avoiding state "D"
- $N = 2$ state "A" is directly enforced when "D" occurs:
  "DA"
- $N = 3$ state "A" is enforced when "D" occurs:
  directly "DA" or postponed "DB*A"
- $N = 4$ state "A" is enforced when "D" occurs:
  directly "DA" or postponed "DB*A" resp., "DB*B**A"

The corresponding state machines are illustrated in Fig. 5.5. Analogously to the previous section the adjacency matrix can be found, e.g., for $N = 1$ and $N = 2$ by considering directional connections of the state machines in Fig. 5.5

$$\mathbf{D}_1 = \begin{bmatrix} 1 & 1 & 1 \\ 1 & 1 & 1 \\ 1 & 1 & 1 \end{bmatrix}, \quad \mathbf{D}_2 = \begin{bmatrix} 1 & 1 & 1 & 1 \\ 1 & 1 & 1 & 1 \\ 1 & 1 & 1 & 1 \\ 1 & 0 & 0 & 0 \end{bmatrix}. \tag{5.13}$$

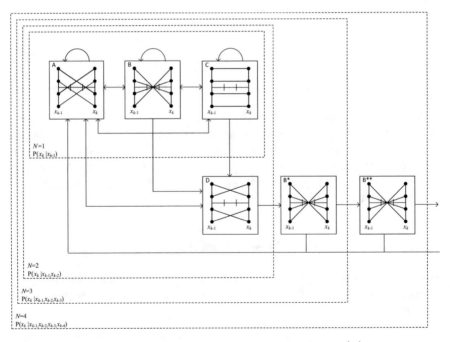

**Fig. 5.5** State machine describing reconstructable sequences with $P(x_k | x_{k-N}^{k-1})$ property, some connections are unidirectional

**Table 5.1** Source entropy rates of reconstructable sequences and independent uniformly distributed symbols

| Sequence property | $N = 1$ | $N = 2$ | $N = 3$ | $N = 4$ | $N = \infty$ | i.u.d. |
|---|---|---|---|---|---|---|
| $H(X)$/[bit per symbol] | 1.585 | 1.7237 | 1.7583 | 1.7678 | 1.7716 | 2 |

**Table 5.2** Number of reconstructable 4-ASK sequences and source entropy

| Super-symbol length | $S = 2$ | $S = 3$ | $S = 4$ | $S = 5$ | $S = 6$ | $S = 7$ | $S = 8$ | $S = 9$ |
|---|---|---|---|---|---|---|---|---|
| Valid super-symbols | 8 | 28 | 96 | 328 | 1120 | 3824 | 13056 | 44756 |
| Source entropy rate w/o bit-mapping | 1.5 | 1.603 | 1.646 | 1.672 | 1.688 | 1.700 | 1.709 | 1.716 |
| Source entropy rate w/direct bit-mapping | 1.5 | 1.333 | 1.5 | 1.6 | 1.667 | 1.571 | 1.625 | 1.667 |

It can be seen that small values for $N$ provide already good approximations with an entropy rate close to the state machine in Fig. 5.4. Therefore we do not consider larger values of $N$. When translating the proposed state machine into a process where the state is given by the number of previous symbols $s_{k-1} = x_{k-N}^{k-1}$ an equivalent adjacency matrix $\tilde{\mathbf{D}}$ can be found. This property allows for a straightforward computation of the achievable rate in Sect. 5.4. Applying (5.11) to the adjacency matrix $\tilde{\mathbf{D}}$ leads to the transition probabilities $P_{i,j} = P(s_k|s_{k-1}) = P(x_k|x_{k-N}^{k-1})$ corresponding to the maxentropic source for the given state machine (Table 5.1).

### 5.3.1.4 Reconstructable Sequences with Fixed Length

A simple method for utilizing reconstructable sequences for transmission of random information bits is given by the consideration of direct bit-mapping to sequences with fixed length [4, 12]. Table 5.2 summarizes the losses in terms of source entropy. The fixed length sequences are termed super-symbols. On one hand there is an entropy loss caused by the fixed length sequences. Furthermore there might be entropy loss considering a direct mapping from information bits to sequences. This case occurs, when the binary logarithm of the number of valid sequences is not an integer value.

## 5.3.2   Run-Length Limited Sequences

An alternative way for obtaining an increase in achievable rate is given by directly increasing the signaling rate of binary input signals [3]. However, when keeping the same channel characteristic fixed as (5.8) or later (5.22) the level of intersymbol interference in terms of number of corrupted symbols rises when increasing the signaling rate. To counteract this limitation the utilization of so-called run-length limited sequences introduces appropriate redundancy which finally enables signal reconstruction. The run-length limited sequences are well established for magnetic recordings and a broad overview can be found in [9] which is partially repeated in this section. Conventionally, a so-called $(d, k)$ sequence is considered before translating it into a run-length limited code. A $(d, k)$ sequence is a binary sequence which is characterized by at least $d$ but at a max of $k$ zeros following a one. The $(d, k)$ sequence is converted into the run-length limited sequences by flipping the sign of a binary antipodal sequence whenever the current symbol of the $(d, k)$ sequence is one. The illustrated sequence fulfills the constraint for $d = 1$

$$(d, k) \text{ sequence} = [\ldots, 1, 0, 1, 0, 0, 0, 1, 0, 1, 0, 0, \ldots]$$

$$\text{run-length limited sequence} = [\ldots, 1, 1, -1, -1, -1, -1, 1, 1, -1, -1, -1, \ldots].$$

The $k$ parameter is introduced for practical purposes like data based clock recovery. Designing chip-to-chip communication such a function is not necessarily required and for this purpose $k = \infty$ is considered. The maximum entropy rate of a $d$-constrained source ($d$ sequence) can be computed by considering the number of valid run-length limited sequences having a defined length fulfilling the $d$-constraint. Alternatively the maximum entropy rate can be computed by considering the state diagram of the Markov source model (Fig. 5.6) and its adjacency matrix. The adjacency matrix, which describes the directional connections of the state machine generating $d$-constrained sequences is given by

$$\mathbf{D}_{d=1} = \begin{bmatrix} 0 & 1 \\ 1 & 1 \end{bmatrix}, \quad \mathbf{D}_{d=2} = \begin{bmatrix} 0 & 1 & 0 \\ 0 & 0 & 1 \\ 1 & 0 & 1 \end{bmatrix}. \tag{5.14}$$

As in the previous example the maximum entropy rate can be directly computed by (5.10). Furthermore the corresponding transition probabilities can be found by applying (5.11). The maximum entropy rates according to a given $d$-constraint are

**Fig. 5.6** State diagram for a $d$-constrained input sequence

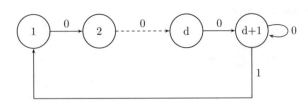

**Table 5.3** Maximum entropy of $d$-constrained sources

| Run-length constraint | $d = 1$ | $d = 2$ | $d = 3$ | $d = 4$ | $d = 5$ | $d = 6$ |
|---|---|---|---|---|---|---|
| Maximum entropy [bits per symbol] | 0.6942 | 0.5515 | 0.4650 | 0.4057 | 0.3620 | 0.3282 |

listed in Table 5.3. Obviously the $d$-constraint introduces redundancy. However, the $d$-constraint enables higher signaling rates which can finally result in increased information per time instant. When it is aimed to transmit binary information (bits) a sophisticated mapping can be applied. Several approaches such as the utilization of fixed length sequences, where a fixed number of bits are mapped to a fixed length run-length limited sequence exist. A popular example is given by the 3 Position Modulation (3PM) Code. Furthermore there exist approaches for variable length mapping where the run-length limited sequence *and* the information bits are not fixed [9].

## 5.4 Achievable Rate

The achievable rate is the theoretical throughput limit for a given channel with a given input distribution. While a survey of methods for computing bounds on the achievable rate for the class of the considered channel can be found in [6] this chapter only reviews a method which has been found out to be a good compromise between accuracy and complexity. The achievable rate of a memory channel can be computed by the methods presented in [1] and [8] employing the Shannon–McMillan–Breiman theorem

$$\lim_{n \to \infty} \frac{1}{n} \hat{I}(X^n; Y^n)$$

$$= \frac{1}{n}(-\log P(\mathbf{y}^n) + \log P(\mathbf{y}^n|x^n)P(x^n) - \log P(x^n)). \qquad (5.15)$$

The probabilities can be approximated by a recursive computation which requires decomposability of the channel. Especially when considering correlated noise samples and the quantization of the channel output the decomposition becomes difficult. To overcome this issue an auxiliary channel model can be introduced which finally corresponds to a lower bound on the achievable rate. The auxiliary channel model is denoted as $W(\cdot)$ with the properties $W(\mathbf{y}^n) \approx P(\mathbf{y}^n)$ and $W(\mathbf{y}^n|x^n) \approx P(\mathbf{y}^n|x^n)$. The auxiliary channel lower bound can be computed with

$$\lim_{n \to \infty} \frac{1}{n} I(X^n; Y^n)$$

$$\geq \frac{1}{n}(-\log W(\mathbf{y}^n) + \log W(\mathbf{y}^n|x^n)P(x^n) - \log P(x^n)), \qquad (5.16)$$

where $W(\cdot) > 0$ holds whenever $P(\cdot) > 0$. In this study the auxiliary channel $W(\cdot)$ is constructed based on the following assumption

$$P(\mathbf{y}_k|\mathbf{y}^{k-1}, x^k) \approx P(\mathbf{y}_k|x_{k-L}^k), \tag{5.17}$$

which becomes for rising signal-to-noise ratio (SNR) values an asymptotically close approximation. In fact this is the region of interest for ultra short range communication such as given for chip-to-chip. This assumption on the channel will be used to compute the probabilities in (5.16) with the forward recursion of the BCJR (Bahl, J. Cocke, F. Jelinek, and J. Raviv) algorithm [2]

$$P(\mathbf{y}^k) \geq W(\mathbf{y}^k) = \sum_{s_k} W(\mathbf{y}^k, s_k) = \sum_{s_k} \mu_k(s_k),$$

$$P(\mathbf{y}^k|x^n) \geq W(\mathbf{y}^k|x^n) = \tilde{\mu}_k = P(\mathbf{y}_k|s_k, s_{k-1})\tilde{\mu}_{k-1},$$

$$P(x^k) = P(s_k|s_{k-1})P(s^{k-1}), \tag{5.18}$$

where $s_k = x_{k-N+1}^k$ and state metrics are denoted as $\mu_k(s_k)$, respectively, $\tilde{\mu}_k$. Each recursion implies

$$\mu_k(s_k) = \sum_{x_{k-N}^{k-1}} P(\mathbf{y}_k|x_{k-N}^k)P(x_k|x_{k-N}^{k-1})\mu_{k-1}(s_{k-1})$$

$$\tilde{\mu}_k = P(\mathbf{y}_k|x_{k-N}^k)\tilde{\mu}_{k-1}, \tag{5.19}$$

where $P(\mathbf{y}_k|x_{k-N}^k) = P(\mathbf{y}_k|x_{k-L}^k)$ for $L < N$.

### 5.4.1 Transition Probabilities

The probability density function of the unquantized received signal is given by

$$p(\mathbf{z}_k|x_{k-L}^k) = \frac{1}{(2\pi)^{\frac{M}{2}}|\mathbf{R}|^{\frac{1}{2}}} \exp\left(-\frac{1}{2}(\mathbf{z}_k - \mathrm{E}\{\mathbf{z}_k\})^T \mathbf{R}^{-1}(\mathbf{z}_k - \mathrm{E}\{\mathbf{z}_k\})\right), \tag{5.20}$$

where $\mathrm{E}\{\mathbf{z}_k\} = \mathbf{V}\mathbf{U}x_{k-L}^k$, the covariance $\mathbf{R} = \mathrm{E}\{\mathbf{G}\mathbf{n}_{k-\xi}^k(\mathbf{n}_{k-\xi}^k)^T\mathbf{G}^T\}$, and $|\cdot|$ denotes the determinant.

The transition probability for the quantized signal can be found with the integration over the quantization interval given by $\mathbf{y}_k$

$$P(\mathbf{y}_k|x_{k-L}^k) = \int_{\mathbf{z}_k \in \mathbb{Y}_k} p(\mathbf{z}_k|x_{k-L}^k)d\mathbf{z}_k, \tag{5.21}$$

where $\mathbb{Y}_k = \{\mathbf{z}_k|Q_1\{\mathbf{z}_k\} = \mathbf{y}_k\}$.

## 5.4.2   Numerical Results

The simulation based computation of the achievable rate is carried out based on sequences with a length of $n = 10^5$ symbols. The shown rates are lower bounds on the true achievable rate but according to the auxiliary channel model the precision is very good for high SNR. It is assumed that TSV channel has low-pass behavior. As an example, the simplified TSV channel with its memory is represented as a truncated raised cosine function

$$h(\tau) = \begin{cases} \sqrt{\frac{2}{3T_s}} \left(1 - \cos\left(2\pi \frac{1}{2T_s}\tau\right)\right), & 0 \leq \tau < 2T_s \\ 0, & \text{else.} \end{cases} \tag{5.22}$$

At the same time, it is assumed, that the thermal noise process is also low-pass filtered by the TSV channel. In this regard, the noise correlation function is given by

$$c_{nn}(\tau) = \sigma_n^2 \int_{-\infty}^{\infty} h(u)h^*(u + \tau)du. \tag{5.23}$$

The SNR is given by the ratio between transmit signal energy in the time interval $T_s$ and noise power density

$$\text{SNR} = \frac{\mathrm{E}\left\{\int_{T_s} |x(t)^2| \, dt\right\}}{\sigma_n^2}. \tag{5.24}$$

Because the channel which filters the noise has unit energy, the noise variance has the same value as the noise power density $\sigma_n^2 = N_0$ of the unfiltered white noise.

### 5.4.2.1   ASK Sequences

The results on the achievable rate in Fig. 5.7 evaluate the construction rules for the design of reconstructable ASK sequences. The reconstruction property can be confirmed by the fact that the achievable rate asymptotically reaches the source entropy rate for increasing SNR. Indeed, besides the special designed sequences, also communication based on identically uniformly drawn input symbols benefits from oversampling in terms of a significant increase in achievable rate. However, so far it is not known how conventional methods with limited complexity can approach these rates when considering the i.u.d. input symbols. The designed sequences proposed in this chapter are therefore the preferred alternative.

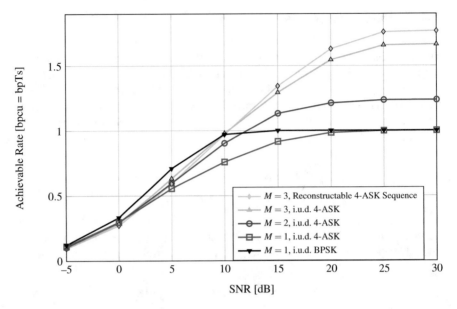

**Fig. 5.7** Achievable rate versus signal-to-noise ratio, bit per channel use is equivalent to bits per $T_s$

### 5.4.2.2 Run-Length Limited Sequences

To illustrate the principle and advantage of run-length limited sequences a special case has been considered where the sampling rate is equivalent to the signaling rate ($M = 1$). Different signaling rates $\frac{M_{Tx}}{T_s}$ have been used for computing the achievable rate. Considering bits per channel use, Fig. 5.8 shows that a larger value for $d$, where $d + 1$ defines the minimum number of equivalent consecutive symbols, reduces the entropy rate and with this the maximum achievable rate.

However, when considering the achievable rate per time interval in Fig. 5.9 there is a significant advantage in terms of achievable rate for a higher signaling rate in combination with a sophisticated $d$ sequence.

## 5.5 Conclusion

In this investigation communication designs are presented which are dedicated for receivers having only sign information of the sampled signal. The coarse quantization results in a loss of achievable rate which is balanced by an increased sampling rate. Two degrees of freedom are exploited, namely, amplitude modulation and signaling rate in terms of faster-than-Nyquist signaling. Both, reconstructability of transmitted signals and increased achievable rate can be obtained by considering a sophisticated sequence design.

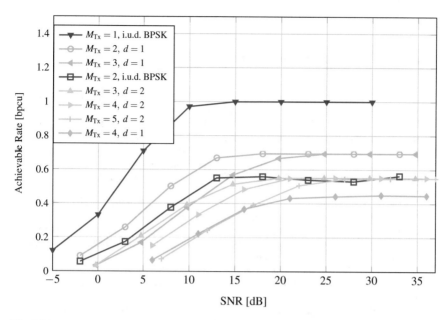

**Fig. 5.8** Achievable rate versus signal-to-noise ratio, bit per channel use

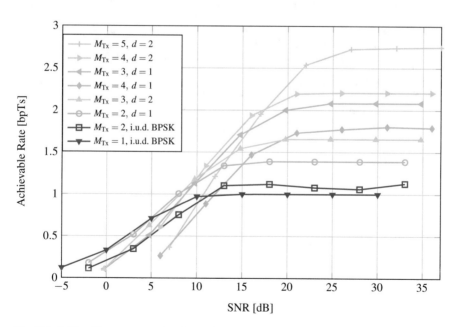

**Fig. 5.9** Achievable rate versus signal-to-noise ratio, bits per $T_s$, for each signaling rate in terms of $M_{Tx}$ an optimal $d$-constraint exists

# References

1. D.M. Arnold, H.A. Loeliger, P.O. Vontobel, A. Kavcic, W. Zeng, Simulation-based computation of information rates for channels with memory. IEEE Trans. Inf. Theory **52**(8), 3498–3508 (2006)
2. L. Bahl, J. Cocke, F. Jelinek, J. Raviv, Optimal decoding of linear codes for minimizing symbol error rate. IEEE Trans. Inf. Theory **20**(2), 284–287 (1974)
3. T. Hälsig, L. Landau, G. Fettweis, Information rates for faster-than-Nyquist signaling with 1-bit quantization and oversampling at the receiver, in *79th IEEE Vehicular Technology Conference (VTC Spring)*, 2014
4. T. Hälsig, L. Landau, G. Fettweis, Spectral efficient communications employing 1-bit quantization and oversampling at the receiver, in *80th IEEE Vehicular Technology Conference (VTC Fall)*, 2014
5. C. Jiangxin, P.H. Siegel, Markov processes asymptotically achieve the capacity of finite-state intersymbol interference channels. IEEE Trans. Inf. Theory **54**(3), 1295–1303 (2008)
6. L. Landau, G. Fettweis, Information rates employing 1-bit quantization and oversampling at the receiver, in *IEEE 15th International Workshop on Signal Processing Advances in Wireless Communications (SPAWC)* (2014), pp. 219–223
7. L. Landau, G. Fettweis, On reconstructable ask-sequences for receivers employing 1-bit quantization and oversampling, in *IEEE International Conference on Ultra-WideBand (ICUWB)* (2014), pp. 180–184
8. H.D. Pfister, J.B. Soriaga, P.H. Siegel, On the achievable information rates of finite state ISI channels, in *Global Telecommunications Conference, 2001. GLOBECOM '01. IEEE* (2001), pp. 2992–2996
9. K.A. Schouhamer Immink, Runlength-limited sequences. Proc. IEEE **78**(78), 1745–1759 (1990)
10. S. Shamai (Shitz), Information rates by oversampling the sign of a bandlimited process. IEEE Trans. Inf. Theory **40**(4), 1230–1236 (1994)
11. C.E. Shannon, A mathematical theory of communications. Bell Syst. Tech. J. **27**, 379–423 (1948)
12. G. Singh, L. Landau, G. Fettweis, Finite length reconstructible ASK-sequences received with 1-bit quantization and oversampling, in *Proceedings of the 10th International ITG Conference on Systems, Communications and Coding (SCC)*, 2015
13. R.B. Staszewski, Digitally intensive wireless transceivers. IEEE Des. Test Comput. **29**(6), 7–18 (2012)

# Chapter 6
# Clock Generators for Heterogeneous MPSoCs Within 3D Chip Stacks

Sebastian Höppner, Dennis Walter, and René Schüffny

## 6.1  Introduction

Modern 3D chip stack systems contain various system components for data processing, storage, and on-chip transmission, which are operated using a large variety of clock signals. This especially holds for heterogeneous multi-processor systems-on-chip (MPSoCs), where different compute cores, hardware accelerators, and I/O interfaces have to be clocked with individual requirements and constraints. Therefore, a globally asynchronous locally synchronous (GALS) clocking scheme, where each system core has an individual clock generator with no defined relations of clock frequency and phase to other cores. However, modern GALS MPSoCs [23, 30] require local clock generators to fulfill some specific requirements:

- A wide range of clock frequencies must be provided for fine-grained power management techniques, such as dynamic voltage and frequency scaling (DVFS) [13].

S. Höppner (✉) • R. Schüffny
Technische Universität Dresden, Chair of Highly-Parallel VLSI-Systems and
Neuro-Microelectronics, Dresden, Germany
e-mail: Sebastian.Hoeppner@tu-dresden.de; Rene.Schueffny@tu-dresden.de

D. Walter
Technische Universität Dresden, Chair of Highly-Parallel VLSI-Systems and
Neuro-Microelectronics, Dresden, Germany
e-mail: Dennis.Walter@tu-dresden.de

© Springer International Publishing Switzerland 2016
I.M. Elfadel, G. Fettweis (eds.), *3D Stacked Chips*,
DOI 10.1007/978-3-319-20481-9_6

101

The switching times between these output frequencies must be as small as possible to reduce idle times when changing the performance level of MPSoC cores.

- Special purpose clocks are required for interface clocking (e.g., to external DRAM), high-speed network-on-chip links [17, 28], and network-in-3D-chip-stack data links [9]. For high data rate applications, clock jitter specifications must be fulfilled.
- The clock generator must be disabled for power gated cycles, with minimum static current consumption. The restart time must be minimized as well.
- Low power consumption is mandatory to reduce the energy overhead of local clocking and to benefit from advanced power management (e.g., DVFS, AVFS).
- The clock generator should be usable in special purpose clocking applications, such as on-chip measurement circuits [6] or power supply circuits [16].
- Small chip area is mandatory for per-core instantiation of the clock generator.
- The clock generator should be easily portable to another semiconductor technology node, to reduce design implementation time. The number of custom-designed circuit blocks should be minimized and to the extent possible content should be realized as digital circuit to benefit from both the fast digital RTL-to-GDS implementation flow and the excellent scaling of digital logic cells in smaller technology nodes.

Ideally, one clock generator circuit should be capable to fulfill all requirements mentioned above. Previously published clock generators are only capable to fulfill parts of these requirements. In [20, 26] simple ring oscillator clock generators are used. They are not suitable for DVFS frequency switching and do not generate low jitter clocks at defined frequencies. The locally calibrated clock generators based on controlled delay lines in [22] do not allow fast switching between clock frequencies. The flying adder frequency synthesizer presented in [32] can generate a wide range of frequencies with low jitter but requires large chip area.

Therefore, the work described in this chapter aims at realizing a clock generator for heterogeneous GALS MPSoCs within 3D chip stack systems, which can fulfill the requirements and specifications. It should be instantiated *per-core*. Thereby, low power consumption and small chip area are the key aspects to be optimized. The required output frequencies are in the range from below 100 MHz (for processor cores at low performance levels) up to few GHz (for high-speed network-on-chip and 3D TSV signaling). These frequencies have to be generated from a global reference clock which is typically in the range of some 10 MHz. Figure 6.1 shows the block level schematics of the local clock generator and the corresponding frequency plan is shown in Fig. 6.2. The ADPLL, as presented in Sect. 6.3, multiplies the reference clock frequency by $N$ and provides a multi-phase output signal with the period $T_0 = T_{ref}/N$. The target period of $T_0$ is chosen such that a robust circuit implementation of the oscillator and the ADPLL loop divider can be robustly realized in the target 28 nm CMOS technology in the presence of PVT variations. From this ADPLL output clock lower core frequencies are generated by open-loop frequency division, as explained in Sect. 6.4. The advantage of this technique is

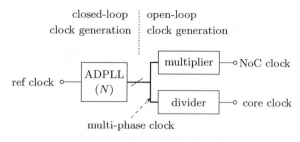

**Fig. 6.1** Local clock generator block level schematic

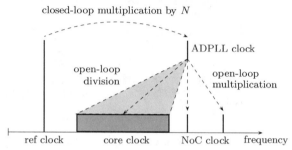

**Fig. 6.2** Local clock generation frequency plan

that changes of the output frequencies can be realized by re-programming the open-loop clock generator without time-consuming re-locking of the closed-loop ADPLL [14, 15].

The proposed concepts have been integrated in an MPSoC clock generator which has been prototyped in 28 nm CMOS technology. Figure 6.3 shows the layout and partial chip photo of the testchip. The measurement results described in the following subsections have been acquired from this silicon prototype.

## 6.2  A Multi-Phase DCO in 28 nm CMOS Technology

Digitally controllable oscillators (DCOs) are the key component of ADPLL based frequency synthesizers. The DCO for the target application has to be locked to $T_0$ and must provide its output clock in multiple clock phases for the open-loop output clock generation methods as presented in Sect. 6.4.

### 6.2.1  Circuit Architecture

The DCO core architecture is shown in Fig. 6.4a. It consists of tri-state inverter cells (Fig. 6.4b), which can be disabled during power down. The cross-coupled inverters in each stage are not completely disabled, but the pull-up (pfet on) and pull-down paths (nfet on) remain on during power down, realizing a reset scheme, where the

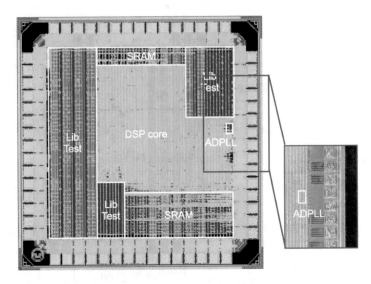

**Fig. 6.3** 28 nm clock generator testchip layout and partial chip photo

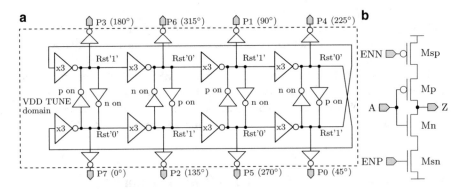

**Fig. 6.4** 28 nm DCO core schematic. (**a**) Ring oscillator, (**b**) core cell

differential ring oscillator nodes keep their differential state all times. This ensures safe and defined startup conditions for instantaneous DCO clock enable.

The tuning circuit is shown in Fig. 6.5. The supply tuning resistance is realized by parallel connection of PMOS devices, which are encapsulated into tune switch cells. Each tune switch cell contains a HVT PMOS with $4 \cdot L_{min}$ gate length, to achieve a higher on-resistance for finer tuning step size. It is enabled by a standard CMOS inverter structure, which additionally serves the purpose of decoupling the digital supply voltage domain (where the ADPLL controller is located) from the analog DCO supply voltage domain, to reduce supply noise on the analog domain. The tune switch cell layout is compatible with the standard cell grid. The remaining layout space for NMOS due to the PMOS switch with increased length is filled with a capacitor device, which reduces ripple on the DCO tuning voltage. Each single

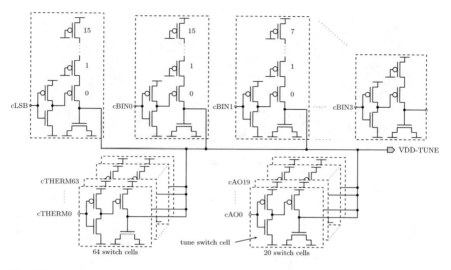

**Fig. 6.5** Tuning circuit schematic with resistance in the supply path

tune switch has the on conductance of $G_0 = 1/R_{\mathrm{on,HVT,PMOS}}$. The tune switch cells are clustered for different components of the total tuning signal $c$ of 10-bit width. The main group has 64 tune switch cells which are controlled by 64 thermometer coded signals cTHERM0 to cTHERM63, representing the upper 6 bits of $c$. The four LSBs of $c$ are directly applied to 4 tune switch cell with binary switched on-resistances using series connection of the HVT PMOS devices, realizing on conductances of $G_0/2$, $G_0/4$, $G_0/8$, and $G_0/16$, respectively. Another switch with $G_0/16$ is added for the delta-sigma modulated LSB tune bit $c_{\mathrm{LSB}} \in [0; 1]$. In order to achieve a small tuning step size and to allow oscillation of the DCO for all possible values of $c_{\mathrm{tune}}$, a set of 20 always-on switch cells is added. The number of the activated always-on cells is selectable by $c_{\mathrm{ao}}$, which is not changed during ADPLL control of this DCO. The total tune conductance in the supply path reads

$$G_{\mathrm{tune}} = c_{\mathrm{ao}} \cdot G_0 + c \cdot \frac{G_0}{16} + c_{\mathrm{DSM}} \cdot \frac{G_0}{16}, \qquad (6.1)$$

where $c_{\mathrm{DSM}}$ is an additional LSB which is used for fractional tuning using a delta-sigma modulator. This purely digital tuning scheme without analog voltage or current processing elements benefits well from technology scaling in terms of area, does not rely on good analog MOS device properties, and allows for easy design implementation in the target 28 nm CMOS technology.

**Fig. 6.6** Layout of the DCO in 28 nm CMOS technology, $28.0 \times 8.6\,\mu m$

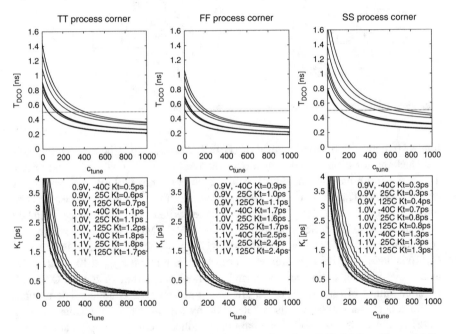

**Fig. 6.7** DCO tuning characteristics and tuning gain, $-40° \leq \theta \leq 125°, 0.9\,V \leq V_{DD} \leq 1.1\,V$

### 6.2.2 Implementation Results

Figure 6.6 shows the layout of the DCO in 28 nm CMOS technology. All components are realized within a standard cell grid. The VDD supply rail of the oscillator core cells is connected to the VDD_TUNE net, whereas the supply rails of the tuning switch array and the output buffers are operating on VDD. Therefore adapter cells are inserted left and right of the ring oscillator core, which also realize required spacing of the *n*-wells on VDD and VDD_TUNE, respectively.

Figure 6.7 shows the simulated tuning characteristics of the DCO within the temperature and supply voltage ranges ($-40° \leq \theta \leq 125°, 0.9\,V \leq V_{DD} \leq 1.1\,V$) for three process corners. The target period of $T_0 = 0.5\,ns$ can always be achieved. The tuning gain $K_t = dT_{DCO}/dc_{tune}$ shows strong dependency on the PVT

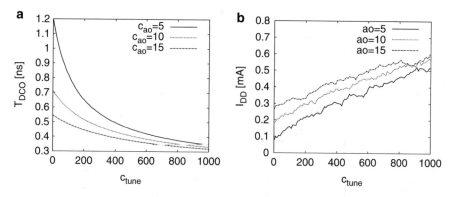

**Fig. 6.8** 28 nm DCO measurement results, $V_{DD} = 1.0$ V, room temperature. (**a**) Tune, (**b**) current consumption

conditions. In the lower plots of Fig. 6.7 the target $K_t$ at $T_{DCO} = T_0$ is printed. It shows variations from 0.3 to 2.4 ps considering all possible PVT variations within the specified ranges. Figure 6.8 shows measurement results of $T_{DCO}$ and the supply current $I_{DD}$ for different numbers of activated always-on switches.

## 6.3   ADPLL Clock Generator in 28 nm CMOS

ADPLLs are suitable circuit architectures for frequency multiplication, especially when being implemented in nanometer CMOS technologies, where area hungry analog filter components of analog PLLs can be replaced by efficient and compact digital filter structures. In addition, digital ADPLL controller implementations enable additional features for lock-in acceleration, as explained in this section.

### 6.3.1   Circuit Architecture

Figure 6.9 shows the block level schematic of the ADPLL. Its custom design part includes the multi-phase DCO as presented in Sect. 6.2 and the ADPLL loop frequency divider with ratio $N$ and a synchronizer block to realize fast lock-in. The loop divider provides two output clocks with periods of $T_{DCO} \cdot N/4$ and $T_{DCO} \cdot N$, respectively, which are employed to realize a digital loop filter being operated at faster frequency than the reference clock. This enables a fast, low latency filter response which results in lower accumulated ADPLL jitter. The clock timing diagram of the controller is shown in Fig. 6.10.

The digital part of the ADPLL contains a main finite state machine (FSM) and a loop filter with lock detection. The state sequence for fast lock-in and operation

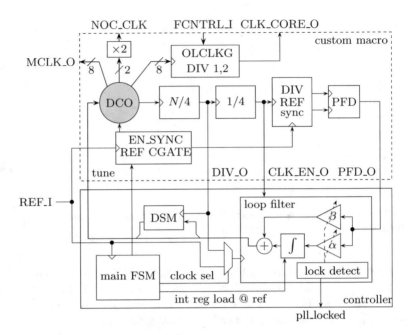

**Fig. 6.9** Block level schematic of the ADPLL in 28 nm CMOS

**Fig. 6.10** Controller timing diagram

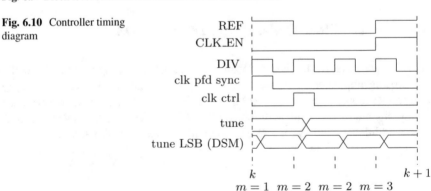

is shown in Fig. 6.11. The main FSM controls the lock-in sequence. Therefore, it runs with the reference clock being available when the DCO is disabled. When running the coarse lock-in sequence as presented in Sect. 6.3.2, the main FSM can directly control the integrator register of the loop filter to set the tuning word $c_{tune}$ of the DCO. Therefore the loop filter can be operated with the reference clock as well. Before changing to closed-loop ADPLL operation, the main FSM switches the clock source of the loop filter to the divider clock. At this time, the DCO is disabled, which prevents the generation of glitches when switching the clocks. The DCO is enabled by the main FSM and the loop filter operates in closed loop. The lock-in detector monitors the transitions of the phase frequency detector (PFD) signal to

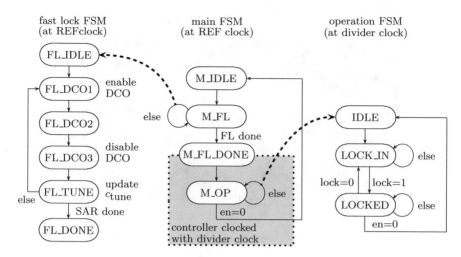

**Fig. 6.11** ADPLL controller state sequence

indicate lock of the ADPLL. A different set of loop filter coefficients $\alpha$ and $\beta$ with higher filter bandwidth is used for LOCK_IN and LOCKED state to speed up fine lock-in.

### 6.3.2  Fast Phase-Lock Architecture

This ADPLL realization features mechanisms for fast lock-in. Generally, the lock-in of an ADPLL includes phase and frequency lock. Both conditions

$$\Delta t_{\mathrm{PFD}} = t_{\mathrm{ref}} - t_{\mathrm{div}} = 0 \tag{6.2}$$

$$N \cdot T_{\mathrm{DCO}} = T_{\mathrm{ref}} \tag{6.3}$$

must be fulfilled. Figure 6.12 shows the simulated lock-in time for closed-loop ADPLL operation when it is started at different initial phase and frequency conditions. For minimized lock-in time, the ADPLL would have to be started in the target lock point of its state-space plane.

In contrast to the conventional ADPLL scheme as shown in Fig. 6.13a, where the phase-lock condition is a result from closed-loop operation, an active single-shot phase synchronizer is proposed in this work as shown in Fig. 6.13b. By means of a configurable delay chain with signal edge detection capability, it resets the phase difference at the PFD input to zero with the first reference clock edge after starting the ADPLL. After this it keeps its delays from the divider output and the reference input to the PFD static during closed-loop operation to allow phase tracking for jitter compensation by the ADPLL. This concept is applicable in this work where

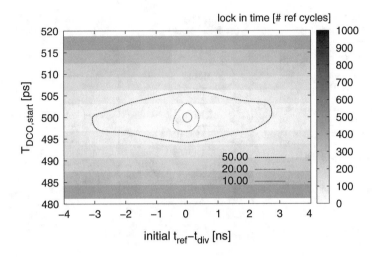

**Fig. 6.12** Simulated ADPLL lock-in time

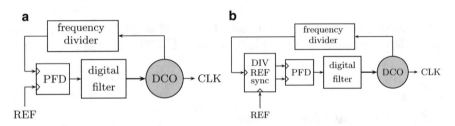

**Fig. 6.13** ADPLL architectures. (**a**) Conventional, (**b**) with single-shot phase synchronizer

the ADPLL is used as *frequency* multiplier for MPSoC clocking. No defined phase relation between the clock generator output and the reference clock is required. A similar fast lock-in architecture based on phase synchronization between the DCO and the reference clock signal has been proposed in [31] where the additional delays of asynchronous loop frequency divider stages are not considered. In contrast, this work provides a versatile phase synchronization solution which is applicable to a wide range of ADPLL architectures by the addition of the single-shot phase synchronizer.

Figure 6.14 shows the schematic of the single-shot phase synchronizer. It synchronizes the rising edges of the divider output clock and the reference clock. Therefore, the divider clock (DIV_I) is fed to a main delay line. After each delay element the signal is selected and fed to the output tri-state bus. The selection bit is stored in a D-latch, which is transparent with the low clock phase. They are transparent after reset. During the capture cycle, the latches store the value of the delayed divider signal in the main delay line. In the acquisition cycle the output of the capture signal flip-flop changes from 0 to 1 with the rising edge of the reference clock. This disables the latches in the main delay line. Thereby the position in the

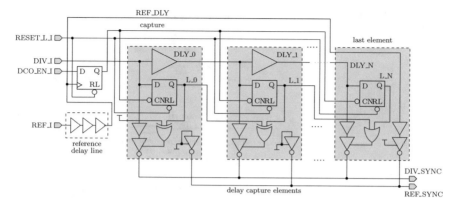

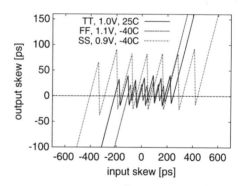

**Fig. 6.14** Single-shot phase synchronizer schematic, bypass circuits for disabling the synchronization are not shown

**Fig. 6.15** Single-shot phase synchronizer simulation results, post-layout

main delay line, where the state of the latches changes from 1 to 0 between two adjacent stages, denotes the point where the rising edges of the divider signal and the reference clock occurred at the same time. It is detected by XOR gates (OR in the last stage), which enable the output tri-state drivers for this particular stage. The un-gated reference clock is fed to the output of a replica tri-state stage for symmetry reasons. In the subsequent clock cycles, all latches are opaque because the gated reference clock is static 1. The phase synchronizer remains in its captured state. All delay variations between the divider signal and the reference clock that occur during ADPLL operation are fed directly to the PFD input. Thereby the phase synchronizer does not disturb the operation of the closed-loop ADPLL.

Figure 6.15 shows the post-layout simulation results of the single-shot phase synchronizer for typical, worst-case and best-case corners. The nominal timing error is $|t_{offset,sync}| < 80$ ps for an input skew within $\pm 200$ ps. The phase synchronizer needs to compensate the timing variability from synchronous DCO startup to the first rising clock edges at the PFD input where this variability is mainly caused by the delay through the frequency divider stages. Therefore, the compensation range of $\pm 200$ ps is sufficient for the application in this work. However, this range can be extended by the addition of more delay stages at cost of higher power consumption and larger chip area.

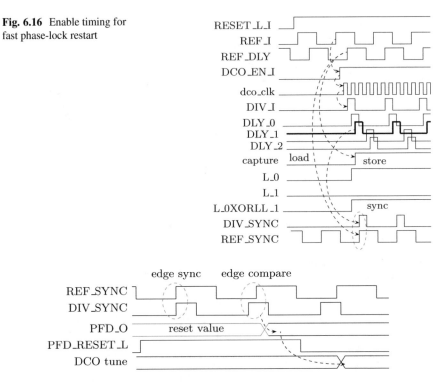

**Fig. 6.16** Enable timing for fast phase-lock restart

**Fig. 6.17** PFD waveform for binary frequency detection

Figure 6.16 visualizes the signal timings for enabling the ADPLL with single-shot phase synchronization. First, the loop dividers are released from reset and the DCO is enabled with the rising edge of the reference clock, triggered by the main FSM of the ADPLL. This results in a first rising edge at the divider output pulse clock (CLK_EN_O), which is fed to the delay line of the single-shot phase synchronizer, where it is synchronized to a delayed copy of the reference clock.

It is proposed to use the phase synchronization effectively for binary frequency detection for fast frequency lock to fulfill the condition of Eq. (6.3). When the first signal edge after the reset is synchronized to the reference edge by the single-shot phase synchronizer, the relative position of the following edges is a measure for the relative period of the reference clock and the divided DCO clock, as illustrated in Fig. 6.17. Therefore the PFD combined with the proposed synchronizer can be used as binary frequency detector with low hardware effort. When the phase is synchronized, the initial phase difference is $t_{offset}$, which is expected to be ideally zero. It includes the offset time of the PFD itself and the remaining timing difference after the enable synchronizer logic $t_{offset,sync}$. The timing difference of the following $n$-th edge at the PFD input can be calculated by

$$\Delta t_{\text{PFD}}(n) = t_{\text{offset}} + \sum_{i=1}^{i+N \cdot n} T_{\text{DCO},i} - \sum_{j=1}^{n} T_{\text{ref},j}, \tag{6.4}$$

where $N$ is the loop divider ratio. From Eq. (6.4) the average value and the standard deviation of $\Delta t_{\text{PFD}}$ are

$$\overline{\Delta t_{\text{PFD}}}(n) = t_{\text{offset}} + n \cdot (N \cdot T_{\text{DCO}} - T_{\text{ref}}), \tag{6.5}$$

$$\sigma_{t_{\text{PFD}}}^2(n) = n \cdot (N \cdot \sigma_{T_{\text{DCO}}}^2 + \sigma_{T_{\text{ref}}}^2). \tag{6.6}$$

To determine the number of signal edges $n$ for a required resolution for the DCO period $\Delta T_{\text{DCO}}$, first the systematic offset $t_{\text{offset}}$ of $\overline{\Delta t_{\text{PFD}}}(n)$ is considered. From Eq. (6.5) it can be concluded that the systematic deviation of the DCO period is

$$|\Delta T_{\text{DCO,systematic}}| = \frac{|t_{\text{offset}}|}{n \cdot N}, \tag{6.7}$$

for constant $T_{\text{DCO}}$ and $T_{\text{ref}}$ (neglecting jitter). Considering $k$-sigma accuracy for determination of the DCO period with respect to random jitter, it is

$$k \cdot \sigma_{t_{\text{PFD}}} < n \cdot N \cdot (|\Delta T_{\text{DCO,max}}| - |\Delta T_{\text{DCO,systematic}}|), \tag{6.8}$$

$$k \cdot \sqrt{n \cdot (N \cdot \sigma_{T_{\text{DCO}}}^2 + \sigma_{T_{\text{ref}}}^2)} < n \cdot N \cdot |\Delta T_{\text{DCO,max}}| - |t_{\text{offset}}|, \tag{6.9}$$

where $\Delta T_{\text{DCO,max}}$ is the maximum allowed deviation from the DCO period. So it is

$$|\Delta T_{\text{DCO,max}}| = \frac{1}{n \cdot N} \cdot \left( |t_{\text{offset}}| + k \cdot \sqrt{n} \cdot \sqrt{N \cdot \sigma_{T_{\text{DCO}}}^2 + \sigma_{T_{\text{ref}}}^2} \right). \tag{6.10}$$

Solving Eq. (6.10) for $n$ leads to the minimum number of measurement edges for a maximum DCO period variation within $k$-sigma accuracy

$$n_{\min} = \left( \frac{1}{2N|\Delta T_{\text{DCO,max}}|} \cdot \left( k\sigma_{\text{jitter}} + \sqrt{k^2 \sigma_{\text{jitter}}^2 + 2N|\Delta T_{\text{DCO,max}}| \cdot |t_{\text{offset}}|} \right) \right)^2 \tag{6.11}$$

with $\sigma_{\text{jitter}}^2 = N \cdot \sigma_{T_{\text{DCO}}}^2 + \sigma_{T_{\text{ref}}}^2$. Figure 6.18 visualizes the results from Eq. (6.10). There exists a trade-off between lock-in time and accuracy of the binary frequency search. For selection of a suitable value of $n$, the ADPLL analysis results from Fig. 6.12 are used. From this, a DCO period resolution of $|\Delta T_{\text{DCO,max}}| < 3\,\text{ps}$ is constrained, with the target of achieving ADPLL lock within 20 reference clock cycles. So a value of $n = 2$ is chosen here.

This binary frequency detection scheme is used for frequency lock of the ADPLL. The divided-by-$N$ DCO period is compared with the reference period $T_{\text{ref}}$. The output of the PFD indicates the sign of the period difference. Based on this

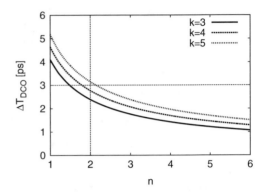

**Fig. 6.18** Maximum DCO period estimation error versus number of PFD frequency compare cycles $n$, $\sigma_{T_{DCO}} = 3$ ps, $\sigma_{T_{ref}} = 10$ ps, $N = 40$, $t_{offset} = 100$ ps, and different statistical safety margins $k$

a successive approximation algorithm is used, where the tuning word setting of $m$-bit accuracy is determined within $m$ measurements. Additionally, the number of successive approximation steps can be reduced if the condition $t_{step} < \sigma_{T_{DCO}}$ is fulfilled, because the result of the fast lock-in sequence will be in the noise floor of the DCO jitter. Therefore, lock-in time can be reduced without significant impact on jitter performance.

Figure 6.19 shows the waveforms from RTL simulation of the ADPLL. First, the binary frequency search is performed in nine binary frequency detection cycles corresponding to the 10-bit tuning word. Then the closed-loop ADPLL is activated with a single-shot phase synchronization of the first rising edges at the PFD input. The lock detection operates by counting the number of PFD transitions in a given time frame. When not locked yet, a wider ADPLL filter bandwidth is achieved by shifting (logic shift left) the loop filter coefficients $\alpha$ and $\beta$ in a gear shifting filter scheme. When lock is detected, they are set back to their nominal value.

### 6.3.3 Implementation Results

Figure 6.20 shows the compact ADPLL layout in 28 nm CMOS technology. The measured 2 GHz output waveform with period jitter histogram of the closed-loop ADPLL operation is visualized in Fig. 6.21a, b shows an oscilloscope waveform of the long-term accumulated jitter measurement (31 ps rms jitter over 2000 accumulated cycles).

The fast lock-in functionality is measured by capturing the DCO period over time using a digital sampling oscilloscope. The time 0 of the measured waveforms corresponds to the first rising clock edge of the DCO signal after turning on the ADPLL. Figure 6.22a shows the measured lock-in waveform when both binary frequency search and phase lock with gear shifted loop filter coefficients (by factor 8) are enabled. Here, the lock condition is valid after $0.8\,\mu$s and detected after $1.5\,\mu$s. Figure 6.22b shows the lock-in wave if the lock signal flag is directly asserted after the binary frequency search and no gear shifting filter operation is used. It

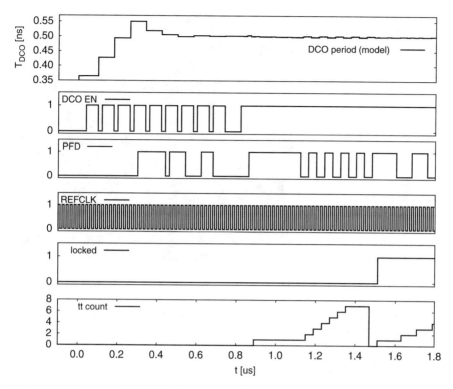

**Fig. 6.19** Fast lock-in ADPLL RTL simulation results

can be seen that the target oscillation period of $T_0 = 500$ ps is directly hit by the frequency search after $0.8\,\mu s$ and does not change during closed-loop operation due to the initial phase synchronization.

Considering this, the proposed phase synchronization method allows to restart the ADPLL immediately, if the previous integrator value of the loop filter (corresponding to the DCO period) is stored. This can be valid in an application scenario where the ADPLL is switched off when the MPSoC core is in idle state. When restarting after a short time (with respect to temperature drifts in the system), the frequency tuning value is still valid. Figure 6.23a shows this measured instantaneous restart capability. For comparison Fig. 6.23b shows the restart waveforms *without* phase synchronization. Although started at the correct frequency, the phase lock needs to be acquired by shifting the DCO period. This leads to increased jitter and prevents using the ADPLL output signal for clocking applications right after startup.

The performances of recently published ADPLL clock generators in sub-100 nm CMOS technologies are summarized in Table 6.1 and compared to the ADPLL implementations in this work. The standard deviation of the period jitter $\sigma_T$ is chosen as the main criterion for output clock quality. Commonly the DCO frequency is much higher than the reference frequency of the ADPLL (clock frequency

**Fig. 6.20** Layout of the
ADPLL in 28 nm CMOS
technology, 52 × 45 μm

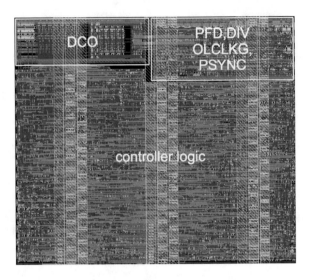

multiplication). Uncorrelated thermal noise as main contributor to DCO period jitter
is not filtered by the closed ADPLL loop. Therefore, in contrast to [8], only the DCO
noise is considered for benchmarking, assuming that it is mainly caused by thermal
noise and that the DCO is the main contributor to power in an ADPLL optimized
for low *period jitter*. So the figure of merit is defined as

$$\text{FOM}_J = \log_{10}\left(\frac{\sigma_T^2}{\text{ps}^2} \cdot \frac{f}{\text{MHz}} \cdot \frac{P}{\text{mW}}\right). \tag{6.12}$$

The ADPLL realization of this work with low power consumption achieves a similar
$\text{FOM}_J$ compared to previous results. It provides a wide range of output frequencies
using the open-loop clock generation method presented in the next section.

## 6.4 Open-Loop Core Clock Generation

An open-loop clock generator for core frequency generation from the multi-phase
ADPLL clock signal is designed. It is based on a reverse phase switching scheme
combined with programmable frequency dividers and thereby realizes sub-integer
division ratios without fractional spurs in the output signal. It provides a wide range
of division ratios with a low minimum ratio of 3, which is in contrast to previously
published fractional-$N$ PLL loop dividers which are designed for realization of
consecutive division ratios. Furthermore, this open-loop clock generator provides
50 % output duty cycle, which is essential for the targeted application within heterogeneous MPSoCs, e.g., for DDE memory interface clocking. It allows instantaneous
changes of the division ratio within a single output clock cycle to realize fast core
frequency changes for DVFS. Its functionality is explained in the following circuit
description and was first presented in [12].

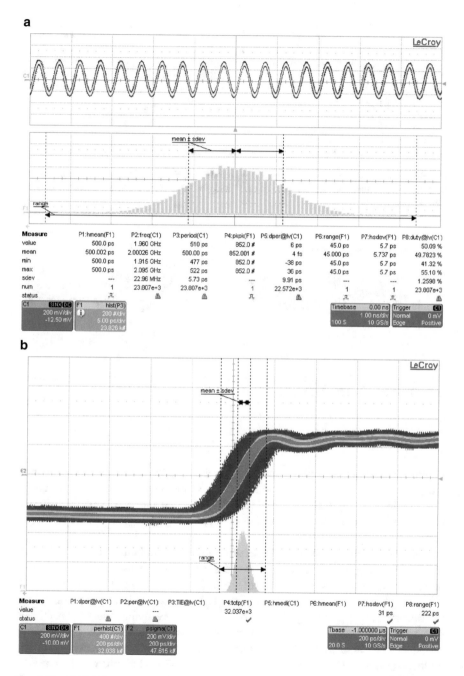

**Fig. 6.21** ADPLL measurement results. (**a**) ADPLL clock signal and period jitter histogram at 2 GHz, (**b**) long-term jitter histogram

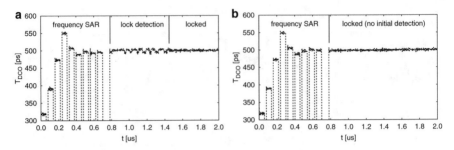

**Fig. 6.22** Measured 28 nm ADPLL lock-in waveform. (**a**) Fast lock-in with lock detection and filter bandwidth adaption by factor 8, (**b**) fast lock-in without initial lock detection, no filter bandwidth adaption

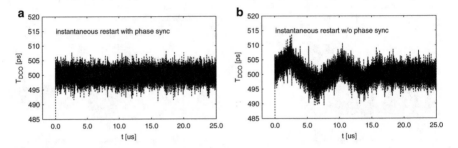

**Fig. 6.23** Measured 28 nm ADPLL instantaneous restart waveform. (**a**) With phase synchronization, (**b**) without phase synchronization

### 6.4.1 Circuit Architecture

Figure 6.24a shows the toplevel schematic of the proposed open-loop clock generator. A phase multiplexer selects 1 out of $M = 8$ input clock phases with period $T_0$. This multiplexer output signal is divided by $N_{23} \in [2, 3]$ and by $N_{sync} \in [2, 4, 6, 8]$ in the output divider. So the base division ratio reads $N_{23} \cdot N_{sync}$ if no phase switching occurs.

The phases are switched in a glitch-free reverse switching scheme [11, 25] as illustrated in Fig. 6.24b. Each time the phase is switched the multiplexer output period is shortened. Following the theory in [7] which suggests a switch step of $|n_{step}| \leq N/4 = 2$, we chose $n_{step} \in [1, 2]$, which leads to reduction of the multiplexer output period by $1/8 \cdot T_0$ or $2/8 \cdot T_0$ per switching event, respectively. The multiplexer select signals are generated by a rotator that acts like a +1 or +2 adder compared to the flying adder frequency synthesis approach [33]. The multiplexer output clock CM is fed to the divide-by-2-or-3 circuit which generates the clock C23 and the enable pulse for the rotator. From $n_{sw} = 0$ up to $n_{sw} = 3$, phase switchings can occur per C23 cycle. The clock C23 is fed to the synchronous frequency divider with *even* division ratios $N_{sync}$ that ensure 50 % duty cycle of the output clock. In summary, the output core period reads

**Table 6.1** Performance comparison of PLL clock generators in sub-100 nm CMOS technologies

| Ref | Tech (nm) | Type | $f_{min}$ (MHz) | $f_{max}$ (MHz) | $\sigma_T$ [ps] | At $f$ (MHz) | $P$ (mW) | $FOM_J$ | $A$ (mm²) |
|---|---|---|---|---|---|---|---|---|---|
| [32] | 90 | FA | 2 | 250 | 9.0 | 148 | 10.0 | 5.1 | 0.1512 |
| [35] | 90 | ADPLL | 700 | 3500 | 1.6 | 2500 | 1.6 | 4.0 | 0.3600 |
| [36] | 90 | ADPLL | 180 | 530 | n.a. | 480 | 0.466 | 5.0 | 0.0086 |
| [27] | 65 | ADPLL | 500 | 8000 | 0.7 | 4000 | 33.6 | 4.8 | 0.0300 |
| [34] | 65 | PLL | 1600 | 3200 | 3.1 | 1600 | 1.62 | 4.4 | 0.0400 |
| [18] | 65 | ADPLL | 3.5 | 1800 | 2.6 | 1600 | 220.0 | 6.4 | 0.5600 |
| [19] | 65 | PLL | 900 | 1000 | 3.1 | 900 | 10.0 | 4.9 | 0.1400 |
| [3] | 65 | PLL | 850 | 1100 | 4.5 | 1000 | 8.4 | 5.2 | 0.3200 |
| [2] | 65 | PLL | 1200 | 1800 | 5.4 | 1500 | 17.0 | 5.9 | 0.2000 |
| [5] | 65 | ADPLL | 600 | 800 | 22.0 | 400 | 3.2 | 5.8 | 0.0270 |
| [10] | 65 | ADPLL | 190 | 4270 | 1.4 | 3000 | 11.8 | 4.8 | 0.0400 |
| [24] | 45 | ADPLL | 840 | 13,300 | 1.1 | 3800 | 16.5 | 4.9 | 0.0280 |
| [21] | 22 | ADPLL | 600 | 3600 | n.a. | n.a. | 18.4 | n.a. | 0.0296 |
| This | 28 | ADPLL | 83 | 2000 | 3.0 | 2000 | 0.64 | 4.1 | 0.00234 |

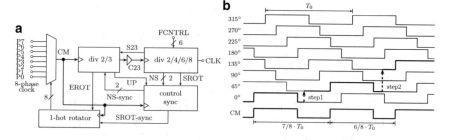

**Fig. 6.24** Open-loop clock generator for frequency division based on reverse phase switching, [12]. (**a**) Block schematic, (**b**) reverse phase switching waveforms with eight-phase clock signal

$$T_{core} = T_0 \cdot N_{sync} \cdot \left( N_{23} - \frac{n_{sw} \cdot n_{step}}{8} \right). \qquad (6.13)$$

The open-loop clock generator is controlled by a 6-bit signal FCNTRL, summarized in Table 6.2. For the 64 control words, 33 different division ratios in the range from 3 to 24 can be realized as shown in Fig. 6.25. Only the integer primes 13, 17, and 19 are missing. For $T_0 = 500$ ps the available output frequencies include 100, 133, 166, 200, 266, 333, 400, 533, and 666 MHz with 50 % duty cycle for DDR memory interfaces. In the following calculations, we assume $T_0 = 500$ ps as nominal operation frequency.

Figure 6.26 shows the schematic of the divide-by-2/3 circuit which is realized as a 2-bit state machine, driven with the clock CM. The state transfer equations are

**Table 6.2** Open-loop clock generator control signal definition

| FCNTRL | Comment | Values |
|---|---|---|
| 5:4 | Output division ratio | 00: $N_{sync} = 0$; 01: $N_{sync} = 1$; 10: $N_{sync} = 2$; 11: $N_{sync} = 3$ |
| 3 | Divide-by-2/3 select signal ($\overline{S23}$) | 0: $N_{23} = 2$; 1: $N_{23} = 3$ |
| 2:1 | Switchings per C23 cycle (NS[1:0]) | 00: $n_{sw} = 3$; 01: $n_{sw} = 2$; 10: $n_{sw} = 1$; 11: $n_{sw} = 0$ |
| 0 | Rotate step (SROT) | 0: $n_{step} = 2$; 1: $n_{step} = 1$ |

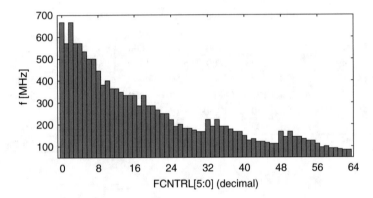

**Fig. 6.25** Available clock generator output frequencies for $T_0 = 500\,ps$ [12]

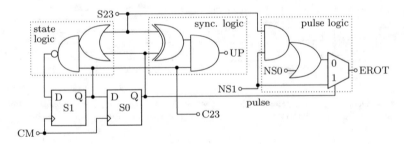

**Fig. 6.26** Divider by 2 and 3 and rotate pulse generation logic [12]

$$S0' = S1 \tag{6.14}$$

$$S1' = \overline{(S23 + S0) \cdot S1} \tag{6.15}$$

leading to a divide-by-2 state sequence of $10 \rightarrow 01 \rightarrow 10\ldots$ if $S23 = 1$ and a divide-by-3 state sequence of $10 \rightarrow 11 \rightarrow 01 \rightarrow 10\ldots$ if $S23 = 0$. The output clock C23 is directly derived from S1 (C23 = S1). This circuit generates enable pulses EROT for the phase rotator depending on $n_{sw}$ (NS1 and NS2) as shown in Table 6.2. During one C23 cycle, 0–3 enable pulses can be generated. EROT is defined by

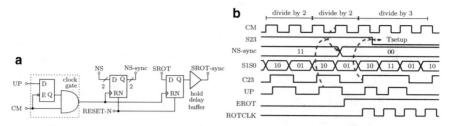

**Fig. 6.27** Synchronization of frequency division control signals [12]. (**a**) Control signal synchro nizer. (**b**) State sequence of Div 23, Change of S23 and NS1 and NS2

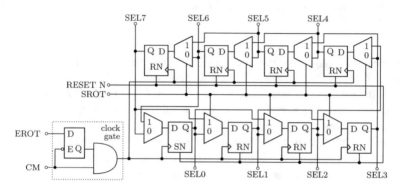

**Fig. 6.28** One-hot rotator schematic [12]

$$EROT = S0 \cdot NS1 + \overline{S0} \cdot (NS0 + NS1 + S23). \qquad (6.16)$$

Additionally, this state machine synchronizes the control signals for the rotator step size and the number of phase switchings by generation of an update signal UP. The rising edge of C23 occurs two cycles after UP is set to 1, with respect to the selected division ratio S23

$$UP = S23 \cdot \overline{S0} \cdot S1 + \overline{S23} \cdot S0 \cdot S1 \qquad (6.17)$$

which ensures in combination with the control synchronizer shown in Fig. 6.27a that SROT_sync and NS_sync are constant within each C23 cycle. Figure 6.27b shows an example state sequence where the division ratio is changed from 2 to 3 (on rising edge C23) and the number of switchings from 0 to 3 (on rising edge CM, enabled by UP). The critical timing constraint of this circuit is the setup time of S23, which is generated in the synchronous output divider with rising edge of C23, that must be settled before the next rising edge of CM.

Figure 6.28 shows the schematic of the 1-hot rotator that generates the phase multiplexer select signals. It consists of a closed-loop shift register, of which one flip-flop has a reset state 1 and all others 0. This approach ensures synchronous select signals with minimum skew and avoids glitches at the phase multiplexer. The

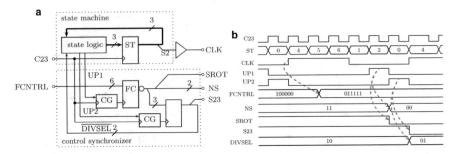

**Fig. 6.29** Synchronous output divider and control synchronizer [12]. (**a**) Schematic, (**b**) state sequence example (div6)

enable signal EROT activates a clock gate. If no phase rotation occurs, the shift register is not clocked which reduces power consumption. Multiplexers are used to control the rotation step which can be 1 or 2 depending on the control signal SROT. This realization enables minimum combinational logic between the flip-flops which is mandatory concerning its worst-case CM clock period of 375 ps.

Figure 6.29a shows the schematic of the synchronous frequency divider. It provides division ratios of 2, 4, 6, and 8 with 50 % output duty cycle using a 3-bit state machine running at clock C23. The state sequences are

div2: $0 \rightarrow 4 \rightarrow 0 \dots$
div4: $0 \rightarrow 4 \rightarrow 6 \rightarrow 2 \rightarrow 0 \dots$
div6: $0 \rightarrow 4 \rightarrow 5 \rightarrow 6 \rightarrow 1 \rightarrow 2 \rightarrow 0 \dots$
div8: $0 \rightarrow 4 \rightarrow 5 \rightarrow 6 \rightarrow 7 \rightarrow 1 \rightarrow 2 \rightarrow 3 \rightarrow 0 \dots$

The output clock CLK is directly derived from the state MSB. Its rising edge occurs at the state transition $0 \rightarrow 4$, independent from the selected division ratio DIVSEL. Additionally, this circuit synchronizes the internal control signals. The frequency control input FCNTRL is captured in register FC at state 0, enabled by the update signal UP1. The S23 control signal is delayed by 1 C23 cycle to be updated with the rising edge of the output clock (state 4) enabled by UP2. This ensures (together with the control signal synchronizer shown in Fig. 6.27a) that the internal control signals remain constant within one output clock cycle of the open-loop clock generator. Figure 6.29b shows an example state sequence of the synchronous output divider with the corresponding control signal updates. When the frequency control input FCNTRL is updated with the rising edge of CLK (state 4) by an external register, it must be settled before state 0 with a certain setup margin of the sampling register FC.

This constraint is modeled in the Liberty (.lib) file of the open-loop clock generator macro. Based on the functional circuit timing (Fig. 6.30a), the black box .lib timing model is defined as shown in Fig. 6.30b, c. An internal virtual clock signal (clk_int) is generated as clock root pin. From this, the virtual FC flip-flop clock is generated with a delay $t_{D,CG}$, modeling the clock gate delay, and the output clock

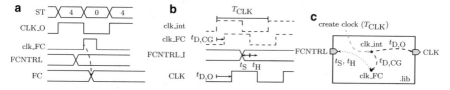

**Fig. 6.30** Timing model of the open-loop core clock generator. (**a**) Functional timing, (**b**) .lib timing, (**c**) .lib timing black box

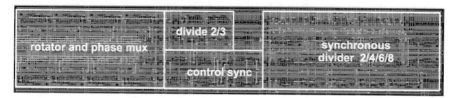

**Fig. 6.31** Open-loop clock generator layout, 28 nm, 29.8 × 5.6 μm

CLK is generated with the output delay $t_{D,O}$. The worst-case timing conditions with respect to the frequency control input FCNTRL are:

- The synchronous output divider has a division ratio of 2, where the FC capture clock (in state 0) corresponds to the falling edge of the output clock.
- The output clock has its shortest period $T_{CLK} = 1.5$ ns.

Therefore, in the .lib file model the FCNTRL inputs are constrained with setup ($t_S$) and hold ($t_H$) with respect to the *falling* edge of the internal pin clk_FC, where $t_S$ and $t_H$ are the setup and hold times of the standard cell flip-flops in the FC register.

### 6.4.2  Implementation Results

The open-loop clock generator has been implemented in 28 nm CMOS technology. Its layout is shown in Fig. 6.31. A high-speed standard cell library is used for circuit implementation. Special customized cells (e.g., fast tri-state driver) are used for timing critical paths. To achieve good phase symmetry and reduce mismatch-induced jitter at the output, the rotator is merged with the phase multiplexer. Thereby, each rotator flip-flop is located next to the corresponding tri-state driver to minimize delay on the select signal line and to ensure symmetry of the 8 clock phase inputs.

Figure 6.32 shows the simulated power consumption of the clock generator for different PVT corners and measurement results. The power consumption increases with the number of phase switchings due to the phase rotator activity. For some frequencies, as shown in Fig. 6.25, different control signal realizations exist with different power consumption. This must be considered when selecting the desired

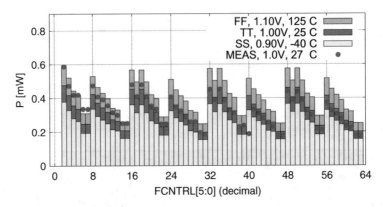

**Fig. 6.32** Simulated and measured power consumption of the open-loop clock generator, $T_0 =$ 500 ps, 28 nm (FCNTRL 0 and 1 not used)

output frequency in the system application. To demonstrate the capability of arbitrary, instantaneous frequency changes, Fig. 6.33 shows instantaneous output frequency changes of the 28 nm testchip implementation.

As presented in [12], Table 6.3 summarizes the performances of recently published frequency dividers and open-loop clock generators in CMOS technologies ≤180 nm for comparison with this work. The Flying Adder (fa) synthesizer reported in [32] provides many different output frequencies but the maximum input frequency and therefore the maximum output frequency is limited due to the arithmetic logic required. The sequential (seq) divider in [1] has a high maximum input frequency but also only high integer division ratios. Wang et al. [29] present a sequential divider with wide division ratio range but only integer division ratios. Chau and Chen [4] use a wide phase multiplexer (32 inputs) and a modulus counter to achieve sub-integer division ratios but allows only small maximum input frequencies.

The proposed design enables high maximum input frequency, as usually achieved in sequential or phase rotating (pr) dividers, combined with a small minimum division ratio as achieved by the flying adder approach. It shows low power consumption and requires only small chip area. By its purely static CMOS logic implementation it scales well with the shrinking of semiconductor technologies.

## 6.5 Conclusion

A versatile clock generator solution for heterogeneous MPSoCs embedded in 3D chip stack systems has been presented. A closed-loop ADPLL clock frequency multiplier is used to generate a fixed frequency high-speed clock, which is available in 8 equally spaced clock phase outputs. From this, a wide range of processor core clock frequencies is generated by an open-loop method using phase rotation and clock

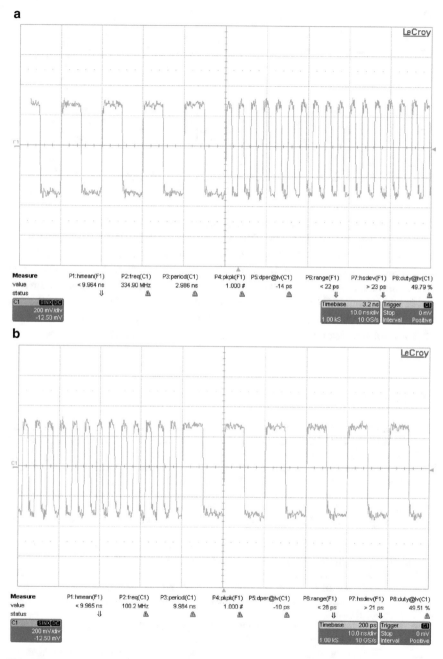

**Fig. 6.33** Measured instantaneous output period changes of the open-loop clock generator, 28 nm realization, "Cool28SoC" testchip. (**a**) Change from 10 to 3 ns output period. (**b**) Change from 3 to 10 ns output period

**Table 6.3** Open-loop clock generator design comparison

| Ref | Type | Tech. (nm) | $f_{in}$ (MHz) | $f_{out,min}$ (MHz) | $f_{out,max}$ (MHz) | $N_{min}$ $N_{max}$ | #$N$ | ($\mu m^2$) | Area (mW) | Power |
|---|---|---|---|---|---|---|---|---|---|---|
| [32] | fa | 90 | 1269 | 100 | 250 | n.a. | n.a. | 60 | 151,000 (with PLL) | n.a. |
| [1] | seq | 90 | 3500 | 130 | 146 | 24 | 27 | 4 | 21800 | 4.5 |
| [29] | seq | 180 | 1500 | 5.8 | 750 | 2 | 256 | 255 | n.a. | 1.3 |
| [4] | pr | 180 | 238 | 7.4 | 122.6 | 2 | 32 | frac | 91,000 | 48 |
| This work | pr | 65 | 2000 | 83 | 667 | 3 | 24 | 33 | 744 | 0.62–1.6 |
| This work | pr | 28 | 2000 | 83 | 667 | 3 | 24 | 33 | 167 | 0.20–0.47 |

frequency division. This enables instantaneous frequency changes for ultra-fast dynamic power management. In addition, the high-speed clock signal can directly be employed for network-on-chip and 3D TSV transceiver clocking. The ADPLL clock generator is equipped with a fast lock-in architecture that allows for fast initial lock-in and instantaneous restart capability, which significantly reduces system idle power consumption. Its small chip area and low active power consumption make this clock generator approach ideally suited for per-core instantiation within complex MPSoCs for fine-grained power management within 3D chip stack systems.

# References

1. F. Barale, P. Sen, S. Sarkar, S. Pinel, J. Laskar, Programmable frequency-divider for millimeter-wave PLL frequency synthesizers, in *38th European Microwave Conference, 2008. EuMC 2008* (Oct 2008), pp. 460–463
2. J.-Y. Chang, S.-I. Liu, A 1.5 GHz phase-locked loop with leakage current suppression in 65 nm CMOS. IET Circuits Devices Syst. **3**(6), 350–358 (2009)
3. J.-Y. Chang, S.-I. Liu, A phase-locked loop with background leakage current compensation. IEEE Trans. Circuits Syst. II Express Briefs **57**(9), 666–670 (2010)
4. Y. Chau, C.-F. Chen, High-performance glitch-free digital frequency synthesiser. Electron. Lett. **44**(18), 1063–1064 (2008)
5. M.-W. Chen, D. Su, S. Mehta, A calibration-free 800 MHz fractional-N digital PLL with embedded TDC. IEEE J. Solid-State Circuits **45**(12), 2819–2827 (2010)
6. S. Dietel, S. Hoppner, T. Brauninger, U. Fiedler, H. Eisenreich, G. Ellguth, S. Hanzsche, S. Henker, R. Schüffny, A compact on-chip IR-drop measurement system in 28 nm CMOS technology, in *IEEE International Symposium on Circuits and Systems (ISCAS), 2014* (June 2014), pp. 1219–1222
7. B. Floyd, Sub-integer frequency synthesis using phase-rotating frequency dividers. IEEE Trans. Circuits Syst. I: Regul. Pap. **55**(7), 1823–1833 (2008)
8. X. Gao, E. Klumperink, P. Geraedts, B. Nauta, Jitter analysis and a benchmarking figure-of-merit for phase-locked loops. IEEE Trans. Circuits Syst. II Express Briefs **56**(2), 117–121 (2009)

9. J. Gorner, S. Hoppner, D. Walter, M. Haas, D. Plettemeier, R. Schuffny, An energy efficient multi-bit TSV transmitter using capacitive coupling, in *21st IEEE International Conference on Electronics Circuits and Systems 2014 (ICECS 2014)*, 2014
10. W. Grollitsch, R. Nonis, N. Da Dalt, A 1.4ps rms-period-jitter TDC-less fractional-N digital PLL with digitally controlled ring oscillator in 65nm CMOS, in *IEEE International Solid-State Circuits Conference Digest of Technical Papers (ISSCC), 2010* (Feb 2010), pp. 478–479
11. S. Hoppner, R. Schuffny, M. Nemes, A low-power, robust multi-modulus frequency divider for automotive radio applications, in *MIXDES '09. MIXDES-16th International Conference Mixed Design of Integrated Circuits Systems, 2009* (June 2009), pp. 205–209
12. S. Hoppner, S. Henker, H. Eisenreich, R. Schuffny, An open-loop clock generator for fast frequency scaling in 65 nm CMOS technology, in *Proceedings of the 18th International Conference Mixed Design of Integrated Circuits and Systems (MIXDES), 2011* (June 2011), pp. 264–269
13. S. Hoppner, C. Shao, H. Eisenreich, G. Ellguth, M. Ander, R. Schuffny, A power management architecture for fast per-core DVFS in heterogeneous MPSoCs, in *IEEE International Symposium on Circuits and Systems (ISCAS), 2012* (May 2012), pp. 261–264
14. S. Hoppner, H. Eisenreich, S. Henker, D. Walter, G. Ellguth, R. Schuffny, A compact clock generator for heterogeneous GALS MPSoCs in 65-nm CMOS technology. IEEE Trans. Very Large Scale Integr. Syst. **21**(3), 566–570 (2013)
15. S. Hoppner, S. Haenzsche, G. Ellguth, D. Walter, H. Eisenreich, R. Schuffny, A fast-locking ADPLL with instantaneous restart capability in 28-nm CMOS technology. IEEE Trans. Circuits Syst. II Express Briefs **60**(11), 741–745 (2013)
16. S. Hoppner, S. Haenzsche, S. Scholze, R. Schuffny, An all-digital PWM generator with 62.5ps resolution in 28nm CMOS technology, in *IEEE International Symposium on Circuits and Systems (ISCAS), 2015*, May 2015
17. S. Höppner, D. Walter, T. Hocker, S. Henker, S. Hanzsche, D. Sausner, G. Ellguth, J.-U. Schlussler, H. Eisenreich, R. Schuffny, An Energy Efficient Multi-Gbit/s NoC transceiver architecture with combined AC/DC drivers and stoppable clocking in 65 nm and 28 nm CMOS. IEEE J. Solid-State Circ. **50**(3), 749–762 (2015). doi: 10.1109/JSSC.2014.2381637
18. P.-H. Hsieh, J. Maxey, C.-K.K. Yang, A phase-selecting digital phase-locked loop with bandwidth tracking in 65-nm CMOS technology. IEEE J. Solid-State Circuits **45**(4), 781–792 (2010)
19. C.-C. Hung, S.-I. Liu, A leakage-compensated PLL in 65-nm CMOS technology. IEEE Trans. Circuits Syst. II Express Briefs **56**(7), 525–529 (2009)
20. R. Jipa, Dedicated solution for local clock programming in GALS designs, in *International Semiconductor Conference, 2008. CAS 2008*, vol. 2 (Oct 2008), pp. 393–396
21. Y. Li, C. Ornelas, H.S. Kim, H. Lakdawala, A. Ravi, K. Soumyanath, A reconfigurable distributed all-digital clock generator core with SSC and skew correction in 22nm high-k tri-gate LP CMOS, in *IEEE International Solid-State Circuits Conference Digest of Technical Papers (ISSCC), 2012* (Feb 2012), pp. 70–72
22. S. Moore, G. Taylor, P. Cunningham, R. Mullins, P. Robinson, Self calibrating clocks for globally asynchronous locally synchronous systems, in *Proceedings of the International Conference on Computer Design 2000* (2000), pp. 73–78
23. B. Noethen, O. Arnold, E. Perez Adeva, T. Seifert, E. Fischer, S. Kunze, E. Matus, G. Fettweis, H. Eisenreich, G. Ellguth, S. Hartmann, S. Höppner, J. Schiefer, J.-U. Schlusler, S. Scholze, D. Walter, R. Schüffny, A 105GOPS 36mm2 heterogeneous SDR MPSoC with energy-aware dynamic scheduling and iterative detection-decoding for 4G in 65nm CMOS, in *IEEE International Solid-State Circuits Conference Digest of Technical Papers (ISSCC), 2014* (Feb 2014), pp. 188–189
24. A. Rylyakov, J. Tierno, G. English, M. Sperling, D. Friedman, A wide tuning range (1 GHz-to-15 GHz) fractional-N all-digital PLL in 45nm SOI, in *Custom Integrated Circuits Conference, 2008. CICC 2008*. IEEE (Sept 2008), pp. 431–434
25. K. Shu, E. Sinchez-Sinencio, *CMOS PLL Synthesizers: Analysis and Design* (Springer, Berlin, 2005)

26. A. Sobczyk, A. Luczyk, W. Pleskacz,   Controllable local clock signal generator for deep submicron GALS architectures,   in *11th IEEE Workshop on Design and Diagnostics of Electronic Circuits and Systems, 2008. DDECS 2008* (Apr 2008), pp. 1–4
27. J.A. Tierno, A.V. Rylyakov, D.J. Friedman, A wide power supply range, wide tuning range, all static CMOS all digital PLL in 65 nm SOI. IEEE J. Solid-State Circuits **43**(1), 42–51 (2008)
28. D. Walter, S. Hoppner, H. Eisenreich, G. Ellguth, S. Henker, S. Haenzsche, R. Schuffny, M. Winter, G. Fettweis A source-synchronous 90Gbit/s capacitively driven serial on-chip link over 6mm in 65nm CMOS,   in *IEEE International Solid-State Circuits Conference, 2012. ISSCC 2012. Digest of Technical Papers* , Feb 2012
29. L. Wang, S. Yue, Y. Zhao, L. Fan,   An SEU-tolerant programmable frequency divider,   in *8th International Symposium on Quality Electronic Design, 2007. ISQED '07* (March 2007), pp. 899–904
30. M. Winter, S. Kunze, E. Adeva, B. Mennenga, E. Matus, G. Fettweis, H. Eisenreich, G. Ellguth, S. Höppner, S. Scholze, R. Schüffny, T. Kobori,  A 335Mb/s 3.9mm2 65nm CMOS flexible MIMO detection-decoding engine achieving 4G wireless data rates,  in *IEEE International Solid-State Circuits Conference Digest of Technical Papers (ISSCC), 2012* (Feb 2012), pp. 216–218
31. C.-T. Wu, W.-C. Shen, W. Wang, A.-Y. Wu, A two-cycle lock-in time ADPLL design based on a frequency estimation algorithm. IEEE Trans. Circuits Syst. II Express Briefs **57**(6), 430–434 (2010)
32. L. Xiu,  A flying-adder on-chip frequency generator for complex SoC environment. IEEE Trans. Circuits Syst. II Express Briefs **54**(12), 1067–1071 (2007)
33. L. Xiu, Z. You, A new frequency synthesis method based on "flying-adder" architecture. IEEE Trans. Circuits Syst. II Analog Digit. Signal Process. **50**(3), 130–134 (2003)
34. Y. Yang, L. Yang, Z. Gao, A PVT tolerant sub-mA PLL in 65nm CMOS process, in *15th IEEE International Conference on Electronics, Circuits and Systems, 2008. ICECS 2008* (Sep 2008), pp. 998–1001
35. W. Yin, R. Inti, A. Elshazly, B. Young, P.K. Hanumolu, A 0.7-to-3.5 GHz 0.6-to-2.8 mW highly digital phase-locked loop with bandwidth tracking. IEEE J. Solid-State Circ. **46**(8), 1870–1880 (2011). doi: 10.1109/JSSC.2011.2157259
36. C.-Y. Yu, C.-C. Chung, C.-J. Yu, C.-Y. Lee,  A low-power DCO using interlaced hysteresis delay cells. IEEE Trans. Circuits Syst. II Express Briefs **59**(10), 673–677 (2012)

# Chapter 7
# Two-nanometer Laser Synthesized Si-Nanoparticles for Low Power Memory Applications

Nazek El-Atab, Ali K. Okyay, and Ammar Nayfeh

## 7.1 Introduction

In the past decade, memory chips with low cost and low power consumption have gained more attention due to the growing market of consumer electronic equipment such as digital cameras and cellular phones. Moreover, the demand for memories

Reprinted with permission from El-Atab N, Ozcan A, Alkis S et al. (2014) Low power zinc-oxide based charge trapping memory with embedded silicon nanoparticles via Poole-Frenkel hole emission. Appl. Phys. Lett. 104:013112. Copyright 2014. AIP Publishing LLC.
Reprinted with permission from El-Atab N, Ozcan A, Alkis S et al. (2014) Silicon Nanoparticle Charge Trapping Memory Cell. Phys. Status Solidi RRL 8:629. Copyright (c) 2014. John Wiley and Sons.
Reprinted with permission from El-Atab N, Rizk A, Tekcan B et al. (2015) Memory effect by charging of ultra-small 2-nm laser-synthesized Si-nanoparticles embedded in Si-Al2O3-SiO2 structure. Phys. Status Solidi A 212. Copyright (c) 2015. John Wiley and Sons.

N. El-Atab • A. Nayfeh (✉)
Department of Electrical Engineering and Computer Science (EECS), Institute Center for Microsystems–iMicro, Masdar Institute of Science and Technology, Abu Dhabi, United Arab Emirates
e-mail: nelatab@masdar.ac.ae; anayfeh@masdar.ac.ae

A.K. Okyay
Department of Electrical and Electronics Engineering, UNAM-National Nanotechnology Research Center, and Institute of Materials Science and Nanotechnology, Bilkent University, Ankara 06800, Turkey
e-mail: aokyay@ee.bilkent.edu.tr

© Springer International Publishing Switzerland 2016
I.M. Elfadel, G. Fettweis (eds.), *3D Stacked Chips*,
DOI 10.1007/978-3-319-20481-9_7

129

with higher density is expected to increase. From a device perspective, increasing the density has made the scaling of flash memory devices possible. In fact, flash memories scaled by a factor of 1000 during the period from 1986 to 2006 to reach 65 nm [1]. In the past, it was easier to scale these devices. Currently, scaling the NOR flash memory devices below the 45 nm technology node is becoming more difficult due to different unsolved problems such as short channel effect, drain induced barrier lowering (DIBL), and sub-surface punch-through effect which cause high leakage currents.

As a matter of fact, the scaling of CMOS logic device is far ahead of the semiconductor flash memory. In 2001, the minimum feature size of a single CMOS FET has reduced to 15 nm with an equivalent gate oxide thickness (EOT) of 0.8 nm [2]. However, for semiconductor flash memories, the EOT of the tunnel oxide is still larger than 6–8 nm. Moreover, the required operation voltage of semiconductor flash memory is still higher than 10 V, while the CMOS logic operation voltage has been reduced to around 1 V.

It is essential to scale down the EOT of the gate stack in order to obtain a small memory cell size and extend battery life. In fact, there is a relation between the operation voltage of semiconductor flash memory devices and the gate stack scaling. In order to meet the 10-year data retention requirement in the commercial flash memory chip, a tunnel oxide thickness of more than 8 nm is presently used. The operation voltage can be scaled down from 10 V to about 4 V if the tunnel oxide thickness was made thinner than 2 nm, however, the retention time would be reduced from 10 years to several seconds [3].

Silicon–Oxide–Nitride–Oxide–Silicon (SONOS) memories were then introduced as an improved version of flash memories. Actually, SONOS has a very similar structure to the standard double poly-silicon flash memory but with an $Si_3N_4$ charge-trapping layer replacing the poly-silicon. SONOS offers higher quality storage because $Si_3N_4$ film is homogeneous unlike the ploy-silicon which has small irregularities. Also, SONOS is considered a lower-cost version of the flash memory because its oxide layering can be fabricated easily on the existing lines and can be easily combined with CMOS logic. However, flash memory requires the production of a high-performance insulating barrier on the gate of its transistors, which leads to an additional number of around 9 fabrication steps [4].

Moreover, the charge-trapping layer of the SONOS structure is nonconductive which means that a shorting defect in the oxide will affect only a localized area of charge unlike the flash memory with poly-silicon where the short would be considered as a single point of failure. SONOS memory is also more scalable and has a lower operating voltage of 8–12 V [5]. However, the scaling and operating voltage of SONOS are still not comparable with those of CMOS logic.

Most of the scaling restrictions are now governed by the device structure and materials. Therefore, there is a need to seek novel materials and memory structures. This chapter will investigate several ways to reduce the operation voltage and reduce the tunnel oxide thickness without compromising the retention time by using new materials such as incorporating nanoparticles and novel oxides in the channel, charge-trapping layer, and tunnel oxide. The main goals are to optimize the charge-trapping layer in order to increase the charge-trapping density and therefore the memory window, in addition to optimizing the tunnel oxide layer in order to reduce the operating voltage without degrading the retention characteristic of the memory. It is worth to mention that these optimized memory cells can be stacked in 3D structures in order to further increase the density of current memory chips. Moreover, one of the main challenges that 3D IC stacks face is how to dissipate the thermal heating that is built up within the stacks. This is an inevitable issue as electrical proximity correlates with thermal proximity. Therefore, these optimized memory cells with lower operation voltage can help in mitigating this problem.

In this chapter, the background on flash memory structure and operation is first reviewed. Subsequently, the use of nanoparticles (NPs) for charge storage is discussed and an alternative device structure is suggested where the material used as the channel of the memory device is ZnO. Then, two different ZnO-based memory structures with embedded Si-NPs are fabricated and electrical measurements such as current–voltage (*I–V*) measurements are performed in order to understand the observed memory behavior, including the nature of charging and retention characteristics. Analysis of the device operation is performed by constructing the energy bandgap of the memory structure. Finally, a simpler and lower-cost memory device based on a metal-oxide-semiconductor (MOS) structure is proposed and fabricated. The behavior of the device is electrically characterized by performing capacitance–voltage (C-V) measurements and by analyzing the energy band diagram of the structure.

## 7.2 Basic Memory Cells Overview

A nonvolatile memory (NVM) device is an essential component of almost all modern electronic systems such as cell phones, personal computers, digital cameras, automotive systems, etc. Memory operation consists of a process of storing, retaining, and retrieving data [6]. Storing information is performed by programming/erasing the memory and retrieving is done via reading the state of the device. The building block of current memory devices is the metal-oxide-semiconductor field-effect-transistor (MOSFET).

Recently, transistors on the order of tens of nanometers for memory and logic applications have been fabricated [7]. These advancements in semiconductor processing technology have allowed storage of information capacity of the order of $10^9$ bits. However, greater scalability and higher storage capacity density are still needed, which would require more research on new NVM device materials and structures.

### 7.2.1 Basic Nonvolatile Memory Cell Structure

The structure of a flash memory is shown in Fig. 7.1. The memory structure is exactly the same as the MOSFET transistor structure but with an extra layer

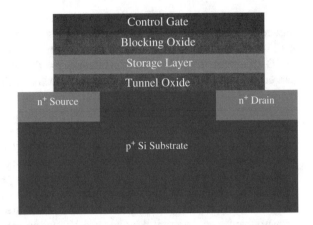

embedded in the oxide for charge storage and retention. In the case of floating-
gate type memory, the storage layer consists of a highly doped poly-silicon, while
in the case of a charge-trapping memory device, the storage layer is conventionally
a nitride layer. The tunnel and blocking oxides are meant to prevent charge leakage
into the substrate or to the control gate, and the material used for these oxides is
conventionally $SiO_2$ due to its excellent interface properties with Si. However, other
high-$\kappa$ dielectrics such as $Al_2O_3$ are being studied as possible replacement with the
scaling of flash memory.

### 7.2.2  Basic Nonvolatile Memory Cell Operation

To understand the mechanism of nonvolatile memory devices, the programming
operation is first illustrated in Fig. 7.2. The memory is programmed by applying a
positive voltage at the gate, electrons in the channel will tunnel through the tunnel
oxide and get trapped in the trapping layer. During the erase operation, a negative
voltage is applied on the gate and the electrons are removed from the storage layer
as shown in Fig. 7.3.

In the retention operation, no voltage is applied on the gate and the electrons
are stored in the trapping layer as shown in Fig. 7.4. The electrons loss during the
retention in conventional memories is due to defects in the tunnel oxide.

The basic nonvolatile memory mechanism is also illustrated in Fig. 7.5 which
shows the drain current ($I_{drain}$) vs. the gate voltage ($V_g$) of the device. The charged
state "0" is achieved when the electrons are trapped in the storage layer, and
the uncharged state "1" is reached when the memory is erased and electrons are
removed from the storage layer. In order to read the state of the memory, a read
voltage $V_{read}$ is applied on the gate and the drain current is sensed. The difference
in the drain current levels between states "0" and "1" is used to identify the state of
the memory [8, 9].

**Fig. 7.2**  Programming of the memory device by applying a positive gate voltage

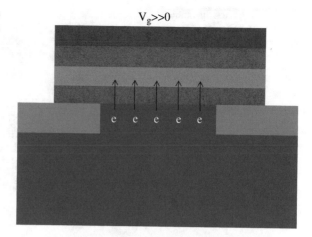

**Fig. 7.3**  Erase operation of the memory device by applying a negative gate voltage

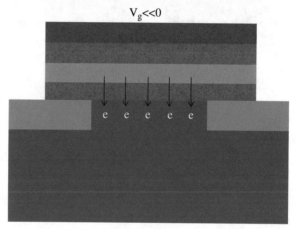

The threshold voltage of the memory is given by (7.1)

$$V_t = V_{FB} - 2\phi_P + \frac{t_{ox}}{\epsilon_{ox}} \sqrt{2\epsilon_{Si} q N_D (2\phi_P)}, \qquad (7.1)$$

where $\phi_P$ is the potential in the Si substrate, $t_{ox}$ is the thickness of the oxide, $N_D$ is the substrate doping, $\epsilon_{ox}, \epsilon_{Si}$ are the oxide and Si dielectric constants, and $V_{FB}$ is the flat-band voltage given by (7.2)

$$V_{FB} = \phi_m - \phi_s + \frac{t_{cntrl} \times Q_s}{\epsilon_{ox}}, \qquad (7.2)$$

**Fig. 7.4** Retention operation
of the memory device with no
applied bias

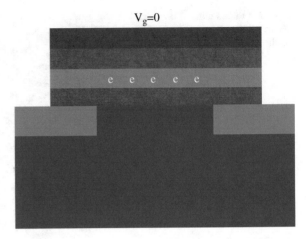

**Fig. 7.5** $I_d - V_g$
characteristic of a memory
device

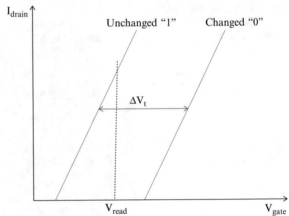

where $\phi_m$ is the metal work-function, $\phi_s$ the MOS channel work-function, $Q_s$ is the
areal density of a sheet charge located at a distance $t_{cntrl}$ from the gate electrode, and
$\epsilon_{ox}$ is the oxide permittivity.

When the memory is programmed, the stored electron charge $Q$ causes the $I_d - V_g$
characteristic of the memory to shift to the right by a voltage $\Delta V_t$ related to $Q$
by (7.3)

$$Q = \frac{C_t \times \Delta V_t}{q},$$                                    (7.3)

where $C_t$ is the capacitance of the MOS memory per unit area, $\Delta V_t$ is the $V_t$
shift, and $q$ is the elementary charge [10]. However, after erasing the memory,
the threshold voltage of the memory shifts back to its original value and so does
the $I_d - V_g$ curve. So, the shift in the threshold voltage represents the program/erase
states.

## 7.3   Nanoparticles for Charge Storage

High-density, low-power consumption, and low-cost memory cells are highly needed because of the scaling of memory devices and the increase in memory array size. The write/erase speed of the floating-gate device can be improved by reducing the thickness of the tunnel oxide below 2.5 nm, which would achieve 100 ns write/erase time at an acceptable programming voltage (<10 V) [3]. However, the retention time would be significantly reduced and the stress-induced leakage current (SILC) will further lower it making the device unreliable. To mitigate these issues, current commercial memory devices have thick tunnel oxides (around 8 nm), and consequently high operating voltages and slow speed, in order to guarantee the 10-year data retention. Nanoparticles embedded in the oxide have been proposed as a solution for this problem [11]. In fact, unlike the floating-gate device which will be totally discharged if a defect is present in the tunnel oxide, nanoparticles based memories will discharge only the stored charge in the nanoparticles present above the tunnel oxide defect. This makes the use of thinner tunnel oxides possible, and consequently, lower operating voltages and higher speed can be achieved without compromising the good retention time (>10 years). In this chapter, we study the effect of embedding 2-nm Si-nanoparticles on the memory device behavior.

### 7.3.1   Si-NPs Electrical Properties

The majority of recent experimental work regarding the use of silicon nanoparticles (Si-NPs) in memory device applications is about SONOS memory devices with >5-nm Si-NPs which exhibit bulk-like trapping characteristics [12, 13]. However, competitive and technologically feasible future devices require nanoparticles of sub-3-nm dimensions; a zero-dimensional regime where important modifications to the silicon electronic structure occur. In fact, as the size of the nanoparticles reduces, their electrical properties change such as their bandgap increases due to quantum confinement in 0-D [14], their dielectric constant decreases [15], their work-function increases [16], and their electron affinity decreases [16]. Additionally, the charging energy is increased as the size of the nanoparticle decreases. The Coulomb charging energy represents the energy needed to add a single electron or hole to the nanoparticle and is given by (7.4)

$$E = \frac{q^2}{C},$$
(7.4)

where $q$ is the Coulomb charge and $C$ is the capacitance of the nanoparticle. For small conductors like nanoparticles, it is convenient to calculate the capacitance $C$ by assuming that the conductor is a sphere of radius $R$. From Gauss's law, the potential at a point with radius $r$ from the center of the sphere ($r > R$) is given by (7.5)

$$V(r) = \frac{Q}{4\pi\epsilon r}, \tag{7.5}$$

where $\epsilon$ is the dielectric constant and $Q$ is the net charge on the sphere. Assuming the potential at infinity is zero, then the potential of the sphere can be calculated using (7.6)

$$V \equiv V(R) = \frac{Q}{4\pi\epsilon R} \tag{7.6}$$

and the capacitance $C$ is calculated using (7.7)

$$C = \frac{Q}{V} = 4\pi\epsilon R. \tag{7.7}$$

When the size of the nanoparticle decreases to the nm range, its charging energy increases and becomes much higher than the room temperature thermal energy ($kT = 25$ meV, where $k$ is Boltzmann constant and $T$ is the temperature). The calculated Coulomb charging energy of Si-NPs of size 2 nm is 1.1 eV.

### 7.3.2   Si-NPs Fabrication

Various techniques have been established to produce Si-nanoparticles including thermal annealing of Si grown by electron-beam evaporation [17, 18], laser abla-tion [19], Si ion implantation [17], atmospheric pressure micro-discharges [20], etc. Alkis et al. showed that it is possible to obtain 2 nm average size Si-NPs through laser ablation and an acid-free sonication and filtration post-treatment method [19]. Firstly, production of Si-NPs was achieved by focusing a femtosecond pulsed laser of $\lambda = 800$ nm with pulse duration of 200 fs, an average output power of 1.6 W at a pulse repetition rate of 1 kHz on a silicon wafer immersed in deionized water. Next, Si-NPs ranging from 1–5.5 nm in size (average size 2 nm) were synthesized by performing sonication at 40 kHz for 200 min then filtration of the NPs colloidal using filters with a pore size of 100 nm [19]. A TEM image of the synthesized non-agglomerate Si-NPs is shown in Fig. 7.6.

## 7.4   ZnO MOSFET-Based Memory Devices

In this section, two different fabricated ZnO-based memory devices with different charge-trapping layers are demonstrated. The memory cells are electrically char-acterized by doing current–voltage ($I$–$V$) measurements in order to understand the observed memory behavior, including the nature of charging. In addition, the effect of the programming voltage on the threshold voltage shift is studied as well as the

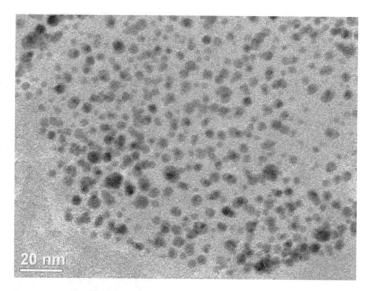

**Fig. 7.6** TEM image of the laser synthesized Si-nanoparticles [19]

retention characteristics of the memory cell. The physics of the device operation are analyzed by constructing the energy bandgap of the memory structure, and by plotting the threshold voltage shift versus a power of the tunnel oxide electric field.

### 7.4.1 Si-NPs Embedded in Zinc-Oxide Charge-Trapping Layer

#### 7.4.1.1 Memory Structure and Fabrication

The channel-last memory cells are fabricated on highly doped ($10^{-18}$ milliohm-cm) p-type (111) Si wafer which is used as a back-gate electrode. First, a 15-nm-thick $Al_2O_3$ blocking oxide is first deposited by ALD using a Savannah 100 system, followed by a 2-nm-thick ZnO charge-trapping layer. Then, Si-NPs are spun on the ZnO at a speed of 700 rpm and an acceleration of 250 rpm/s for 10 s. Again, a 2-nm-thick ZnO charge-trapping layer is ALD deposited so that the Si-NPs are embedded within the charge-trapping ZnO. This is followed by ALD deposition of a 3.6-nm-thick $Al_2O_3$ tunneling oxide and an 11-nm-thick ZnO channel at 250 °C [21]. A solution of 98:2 $H_2O$:$H_2SO_4$ is used for 2 s to etch the channel after patterning by optical lithography (shown in Fig. 7.7). The source and drain contacts are created by depositing 100 nm Al by thermal evaporation followed by liftoff. Using plasma enhanced chemical vapor deposition, a 360-nm-thick $SiO_2$ layer is deposited for device isolation. Finally, rapid thermal annealing in forming gas ($H_2$:$N_2$ 5:95) for

**Fig. 7.7** Mask aligner with nanoimprint lithography at UNAM, Bilkent University, Ankara, Turkey

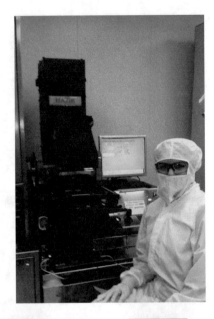

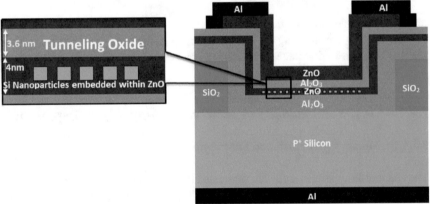

**Fig. 7.8** Cross-section of the fabricated memory cell with embedded Si-nanoparticles

10 min at 400 °C is performed on the samples. Figure 7.8 shows a cross-section of the final device structure with the Si-nanoparticles. Figure 7.9 shows a top view of the fabricated memory devices.

**Fig. 7.9** *Top view* of the fabricated memory cell with embedded Si-nanoparticles

### 7.4.1.2 Characterization

*$I_d - V_g$ Characteristic*

In an attempt to investigate the effect of the Si-NPs on the performance of the memory device, the memory cells are probed using the Agilent-Signatone probe station. Signatone probe station shown in Fig. 7.10 was connected to the Agilent Analyzer shown in Fig. 7.11. Samples are positioned in the middle of the chuck and their location is secured with vacuum. In total darkness, gate voltage is swept in a desired range in order to generate the $I$–$V$ characteristics of a memory cell.

In order to program and erase the memory cell $-10\,V/10\,V$ is applied on the gate for 5 s with the source and drain being grounded. Then, the gate voltage is swept from 0 V up to 20 V with a drain voltage $V_d$ of 10 V and the source being grounded in order to read the state of the cell. It is found that the memory cells are programmed by applying a negative gate voltage and erased by applying a positive gate voltage which suggests that holes are being trapped. The measured $I_{drain} - V_{gate}$ curves of the programmed and erased states of memory devices with and without Si-NPs are depicted in Fig. 7.12 and the $\Delta V_t$ is extracted at a drain current of $4 \times 10^{-5}$ A which is near the extrapolated turn on of the device. The $\Delta V_t$ is increased by an amount of 3.7 V (from 2.6 V) with the Si-NPs. This shows that the Si-NPs behave as charge-trapping centers with a high charge-trapping density within the bandgap of ZnO [24, 25].

In order to see the effect of the programming voltage on the charging of the Si-NPs, the samples are programmed and erased (P/E) at different voltages as shown in Fig. 7.13. As expected, the $V_t$ shift increases with the program and erase voltages. Also, the $V_t$ shift can be as high as 3.4 V at a very low program/erase voltage of

**Fig. 7.10** Signatone probe
station (Model S-1160A) [22]

S-1160A Probe Station

**Fig. 7.11** Agilent
semiconductor analyzer
(Model B1505A) [23]

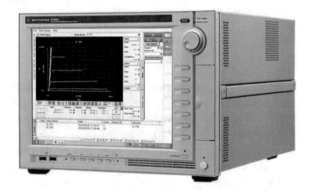

−1 V/1 V due to the Si-NPs which suggests that a mechanism other than tunneling
can cause the hole injection into charge-trapping layer. Figure 7.13 depicts the mean
and standard deviation of the measured $V_t$ shifts. The plot shows that the variation
obtained with Si-NPs is larger. The reason for this larger measured deviation can
be due to the different sizes and numbers of the nanoparticles embedded within
each memory cell. In fact, the fabricated Si-NPs size ranges from 1 to 5.5 nm,
which makes it very difficult to obtain a uniform distribution of NPs in all the
devices. In addition, the deposition method of the Si-NPs can result in a non-uniform
distribution.

*Retention Time*

Furthermore, the retention characteristic of the memory with and without
nanoparticles is analyzed. Figure 7.14 depicts the $V_t$ shift versus time after a single
−10 V programming event The plot shows that 36 % of the stored initial charge is

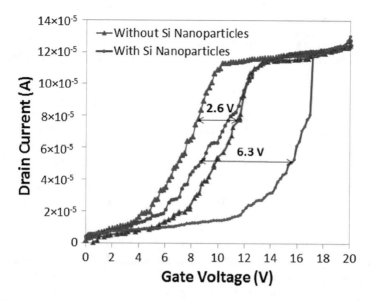

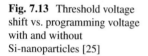

**Fig. 7.12** $I_d - V_g$ showing the obtained $\Delta V_t$ with and without Si-nanoparticles $V_d = 10$ V. The memory is programmed by applying $V_g = -10$ V for 5 s with source and drain being grounded, and erased by applying $V_g = 10$ V for 5 s [24]

**Fig. 7.13** Threshold voltage shift vs. programming voltage with and without Si-nanoparticles [25]

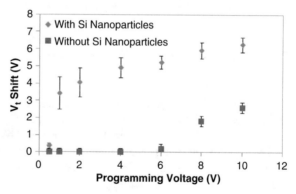

lost in 1 year with Si-NPs while that takes only 70 min in devices without NPs. Also 41 % of the initial stored charge is lost in 10 years with Si-NPs while that only takes 100 min for devices without NPs. The plot indicates that the rate of charge loss is reduced which is revealed by the improved slope of the retention time curve and this is due to Si-NPs better confinement of charges.

The memory with Si-NPs has a large $V_t$ shift of 3.6 V after 10 years while the memory without NPs has a retention time which is much lower than 10 years as shown in Fig. 7.14. The good retention of the memory cell is attributed to the large tunnel oxide thickness, large barrier and good confinement of holes in the Si-NPs making it difficult for holes to leak out without a large reverse electric field.

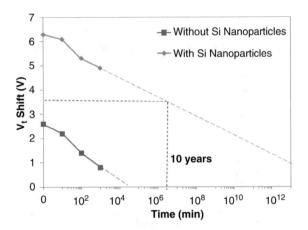

**Fig. 7.14** $V_t$ shift vs. time measured for the memory structures with and without Si-nanoparticles [25]

The charge trap states density can be calculated using (7.3) assuming the threshold voltage shift is mainly due to the stored holes in the charge-trapping layer. With a capacitance of the charge-trapping layer per unit area $C_t = 560\,\text{nF/cm}^2$ and at a $-10\,\text{V}$ programming voltage, the $\Delta V_t$ is 6.3 V which corresponds to a charge trap states density of $1.1 \times 10^{13}\,\text{cm}^{-2}$ or equivalently $1.67 \times 10^{-6}\,\text{C/cm}^{-2}$, and at a programming voltage of $-1$ V, the $\Delta V_t$ is 2.6 V corresponding to a charge trap states density of $5.95 \times 10^{12}\,\text{cm}^{-2}$ or $9.52 \times 10^{-7}\,\text{C/cm}^{-2}$.

### 7.4.1.3 Analysis of Memory Device Operation

*Energy Band Diagram*

To analyze the charge emission mechanism, the energy band diagram of the memory with Si-NPs is plotted in Fig. 7.15. The reported material properties for ZnO, $Al_2O_3$ [21, 26–28], and 2-nm Si-nanoparticles [14–16] are shown in Table 7.1.

The constructed energy band diagram takes into consideration changes in the electronic structure of the Si-NPs due to quantization and Coulomb charging energy. The calculated Coulomb charging energy of Si-NPs of size 2 nm using (7.4) is 1.1 eV and is shown in the energy band diagram plot, where an equal partition in the increase of the energy gap from bulk silicon to 2-nm Si-NPs is added to both the valence band and conduction band. This is a good estimation for 2-nm Si-NPs because the "effective mass" for holes and electrons has been reported to be similar at this size scale [29]. In Fig. 7.15, it is shown that the conduction band offset between channel and tunnel oxide ($\Delta E_c = 1.92\,\text{eV}$) is larger than the valence band offset ($\Delta E_v = 1.36\,\text{eV}$), which makes the holes more prone to overcoming the barrier than electrons. In addition, since the electron affinity of the Si-NPs is small, the conduction band minimum of the Si-NPs is above that of the adjacent ZnO which could inhibit storage of electrons; however, the valence band minimum of the Si-NPs is above that of the adjacent ZnO which forms a quantum well for holes. This analysis supports the observed storage of holes in the memory cell.

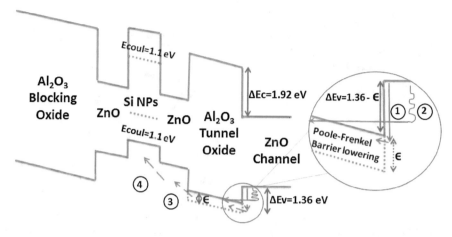

**Fig. 7.15** Energy band diagram of the ZnO memory with Si-nanoparticles with applied negative bias. The changes due to quantization and Coulomb charging energy of the 2-nm Si-nanoparticles are included. (*1*) The Poole–Frenkel effect reduces the barrier for holes allowing them to overcome the potential barrier and be emitted to Al$_2$O$_3$. (*2*) Holes are thermally excited and tunnel via phonon-assisted tunneling (PAT). (*3*) Holes in Al$_2$O$_3$ tunnel oxide drift to the ZnO due to the electric field in the oxide. (*4*) Holes are trapped in the available quantum states in the ZnO bandgap and in the quantum well formed due to valence band offset between the Si-nanoparticles and ZnO trapping layers [25]

**Table 7.1** Material properties for ZnO, Al$_2$O$_3$, and 2-nm Si-nanoparticles

|  | ZnO | Al$_2$O$_3$ | 2-nm Si-NPs |
|---|---|---|---|
| Energy bandgap | 3.37 eV | 6.65 eV | 2.4 eV |
| Relative dielectric constant | 8.75 | 9.5 | 1.78 |
| Electron affinity | 4.5 eV | 2.58 eV | 2.5 eV |

*Holes Transport Mechanism*

In order to determine the holes emission mechanism, the $\Delta V_t$ versus the square root and vs. the square of the electric field are studied and plotted in Figs. 7.16 and 7.17, respectively. Using physics based TCAD simulations, the electric field across the tunnel oxide is calculated [30, 31]. The electric field across the tunnel oxide is 0.36 MV/cm with a 1 V gate voltage and 3.6 MV/cm with a 10 V gate voltage. Tunneling over a potential barrier of 1.36 eV is negligible at an electric field of 1 MV/cm [30, 32, 33]. As a matter of fact, when a very small programming negative gate voltage is applied, the holes in the channel gain enough energy and drift towards the channel/tunnel oxide interface although their energy is not sufficient for tunneling through the 3.6-nm-thick tunnel oxide into the Si-NPs. This is because of the large barrier ($\Delta E_v = 1.36$ eV). However, thermal emission of holes over the barrier is dominant at lower electric fields. This barrier can be further lowered by the electric field in square-root dependence via the Poole–Frenkel effect (PFE) [30].

**Fig. 7.16** $V_t$ shift vs. square root of the electric field across the tunnel oxide. *Linear trend* indicates that the Poole–Frenkel effect is the mechanism for holes emission and capture [25]

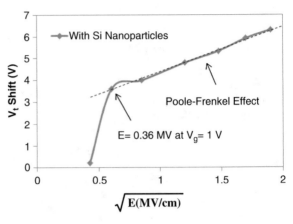

**Fig. 7.17** $V_t$ shift vs. square of the electric field across the tunnel oxide. *Linear trend* indicates that phonon-assisted tunneling (PAT) is the mechanism for holes emission and capture [25]

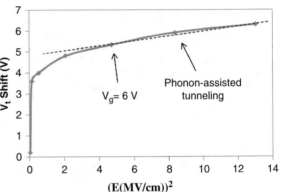

In 1938, Frenkel explained the increase of the carriers thermal emission rate in an external electric field by the barrier lowering associated with the Coulomb potential of the carriers: as the applied field increases, the barrier height decreases further, and due to this barrier lowering, the thermal emission rate of charges exponentially increases [32, 34, 35]. This effect has often been assigned to a donor trap which is neutral when it contains an electron and is positively charged when the electron is absent so that a Coulombic attraction exists. In the fabricated ZnO memory, the ZnO channel is n-type due to native crystallographic defects such as oxygen vacancies and interstitial zinc which behaves as electron donors and the holes are minority carriers. So a Coulombic attraction is available, and when an external electric field exists, the Poole–Frenkel mechanism becomes applicable. In fact, Fig. 7.16 shows a linear dependence of $V_t$ shift on the square root of the electric field. This confirms that PFE is the dominant holes emission mechanism from channel to charge-trapping layers at low electric fields [32]. This also elucidates why large $V_t$ shifts are observed with low program/erase voltages. As a matter of fact, the smaller barrier height for the holes ($\Delta E_v = 1.36\,\text{eV}$) is further lowered in the presence of an electric field by an amount $\epsilon$ given in (7.8) [36, 37] due to PFE:

$$\epsilon = \sqrt{\frac{q^3 E}{\pi \epsilon_0 \epsilon_r}}. \tag{7.8}$$

Where $\epsilon_r$ is the dielectric constant of the tunnel oxide, $q$ is the Coulomb charge, and $E$ is the electric field across the tunnel oxide. The barrier lowering is calculated at a gate voltage $V_g = 1$, 2, and $10\,V$ to be $0.16\,eV$, $0.23\,eV$, and $0.5\,eV$, respectively. The barrier lowering exponentially increases the amount of holes which will overcome the barrier as depicted in Fig. 7.15. Moreover, Fig. 7.17 depicts a linear dependence of $V_t$ shift on the square of the electric field which indicates that phonon-assisted tunneling (PAT) is the dominant hole transmission mechanism at $E > 2.7\,MV/cm$ where holes are thermally excited. This would increase the probability of holes tunneling through the tunnel oxide as shown in Fig. 7.15 [30, 32]. The large electric field allows the holes to drift to the ZnO charge-trapping layer and some holes will be captured by Si-NPs because there is no barrier seen by the holes as depicted in Fig. 7.15. Once there, holes are confined within the nanoparticles or within the available energy states in the quantum well formed by the valence band offset between Si-NPs and adjacent ZnO layers [24].

#### 7.4.1.4  Summary

A low-power, ZnO-based charge-trapping memory with Si-nanoparticles is demonstrated. The memory cells exhibit a much higher $V_t$ shift and a longer retention time (>10 years) with 2-nm Si-NPs. The results show that PFE is the dominant hole-injection mechanism at low electric fields making low programming voltages possible. The large $V_t$ shifts obtained with Si-nanoparticles at low voltages and the excellent retention highlight a promising technology for future ultra-low-power NVM devices.

### 7.4.2  Si-NPs Charge-Trapping Layer

#### 7.4.2.1  Memory Structure and Fabrication

The memory cells are fabricated on a highly doped ($10^{-18}$ milliohm-cm) p-type (111) Si wafer. Firstly, a 360-nm-thick $SiO_2$ layer is deposited using plasma enhanced chemical vapor deposition for device isolation. The $SiO_2$ layer is patterned by optical lithography and etched using the buffered oxide etch. Using a Cambridge Nanotech Savannah 100 ALD system, a 15-nm-thick $Al_2O_3$ is deposited at $250\,°C$ as a blocking oxide. Next, Si-NPs are deposited by spin coating the NPs solution on the sample. A 4-nm-thick $Al_2O_3$ tunneling oxide followed by an 11-nm-thick ZnO channel are deposited by ALD. The ZnO channel is then patterned by optical lithography and etched in a solution of 98:2 $H_2O$:$H_2SO_4$. The source and drain

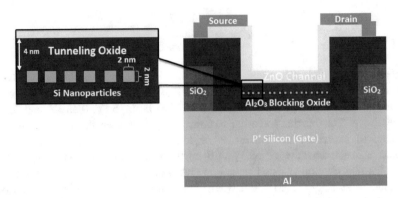

**Fig. 7.18** Cross-sectional illustration of the fabricated charge-trapping memory with 2-nm Si-nanoparticles [36]

contacts are created by thermally evaporating 100 nm Al followed by liftoff. Finally, rapid thermal annealing in forming gas ($H_2$:$N_2$ 5:95) at 400 °C for 10 min is conducted on the devices. The structure of the fabricated memory cell is shown in Fig. 7.18 [37].

### 7.4.2.2 Characterization

*$I_d - V_g$ Characteristic*

By analyzing the $I_{drain} - V_{gate}$ curves of the programmed and erased states of memory devices with and without Si-NPs, the charging effect of the Si-NPs is studied. Using the Agilent-Signatone B1505A semiconductor device analyzer, the memory cells are programmed and erased by applying a gate voltage of −10 V and 10 V, respectively, for 5 s. Next, the gate voltage is swept from 0 V up to 20 V with a drain voltage $V_d$ of 10 V in order to read the state of the cell. In fact, a negative gate voltage is found to program the memory and a positive gate voltage would erase it. This confirms that holes are injected into the Si-NPs. The measured $V_t$ shift is 2.9 V with Si-NPs while it is negligible without NPs as depicted in Fig. 7.19. This proves that $Al_2O_3$ of thickness up to 19 nm has a negligible charge-trapping density which makes this oxide a promising candidate for blocking and tunnel oxides. Also, the charge trap states density can be calculated by adopting (7.3) and assuming the shift in the threshold voltage is mainly due to the trapped charges in the Si-NPs. For a 2.9 $V_t$ shift, the charge trap states density is roughly $7.14 \times 10^{12}$ cm$^{-2}$.

In addition, the erased $I–V$ characteristic of the memory with Si-NPs is shifted to the right with respect to the erased $I–V$ characteristic of the memory without NPs as depicted in Fig. 7.19. This shift of the threshold voltage indicates that the 2-nm Si-NPs are initially negatively charged. The value of the initial negative charge on the Si-NPs can be calculated using (7.3) resulting in a charge density of $7 \times 10^{12}$ cm$^{-2}$. At a negative applied gate voltage, during the programming

**Fig. 7.19** $I_d - V_g$ of the memory showing $V_t$ shift with and without Si-nanoparticles. The memory is programmed by applying $V_g = -10$ V for 5 s [36]

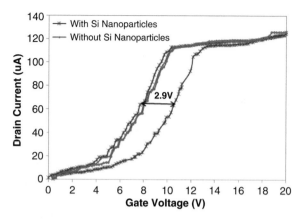

**Fig. 7.20** Threshold voltage shift vs. programming voltage with and without Si-NPs [36]

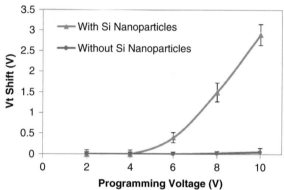

operation, the Si-NPs initial negative charge increases the electric field across the tunnel oxide, therefore enhancing the effect of hole injection. In addition, the mean and standard deviation of the measured $V_t$ shifts at different programming voltages with and without nanoparticles are shown in Fig. 7.20. The $V_t$ shift at a programming voltage of −8 V is around 1.4 V with Si-NPs. Figure 7.20 confirms that Si-NPs act as charge-trapping centers with large charge-trapping density. Also, the standard deviation of the measured $V_t$ shifts with Si-NPs is larger due to the non-uniformity of the distribution of the spin-coated Si-NPs in addition to the different sizes and numbers of the Si-NPs creating the charge-trapping layer of each memory cell.

*Retention Time*

The retention characteristic of the memory is studied by plotting the $V_t$ shift vs. time as shown in Fig. 7.21. The curve is extrapolated to 10 years where the memory cell shows a noticeable $V_t$ shift of 1.25 V, which means a 57 % loss of the initial charge in 10 years. The good retention of the memory cell is due to the good confinement of holes in the Si-NPs.

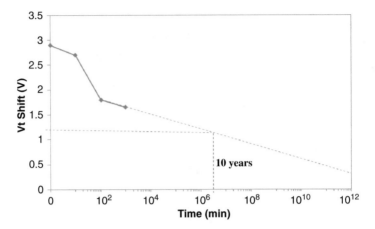

**Fig. 7.21** $V_t$ shift vs. time extrapolated to 10 years with 2-nm Si-nanoparticles [36]

**Fig. 7.22** $V_t$ shift vs. square of the electric field across the tunnel oxide showing a linear dependency, indicating phonon-assisted tunneling (PAT) for emission [36]

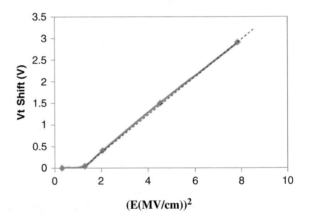

### 7.4.2.3 Analysis of Memory Device Operation

*Holes Transport Mechanism*

In order to determine the hole injection mechanism, $V_t$ shift versus the square of the electric field across the tunnel oxide is plotted in Fig. 7.22. The electric field is calculated using physics based TCAD simulations [30, 31]. The linear relation shown in Fig. 7.22 between $V_t$ shift and the square of the electric field proves that PAT is the main mechanism for hole injection from the channel to Si-NPs at electric fields larger than 1.2 MV/cm. The emission rate in PAT increases exponentially with the square of the electric field intensity according to (7.9) [32–38]:

$$\frac{e(E)}{e(0)} = e^{\frac{E^2}{E_c^2}}, \qquad (7.9)$$

**Fig. 7.23** Plot showing the natural logarithm of the $V_t$ shift times the electric field vs. the reciprocal of the electric field. The *linear trend* indicates that Fowler–Nordheim is the dominant emission mechanism at an oxide electric field of 2 MV/cm [36]

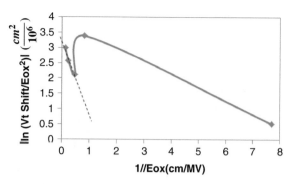

where $e(E)$ and $e(0)$ are the thermal ionization probabilities with and without an electric field E, respectively, and $E_c$ is the characteristic field intensity.

In addition, the natural logarithm of the $V_t$ shift over the square of the electric field across the tunnel oxide versus the inverse of the field is plotted in Fig. 7.23, and the linear trend shows that the Fowler–Nordheim (F–N) tunneling is valid at electric fields larger than 2 MV/cm. However, the dominant mechanism at higher fields ($E > 2$ MV/cm) is F–N tunneling because F–N tunneling has a stronger dependence than PAT on the electric field. In fact, in F–N tunneling, the charges are injected into the tunnel oxide by tunneling into its conduction band through a triangular energy barrier and then are swept by the electric field into the Si-NPs. The emission rate of charges in F–N tunneling follows (7.10) [10]:

$$J = C_1 E_{\mathrm{ox}} e^{\frac{-C_2}{E_{\mathrm{ox}}}},$$ 
(7.10)

where $J$ is the F–N tunneling current, $E_{\mathrm{ox}}$ is the electric field across the tunnel oxide, and $C_1$ and $C_2$ are constants in terms of the effective mass and barrier height. Furthermore, the negatively charged Si-NPs enhance the electric field across the tunnel oxide allowing for the PAT and F–N mechanisms to be applicable at lower electric fields.

*Holes Transport Mechanism*

Using the reported material properties of ZnO, Al$_2$O$_3$, and 2-nm Si-nanoparticles which are listed in Table 7.1, the energy band diagram of the memory cell with Si-NPs is constructed and shown in Fig. 7.24. Figure 7.24 shows that the conduction band offset between tunnel oxide and channel ($\Delta E_c = 1.92$ eV) is larger than the valence band offset ($\Delta E_v = 1.36$ eV), which makes the probability of holes tunneling through the tunnel oxide much higher than electrons tunneling probability. Also, since the electron affinity is lowered as the size of the Si-NPs reduces, this causes the conduction band minimum of the Si-NPs to be above that of the adjacent Al$_2$O$_3$ by 0.08 eV which could prevent electrons storage; however, the valence band minimum of the Si-NPs is above that of the adjacent Al$_2$O$_3$ by 2.13 eV which forms a quantum well where holes can be confined in. The observed hole trapping in the Si-NPs is therefore supported by this analysis. Figure 7.24 illustrates the PAT

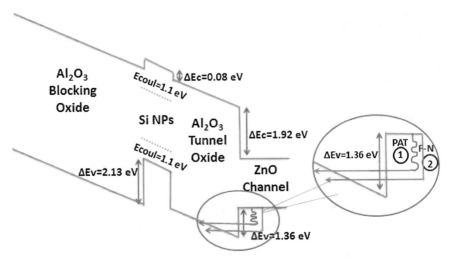

**Fig. 7.24** Energy band diagram of the Si-NPs charge-trapping memory at a negative applied bias. The changes in the electronic structure of the 2-nm Si-nanoparticles due to quantization and Coulomb charging energy are included. (*1*) Holes gain a thermal energy and tunnel through the tunnel oxide. (*2*) Holes are captured by the silicon nanoparticles [36]

and F–N hole-injection mechanisms from the channel into Si-NPs. Lower operating voltage is achieved in Sect. 7.4 (embedded Si-NPs in ZnO) because the mechanism driving the emission is the PFE. The needed electric field is less than 1 MV/cm because the ZnO layer adds available charge trap states to assist the emission [25].

### 7.4.2.4 Summary

An Si-nanoparticle charge-trapping NVM is demonstrated. A 2.9 V $V_t$ shift is obtained at −10 V programming voltage with the Si-NPs. This confirms that the fabricated Si-NPs behave as charge-trapping centers with high charge-trapping density. When 1.2 MV/cm $< E <$ 2 MV/cm, the main mechanism for hole emission is found to be PAT, and when $E >$ 2 MV/cm, F–N tunneling dominates. The negatively charged nature of the Si-NPs increases the electric field across the tunnel oxide during the program function, allowing for an enhanced hole injection through PAT and F–N. The presented results highlight the importance of the charge-injection mechanism on the voltage required to achieve the memory effect. The good retention time of the memory makes further scaling of the tunnel oxide possible without compromising the retention or $V_t$ shift.

## 7.5   MOSCAP-Based Memory Devices

In this section, a MOS memory structure with 2-nm Si-NPs is investigated. The fabricated MOS memories are electrically characterized by plotting the capacitance–voltage (C-V) characteristic and the effect of the gate voltage on the memory hysteresis is studied. In addition, the retention time, endurance, and energy band diagram of the MOS memories are analyzed.

### 7.5.1   MOS Memory Devices

#### 7.5.1.1   Si-NP Charge-Trapping Layer

*Fabrication*

The MOS memory cells are fabricated on a p-type (100) ($10^{-20}$ milliohm-cm) Si wafer. Using a Cambridge Nanotech Savannah 100 ALD system, a 5-nm-thick $Al_2O_3$ oxide is first deposited at 250 °C. Next, Si-NPs—fabricated by Alkis et al.—are delivered by spin coating the Si-NPs solution at a speed of 700 rpm and an acceleration of 250 rpm/s for 10 s. Using a shadow mask, a 10-nm-thick $SiO_2$ gate oxide followed by a 450-nm-thick are sputtered. The use of the shadow mask allowed for patterning the metal gate to a circular shape of diameter 1 mm without the need for lithography or etching [38–45]. Figure 7.25 shows a cross-section of the final device structure with Si-NPs.

*Characterization*

The charging effect of the Si-NPs is analyzed by studying the C-$V_{gate}$ curves of the programmed and erased states of memory devices at high frequency (1 MHz). Using the Agilent B1505A Semiconductor Device Parameter Analyzer, the memory cells gate voltage was first swept from 0 V back to −2 V then forward to 0 V.

**Fig. 7.25** Cross-section schematic of the fabricated MOS memory with Si-nanoparticles

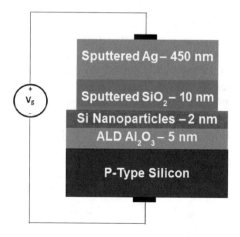

**Fig. 7.26** Measured hysteresis behavior of the $I_{drain} - V_{gate}$ characteristics with gate voltage sweep [40]

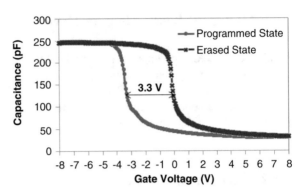

At this low value of applied gate voltage, there was no observed memory hysteresis and the measured C-V curve was the same as the erased state curve shown in Fig. 7.26.

Then, upon sweeping the gate voltage from $-8$ V to $+8$ V with 20 s hold time at $-8$ V, there was a near parallel shift in the measured C-V characteristic in the negative direction (negative $V_t$ shift) from the uncharged state as seen in Fig. 7.26. The value of the shift in the $V_t$ is 3.3 V. The $V_t$ is extracted at a capacitance of 100 pF just before the beginning of the inversion region. The negative shift in the C-V is due to holes storage in the Si-NPs as described by (7.1) and (7.2). Upon reversing the bias sweep from $+8$ V to $-8$ V with 20 s hold time at $+8$ V, there was a reduction in the stored positive charge, and hence the $V_t$ voltage was observed to shift back to the "uncharged" state as shown in Fig. 7.26. As a result, based on these C-V hysteresis measurements, the storage of holes upon negative gate voltage biasing is the observed programming operation, and the removal of stored holes upon positive bias conditions is the observed erase operation of these memory devices with 2-nm Si-NPs. Assuming the threshold voltage shift is mainly due to the trapped charges in the Si-NPs, the charge trap states density of the Si-NPs can be calculated by adopting (7.3). For a 3.8 V $V_t$ shift, and $C_t$ is 796.18 nF/cm$^2$, the charge trap states density is roughly $1.9 \times 10^{13}$ cm$^{-2}$. Moreover, the endurance characteristic of the MOS memory is analyzed by plotting the $V_t$ of both programmed and erased states vs. the number of program/erase cycles as shown in Fig. 7.27. The measurement is made up to $10^5$ cycles where a good memory window (2.9 V) is still present.

In addition, the retention characteristic of the NPs is investigated by measuring the $V_t$ shift vs. time as shown in Fig. 7.28. The curve is extrapolated to 10 years where the memory cell exhibits a noticeable $V_t$ shift of 2 V, which means a loss of 39.4 % of the initial charge in 10 years. The good retention of the memory cell is due to the good confinement of holes in the Si-NPs.

*Analysis of Memory Behavior*

In addition, the energy band diagram of the memory cell with Si-NPs is constructed and shown in Fig. 7.29. The changes in the electronic structure of the

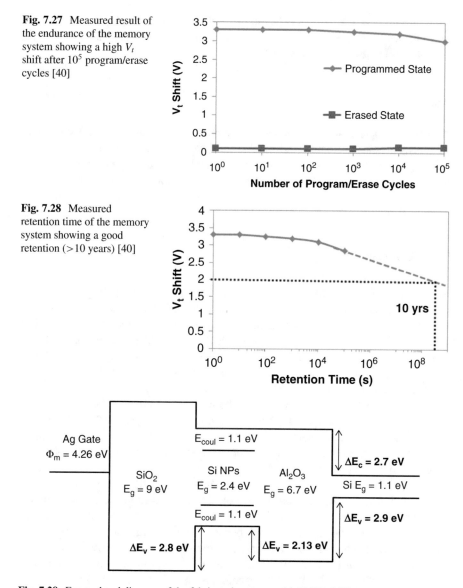

**Fig. 7.27** Measured result of the endurance of the memory system showing a high $V_t$ shift after $10^5$ program/erase cycles [40]

**Fig. 7.28** Measured retention time of the memory system showing a good retention ($>10$ years) [40]

**Fig. 7.29** Energy band diagram of the fabricated memory with Si-NPs [40]

Si-NPs due to quantum confinement in 0-D and to the increased charging energy are taken into account. The Coulomb charging energy of Si-NPs of size 2 nm is calculated to be 1.1 eV using (7.4).

Figure 7.29 shows that there is no conduction band offset between the Si-NPs and the adjacent $Al_2O_3$ which might prevent electrons storage; however, the valence band minimum of the Si-NPs is beyond that of the adjacent $Al_2O_3$ by 2.13 eV so a quantum well is formed where holes can be confined in. This analysis supports the observed hole trapping in the Si-NPs.


#### 7.5.1.2 Summary

A Si-nanoparticle charge-trapping MOS memory is demonstrated. A 3.3 V memory hysteresis is achieved. This confirms that the 2-nm Si-NPs behave as holes charge-trapping centers. The long retention time and the good endurance characteristics of the memory show that Si-NPs are promising in future low-cost nonvolatile memory devices.


## 7.6 Conclusion

Various nonvolatile memory devices with different charge-trapping layers such as 2-nm Si-nanoparticles and 2-nm Si-nanoparticles embedded within ZnO are fabricated and investigated. The memories are based on two structures: MOSFET or MOSCAP. For MOSFET-based memory cells, ZnO is used as the channel material. An alternative MOSCAP-based memory structure with 2-nm Si-nanoparticles is proposed and studied: MOS memories have the lowest number of fabrication steps and thus are the lowest cost memories. The electrical characterization of the various memory devices, calculation of the electric field across the tunnel oxide, the calculation of the charge-trapping density, and the analysis of the energy band diagram allowed the understanding of the behavior of the memory devices based on the properties of the materials used for the charge-trapping layer. Poole–Frenkel mechanism is found to be the dominant mechanism for hole emission at low electric fields (<1 MV/cm) in the Si-nanoparticles embedded in the ZnO memory, while PAT is found to be the mechanism of hole emission at higher electric fields for both ZnO-based memories with Si-nanoparticles. Moreover, the performance of the memory is shown to be highly dependent on the conduction (valence) band offset between the tunnel oxide and charge-trapping layer for effectively storing the electrons (holes) in the trapping layer which would lead to a memory with good retention. In addition, the $V_t$ shift depends on the number of the available states and density of trapping states within the trapping layer. Furthermore, the use of different materials in the charge-trapping layer such as ZnO might have an effect in enhancing the electric field across the tunnel oxide because of its n-type nature, which would lead to a reduced operating voltage memory. The results highlight the importance of the emission mechanism on the magnitude of the voltage needed to achieve the memory effect. Finally, these results show that 2-nm laser synthesized Si-nanoparticles are a good candidate for charge-trapping layers in future low-cost, low-power, nonvolatile memory devices.

# References

1. The International Technology Roadmap for Semiconductors, ITRS (2012)
2. B. Yu, 15nm gate length planar CMOS transistor, in *IEDM*, pp. 257–271 (2001)
3. Y. King, Thin dielectric technology and memory devices. Ph.D dissertation, University of California, Berkeley (1999)
4. K. Ramkumar, Cypress SONOS Technology. Cypress Semiconductor Corporation. Available via DIALOG (2013), http://www.cypress.com/?docID=45736ofsubordinatedocument. Cited 10 Sept 2013
5. M.H. White, D.A. Adams, J. Bu, On the go with SONOS. IEEE Circuits Devices Mag. **16**, 22–31(2000)
6. L. Squire, E. Kandel, *Memory From Mind to Molecule* (Scientific American Library Series, New York, 1999)
7. A. Khakifirooz, D.A. Antoniadis, MOSFET performance – Part I: historical trends. IEEE Trans. Electron Devices **55**, 1391–1400 (2008)
8. P. Pavan, R. Bez, E. Zanoni, Flash memory cells-an overview. Proc. IEEE **85**, 1248–1271 (1997)
9. R. Bez, E. Camerlenghi, A. Modelli, et al., Introduction to flash memory. Proc. IEEE **91**, 489–502 (2003)
10. S.M. Sze, K.K. Ng, *Physics of Semiconductor Devices*, 3rd ed. (Wiley, Hoboken, 2007)
11. S. Tiwari, F. Rana, K. Chan, et al., Volatile and non-volatile memories in silicon with nano-crystal storage, in *IEDM Technical Digest*, p. 521 (1995)
12. W.L. Wilson, P.F. Szajowski, L.E. Brus, Quantum confinement in size-selected, surface-oxidized silicon nanocrystals. Science **262**, 1242–1244 (1993)
13. G. Ledoux, J. Gong, F. Huisken, et al., Photoluminescence of size-separated silicon nanocrystals: confirmation of quantum confinement. Appl. Phys. Lett. **80**, 4834–4836 (2002)
14. J.P. Proot, C. Delerue, G. Allan, Electronic structure and optical properties of silicon crystallites: application to porous silicon. Appl. Phys. Lett. **61**, 1948–1950 (1992)
15. C. Delerue, G. Allan, Effective dielectric constant of nanostructured Si layers. Appl. Phys. Lett. **88**, 173117–173120 (2006)
16. D.V. Melnikov, J.R. Chelikowsky, Electron affinities and ionization energies in Si and Ge nanocrystals. Phys. Rev. B **69**, 113305–113309 (2004)
17. H.H. Hanafi, S. Tiwari, I. Khan, et al., Fast and long retention-time nano-crystal memory. IEEE Trans. Electron Devices **43**, 1553–1558 (1996)
18. Q. Wan, T.H. Wang, M. Zhu, et al., Structural and electrical characteristics of Ge nanoclusters embedded in Al2O3 gate dielectric. Appl. Phys. Lett. **82**, 4708–4710 (2003)
19. S. Alkis, A.K. Okyay, B. Ortac, Post-treatment of silicon nanocrystals produced by ultra-short pulsed laser ablation in liquid: toward blue luminescent nanocrystal generation. J. Phys. Chem. **116**, 3432–3436 (2012)
20. R.M. Sankaran, D. Holunga, R.C. Flagan, et al., Synthesis of blue luminescent Si nanoparticles using atmospheric-pressure microdischarges. Nano Lett. **5**, 537–541 (2005)
21. N. El-Atab, S. Alqatari, F.B. Oruc, et al., Diode behavior in ultra-thin low temperature ALD grown zinc-oxide on silicon. AIP Adv. **3**, 102119 (2013)
22. Signatone, Probe Station. Signatone Corporation. Available via DIALOG (2013), www.signatone.com. of subordinate document. Cited 10 Sept 2013
23. Agilent, Semiconductor Device Analyzer B1505A. Agilent Technologies. Available via DIALOG (2014), www.home.agilent.com of subordinate document. Cited 10 Sept 2013
24. L.W. Lai, C.H. Liu, C.T. Lee, et al., Investigation of silicon nanoclusters embedded in ZnO matrices deposited by cosputtering system. J. Mater. Res. **23**, 2506–2511 (2008)
25. N. El-Atab, A. Ozcan, S. Alkis, et al., Low power zinc-oxide based charge trapping memory with embedded silicon nanoparticles via Poole-Frenkel hole emission. Appl. Phys. Lett. **104**, 013112 (2014)

26. M.L. Huang, Y.C. Chang, C.H. Chang, et al., Energy-band parameters of atomic-layer-deposition heterostructure. Appl. Phys. Lett. **89**, 012903 (2006)
27. G.D. Wilk, R.M. Wallace, J.M. Anthony, et al., High-$\hat{1}^{\circ}$ gate dielectrics: current status and materials properties considerations. J. Appl. Phys. **89**, 5243 (2001)
28. N. El-Atab, A. Rizk, A.K. Okyay, et al., Zinc-oxide charge trapping memory cell with ultra-thin chromium-oxide trapping layer. AIP Adv. **3**, 112116 (2013)
29. O.M. Nayfeh Nonvolatile memory devices with colloidal 1.0 nm silicon nanoparticles: principle of operation, fabrication, measurements, and analysis. Ph.D dissertation, EECS, MIT, Cambridge (2009)
30. A. Alnuaimi, A. Nayfeh, V. Koldyaev, Electric field and temperature dependence on the activation energy associated with gate induced drain leakage (GIDL). J. Appl. Phys. **113**, 044513–044519 (2013)
31. J. Bu, M.H. White, Design considerations in scaled SONOS nonvolatile memory devices. Solid-State Electron. **45**, 113–120 (2001)
32. S.D. Ganichev, E. Ziemann, W. Prettl, et al., Distinction between the Poole-Frenkel and tunneling models of electric-field-stimulated carrier emission from deep levels in semiconductors. Phys. Rev. **61**, 10361 (2000)
33. C.H. Lee, S.H. Hur, Y.C. Shin, et al., Charge-trapping device structure of SiO2/SiN/high-k dielectric Al2O3 for high-density flash memory. Appl. Phys. Lett. **86**, 152908 (2005)
34. Y.M. Chang, W.L. Yang, S.H. Liu, et al., A hot hole-programmed and low-temperature-formed SONOS flash memory. Nanoscale Res. Lett. **8**, 340 (2013)
35. N. Kramer, C.V. Berkel, Reverse current mechanisms in amorphous silicon diodes. Appl. Phys. Lett. **64**, 1129 (1994)
36. N. El-Atab, A. Ozcan, S. Alkis, et al., Silicon nanoparticle charge trapping memory cell. Phys. Status Solidi RRL **8**, 629 (2014)
37. W.R. Harrell, J. Frey, Observation of Poole-Frenkel effect saturation in SiO2 and other insulating films. Thin Solid Films **352**, 95 (1999)
38. N. El-Atab, F. Cimen, S. Alkis, et al., Enhanced memory effect with embedded graphene nanoplatelets in ZnO charge trapping layer. Appl. Phys. Lett. **105**, 033102 (2014)
39. N. El-Atab, F. Cimen, S. Alkis, et al., Future of health insurance. Enhanced memory effect via quantum confinement in 16 nm InN nanoparticles embedded in ZnO charge trapping layer. Appl. Phys. Lett. **104**, 253106 (2014)
40. N. El-Atab, A. Rizk, B. Tekcan, et al., Memory effect by charging of ultra-small 2-nm laser-synthesized Si-nanoparticles embedded in Si-Al2O3-SiO2 structure. Phys. Status Solidi A **212**, 1751–1755 (2015)
41. A. Nayfeh, A.K. Okyay, N. El-Atab, et al., Low power zinc-oxide based charge trapping memory with embedded silicon nanoparticles, in *Invited, 226th ECS Meeting 2014*, vol. 46, pp. 2143–2143, Cancun (2014)
42. A. Nayfeh, A.K. Okyay, N. El-Atab, et al., Transparent graphene nanoplatelets for charge storage in memory devices, *Invited, 226th ECS Meeting 2014*, vol. 37, pp. 1879–1879, Cancun (2014)
43. N. El-Atab, A. Rizk, B. Tekcan, et al., MOS memory using 2-nm silicon nanoparticles charge trapping layer, in *Advances in Materials and Processing Technologies 2014 Conference*, Dubai (2014)
44. N. El-Atab, A. Ozcan, S. Alkis, et al., 2-nm laser-synthesized Si nanoparticles for low-power charge trapping memory devices, in *14th IEEE International Conference on Nanotechnology*, Toronto (2014)
45. N. El-Atab, B. Turgut, A. Okyay, M. Nayfeh, A. Nayfeh, Enhanced non-volatile memory characteristics with Quattro-layer graphene-nanoplatelets vs. 2.85-nm Si-nanoparticles with asymmetric Al2O3/HfO2 tunnel oxide. Nanoscale Res. Lett. **10**, 248 (2015)

# Chapter 8
# Accurate Temperature Measurement for 3D Thermal Management

Sami ur Rehman and Ayman Shabra

## 8.1 Introduction

3D integration introduces a new degree of freedom for the design of electronic systems which offers many potential advantages such as improvements in energy efficiency, increase in speed, and a reduction in area [1, 2]. It also allows for the creation of new classes of microsystems that integrate not only electronics but also components such as sensors and batteries [3]. Yet, some of the known issues in a 2D setting, such as yield, thermal management, and stress management, become much more challenging in a 3D context. In this chapter, we will focus on the thermal management problem and specifically on accurate thermal state sensing.

Although 3D architectural re-partitioning of 2D systems has been demonstrated to improve energy efficiency by reducing losses in an interconnect-dominated environment, it results in an increase in die temperature [1, 2]. To understand this, it can be noted that the pathways for heat generated on the silicon die are impeded by the presence of other die in a 3D stack. Moreover, the reduction in interconnect length enabled by the 3D routing, also results in highly active circuits falling on top of each other resulting in the growth of the intensity and size of hot-spots. If we consider that power densities as high as 100–300 W/cm2 are common in a 2D setting [4], it becomes clear that the thermal state in a 3D stack will be much more severe. These problems can be partially addressed by a careful selection of the die stacking

S. Rehman (✉)
Technische Universität Dresden, Chair for Circuit Design and Network Theory,
01069 Dresden, Germany
e-mail: sami.rehman@seecs.edu.pk

A. Shabra
MediaTek Inc., Woburn, MA, USA
e-mail: shabra@alum.mit.edu

© Springer International Publishing Switzerland 2016
I.M. Elfadel, G. Fettweis (eds.), *3D Stacked Chips*,
DOI 10.1007/978-3-319-20481-9_8

157

arrangement [1], but better solutions are needed, and creative solutions such as the use of micro-fluidic cooling are currently under research [5].

Another aggravating factor in the 3D thermal problem is the continuous increase in system complexity. For example, the trend towards a larger number of processor cores has lead to projections of hundreds of cores by 2020. This will require a rethinking of architectures and strategies for the implementation of power and thermal management (PMT) modules. An important component of this rethinking is an awareness of the thermal state through temperature measurements. If we consider that the AMD x86-64 core has 95 on-chip power monitoring signals per core [6], we can project that the future needs for power monitoring in a massively multi-core setting will be much more critical. This is yet another motivation for this work to develop a compact and accurate 3D temperature sensor.

The design of an accurate sensor involves two main components. The first is the sensor device and the second is an analog to digital converter (ADC). A variety of integrated devices have been used as temperature sensors include bipolar transistors [7, 8], MOS transistors [9], thermistors [10], and thermal diffusivity structures [11]. In this chapter we report on the use of bipolar transistors since they are widely available and used in CMOS technologies and provide good accuracy. The ADC is another key component. We explore the use of successive approximation register (SAR) converters due to their energy efficiency [12]. We also present a technique to improve the overall accuracy of the combination of sensor and ADC.

## 8.2 CMOS Smart Temperature Sensors

Temperature sensors are used in a wide variety of applications including bio-medical, environmental, automotive, manufacturing, and aerospace. Each application places more stringent requirements on some specifications compared with others. In the case of thermal management in 3D integrated systems, we list some of the most critical specifications for temperature sensors.

1. **Area:** Today's high-performance microprocessors can experience elevated temperatures up to 80–100 °C under high load conditions. There can be large temperature gradients across the surface of the chip, and many hotspots are situated at unpredictable locations as we can't determine task allocation profile at design time. It is therefore customary to place many small sized temperature sensors across the plane of the chip for dynamic and accurate throttling and fan speed regulation [13]. In modern microprocessors several on-chip sensors are embedded inside the microprocessor, and many more will be required for 3D thermal management since thermal contours will become worse in 3D stacked architectures where the thermal profile of each die is strongly affected by the die above and below it.

2. **Accuracy:** (the difference between the real and measured temperature) For a microprocessor low-accuracy, such as ±5 °C, is sufficient only when the junction temperature is sufficiently below the upper temperature range. But at full throttle, to minimize performance impact due to back-off, a temperature

accuracy around $\pm 2\,°C$ is desirable. This will reduce the sandbagging necessary in the thermal budget due to the uncertainty in the temperature estimate. In 3D integration, the worsening thermal state makes a temperature accuracy of $\pm 1\,°C$ desirable.

3. **Resolution:** (two consecutive temperature readings that the sensor can reliably distinguish) Microprocessors can experience rapid thermal transients like $0.5\,°C/s$ at the same location. These transients could easily increase to $1\,°C/s$ in 3D stacks. Therefore, resolution of around $0.1\,°C$ is required for thermal monitoring [14].

4. **Response Time:** Typically smart sensors have a response time that is determined by the thermal time constant of the package and is in the range of few hundred ms. But in 3D ICs, quick readings, on the order of few hundreds of $\mu s$, are necessary to control accurately and dynamically the thermal profile at elevated temperatures [13].

## 8.3  Components of a Smart Sensor

A smart sensor typically consists of three components as shown in Fig. 8.1. A bias generator is used to drive the sensor core, whose output is digitized by the digital sensor interface to perform further signal processing. Next, we discuss the sensor core and interface electronics in detail.

## 8.4  Temperature Sensor Core

### 8.4.1  Parasitic BJT-Based Sensors

A typical temperature sensor in a CMOS process uses parasitic BJTs as shown in Fig. 8.2a. Here bias currents $I_{BIAS}$ and $nI_{BIAS}$ applied to nominally equal transistors Q1 and Q2, respectively. The voltages $V_{EB1}$ and $V_{EB2}$ are complementary to absolute temperature (CTAT) while their difference $\Delta V_{EB}$ is proportional to absolute temperature (PTAT) [15]

$$\Delta V_{EB} = V_{EB2} - V_{EB1} \tag{8.1}$$

$$= V_T \ln\left(\frac{nI_{BIAS}}{I_s}\right) - V_T \ln\left(\frac{I_{BIAS}}{I_s}\right) \tag{8.2}$$

$$= \frac{kT}{q} \ln n. \tag{8.3}$$

**Fig. 8.1** Key components of a silicon temperature sensor

| BIAS CURRENT GENERATOR | SENSOR CORE | INTERFACE ELECTRONICS |

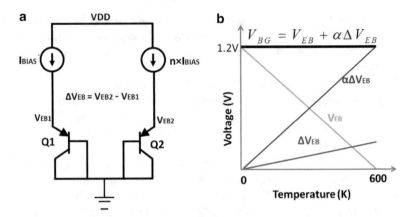

**Fig. 8.2** (a) BJT-based temperature sensor. (b) PTAT, CTAT, and BGR voltages vs temperature

Here $k$ is Boltzmann's constant, $T$ is the absolute temperature and $q$ is the electron charge. The monitoring of $\Delta V_{EB}$ forms the basis for BJT temperature measurements.

If the CTAT voltage is added to an appropriately scaled PTAT voltage, we can generate a bandgap reference (BGR) voltage of 1.2 V given as $V_{BG} = V_{EB} + \alpha \Delta V_{EB}$ [16]. This is illustrated in Fig. 8.2b where $V_{EB}$ has a negative temperature coefficient with an ideal slope of $-2\,\text{mV/}^\circ\text{C}$. The value of $\alpha$ is chosen such that the positive temperature coefficient of $\Delta V_{EB}$ becomes equal in magnitude to the negative temperature coefficient of $V_{EB}$ resulting in a temperature-independent BGR ($V_{BG}$). Since $V_{EB}$ and $\Delta V_{BE}$ are required by the BGR and are available in a BJT temperature sensor core, it should be possible to have self-referenced ADC. This eliminates the requirement of an explicit voltage reference, thereby reducing area, power, and the design overhead of generating an accurate reference.

### 8.4.1.1 Lateral vs Vertical Parasitic pnp BJTs

Because of their availability in bulk CMOS technologies, PNP BJTs are a natural choice for temperature sensing elements. Figure 8.3 shows cross sections of lateral and vertical pnp transistors available in standard CMOS. Lateral pnp BJTs, shown in Fig. 8.3a, have lower 1/f noise but suffer from the presence of parasitic vertical pnp, shown as a dotted BJT. The vertical pnp provides an easy path for the current to flow towards the substrate instead of the collector of the lateral pnp [17]. This current loss results in a non-ideal IV characteristics for the BJT.

A cross section of a vertical pnp (*vpnp*) BJT is shown in Fig. 8.3b. These pnps are biased through their emitters and are typically used as diode connected configurations. Since these pnps are biased through their emitter rather than the collector, the collector current is a function of common emitter current gain, $\beta_F$. The value of $\beta_F$ is typically much smaller than for BJTs fabricated in a bipolar technology, since the wider $n$-well in CMOS serves as the base.

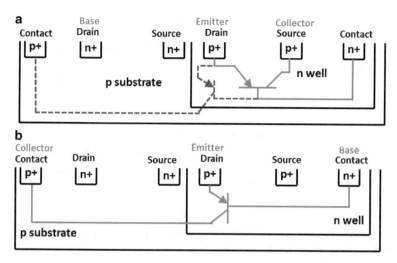

Fig. 8.3 (a) Cross section of lateral pnp. (b) Cross section of vertical pnp

### 8.4.1.2 Principle of Operation

A temperature reading can typically be obtained using a temperature-dependent voltage like $\Delta V_{EB}$, and a temperature-independent reference like BGR voltage. The PTAT voltage $\Delta V_{EB}$, $V_{EB2} - V_{EB1}$, is given by $\frac{kT}{q} \ln(n)$, where $n$ is the difference in current ratios of the two BJTs. This PTAT voltage can serve as an input for the ADC while a BGR can serve as a reference. Since the output of any ADC is the ratio of its input to the reference, [8] gives this ratio for a delta sigma ($\Delta\Sigma$) ADC as:

$$\mu = \frac{\alpha \Delta V_{EB}}{V_{EB} + \alpha \Delta V_{EB}}. \tag{8.4}$$

Here $\alpha$ is a scaling factor that could range from 8 to 20, depending on process and design. An estimate of the temperature can therefore be found as $T_{out} = A \times \mu - 273$ [8], where $A$ is a constant.

## 8.4.2 MOS-Based Sensors

As discussed in the previous section parasitic BJTs are used as sensing elements in conventional CMOS temperature sensors. Typically $V_{EB}$ ranges from 0.7 to 0.8 V over a wide temperature range. With some voltage headroom needed for the MOS current sources, the use of BJT sensors is challenging in sub-1 V nanoscale CMOS design. This problem can be overcome by using a MOSFET in a diode configuration

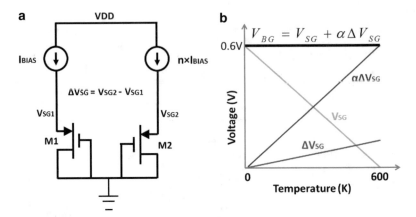

**Fig. 8.4** (**a**) MOS-based sensor operated in weak inversion regime. (**b**) MOS-based PTAT, CTAT, and BGR voltage swept across temperature

as the temperature sensing element as shown in Fig. 8.4. The transistors in Fig. 8.4a are operated in sub-threshold regime, which results in an exponential dependence of $I_D$ on $V_{SG}$, given by

$$I_D \propto \exp \left[ \frac{V_{SG} - V_{TH}}{m \frac{kT}{q}} \right] \tag{8.5}$$

much like $I_C$ dependence on $V_{EB}$. Here $m$ is the MOS slope factor and is equivalent to the ratio of depletion capacitance over oxide capacitance. But unlike BJTs where saturation current, $I_C$, is the only parameter impacted by device mismatch and process spread, while for MOSFET, both threshold voltage and charge mobility are the two parameters, found in Eq. (8.5), which are affected by process variations [9]. As a result, MOS-based sensors have limited accuracy and require at least a two-point trim to achieve the same accuracy as BJT-based sensors.

Similar to $V_{EB}$, $V_{SG}$ is CTAT while the difference between two $V_{SG}$, $\Delta V_{SG}$, is PTAT. The BGR voltage is also generated in a similar manner when CTAT $V_{SG}$ is added in proper weighting with PTAT $\Delta V_{SG}$ shown in Fig. 8.4b. However, the BGR generated by a MOS sensor is approximately 0.6 V [9].

## 8.4.3 Thermal Diffusivity-Based Sensors

Another class of temperature sensors is based on the thermal diffusion constant of silicon, which has a well-defined relationship with temperature. These sensors are realized by measuring the temperature-dependent diffusion of heat in bulk silicon. The construction and the operating principle of such sensors is detailed in [18]. Key components of diffusivity-based sensors are electrothermal filter (ETF), amplifier, phase operated $\Delta\Sigma$ modulator, and VCO. ETF consists of a thermopile and two heaters, with the distance between the thermopile and the filter being $r$. The output of VCO drives the two heaters in anti-phase, which results in temperature difference across thermopile of opposite polarity in the form of a potential difference of a few hundreds of $\mu$Vs. This small difference is first amplified and then digitized to compute the temperature.

The MOS- and BJT-based sensors in CMOS, as discussed earlier, suffer from two critical drawbacks. Firstly, they are heavily prone to mismatch and process spread, which become worse with technology scaling. Secondly, increase in leakage current at elevated temperatures restricts the sensor accuracy. On the other hand, in thermal diffusivity-based sensors the temperature is computed by measuring the phase shift of a filter, which in turn is dependent on $r$ and thermal diffusion constant of Si. The impact of process spread can be reduced by increasing $r$. Since the output of such sensor core is essentially phase shift of a filter, which is independent of device leakage current, such sensors can be used at extremely high temperatures. As a result, such sensors offer better untrimmed accuracy compared to MOS- or BJT-based sensors [19]. One drawback of such sensors is their power inefficiency as the extensive dissipation of heat in the heater could be a few mWs.

### 8.4.4   On-Chip Resistor-Based Sensors

On-chip resistors, $n$-well resistor and polysilicon resistors, have also been used as temperature sensing elements since they have large temperature coefficients. But unlike BJTs and thermal diffusivity-based sensors their temperature dependence is not very well defined and, therefore, more than two-point trim is required to achieve accuracy in comparison to the sensors discussed above [19].

### 8.4.5   Inverter Chain Delay-Based Sensors

The temperature-dependent propagation delay of a chain of inverters can also be measured to estimate the temperature. The propagation delay of a single inverter driving a load capacitance, $C_L$, is a function of VDD, $C_L$, $V_{TH}$, and $\mu$. When VDD and $C_L$ are constant, and VDD is much larger than $V_{TH}$, the propagation delay becomes largely a function of carrier mobility. Although power and area efficient, the accuracy of such sensors is worse than BJT and diffusivity-based sensors.

## 8.5 Interface Electronics for Temperature Sensors

### 8.5.1 Analog-to-Digital Converter

A $\Delta\Sigma$ ADC trades off conversion speed for higher resolution. Therefore, high accuracy sensors [8] typically use $\Delta\Sigma$ modulators along with precision circuit techniques of auto-zeroing, chopping, and dynamic element matching (DEM). Such high-resolution sensors find their applications in industrial, automotive, biomedical, and environmental systems. $\Delta\Sigma$ ADCs, however, tend to occupy a large area, which makes them unsuitable for hotspot thermal monitoring. This is primarily due to the current practice of using a large number of thermal sensors and adjusting their location after experimentally measuring the location of the actual hotspots.

SAR ADCs have also been used [20] and are particularly attractive given the power and area efficiency they have demonstrated to date. The inaccuracy introduced by such ADCs is dependent on device mismatch especially when power consumption must be minimized through the use of the smallest device sizes, satisfying the noise requirements. To address this limitation, pure analog, digital, or mixed calibration of device mismatch must be employed in order to keep the SAR ADC contribution to the sensor error to a minimum.

### 8.5.2 Frequency-to-Digital Converter

Smart sensors that don't employ an ADC typically require an explicit reference voltage [19], which becomes an important contributor to the error budget when high accuracy is desired. A typical BGR circuit is PVT sensitive since it implements the scaling factor $\alpha$, the product $\alpha\Delta V_{EB}$, and the addition $V_{EB}+\alpha\Delta V_{EB}$ in the analog domain. In advanced CMOS nodes, implementing these arithmetic operations in the analog domain is ambitious due to several sub-micron non-idealities, device mismatch, and process spread. Such sensors have been used for microprocessor thermal monitoring and fan speed regulation and have been demonstrated an accuracy of $\pm 1.5\,°C$ after calibration based on measurements at two temperatures.

The voltage reference in a frequency-to-digital converter (FDC) generates $V_{EB}$ and $V_{BG}$ which are converted into their corresponding frequencies using a voltage to frequency converter. The two frequencies are then input to counters. When the counter corresponding to $V_{EB}$ is filled, it stops the counter for $V_{BG}$ and the digital code available in the first counter is the desired ratio $V_{EB}/V_{BG}$ and is the digital estimate of the temperature. A detailed implementation and circuit diagram of an FDC can be found in [21]. A similar digitization approach has been used in [22] for RFID food monitoring.

### 8.5.3 Time-to-Digital Converter

The most power and area efficient sensor can be designed using a time-to-digital converter (TDC). One possible implementation relies on measuring the frequency of two current starved ring oscillators. One of the oscillators is driven by a temperature-independent current source, $I_{REF}$, while the other one by a PTAT current source, $\Delta I_{EB}$. A frequency comparator is then used to compare the two frequencies (or delay of the two pulses) and outputs a pulse with width corresponding to the difference between the two frequencies. The output of a frequency comparator is input to a TDC which consists of a counter whose digital output, the temperature estimate, corresponds to the number of periods of counter frequency it can accommodate within the width of pulse generated by frequency comparator. Mismatch and process spread in the ring oscillator severely limits the accuracy of such sensors and renders it unsuitable for high accuracy thermal management in 3D stacked chips at throttle temperature range. Detailed implementation of such a sensor can be found in [23].

## 8.6 Realization of a Smart Temperature Sensor in 65 nm CMOS

### 8.6.1 Choice of Sensor Core and Interface Electronics

We desire a temperature accuracy of $\pm 1\,°C$, resolution of around $0.1\,°C$, minimum area and power, and a quick response time for managing the thermal profile of 3D integrated IC. Figure 8.5 is taken from [24] and plots the inaccuracy of the sensor versus the number of trims. The figure includes sensors based on BJTs, MOSFETs, resistors (RES), and thermal diffusivity (TD) and provides a useful guide for the sensors selection. It is evident from the figure that the accuracy of MOS and resistor-based sensors is inferior to BJTs. Moreover, MOS, resistors, and delay sensors are sensitive to process variations and present a whole host of challenges for the required accuracy at elevated temperatures. Although thermal diffusivity-based sensors are less sensitive to process spread and provide good accuracy, they are power inefficient and require a precision amplifier at the output of ETF. Parasitic BJTs offer a good choice because they are power and area efficient and can achieve $\pm 0.6\,°C$ error inaccuracy in standard 65 nm CMOS with a single room temperature trim.

Since both FDC- and TDC-based interface electronics require an explicit reference, their accuracy is compromised due to process and mismatch spread in the BGR circuit. $\Delta\Sigma$ modulators, on the other hand, have been reported to achieve low power and high accuracy not only in mature CMOS nodes [8, 9], but also in 65 nm CMOS as reported in [25]. For 3D temperature resolution requirements, $\Delta\Sigma$ modulators are unlikely to be the optimal choice. SAR ADCs can achieve lower power, area, and response time compared to $\Delta\Sigma$ designs.

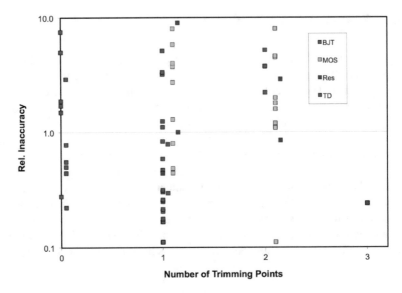

**Fig. 8.5** Temperature sensor inaccuracy vs number of trims [24]

Next, we discuss a temperature sensor in a 65 nm CMOS technology consisting of

1. Vertical *pnp*-based parasitic BJT core driven by
2. PTAT current generator with interface electronics in the form of
3. 12-bit segmented SAR ADC.

## 8.6.2 Vertical PNP BJT Sensor Core

A vertical *pnp* BJT sensor core similar to the one shown in Fig. 8.2a has been implemented. The *vpnp*s in this configuration are biased through their emitters resulting in $V_{EB}$ that is a function of the common emitter current gain $\beta$, which has a typical value of only 1.1 in our process. Spread in $\beta$ can result in erroneous values of $V_{EB}$, thereby compromising sensor accuracy. In the next section, a circuit scheme to compensate for the process spread in $\beta$ will be discussed. Figure 8.6 shows the statistical simulations of estimated temperature error for the *vpnp* core only with single room temperature trim and no curvature correction applied. This result includes the contribution of process variations and device mismatch. Thick red lines in Fig. 8.6 show the average and $\pm 3\sigma$ bounds of error inaccuracy.

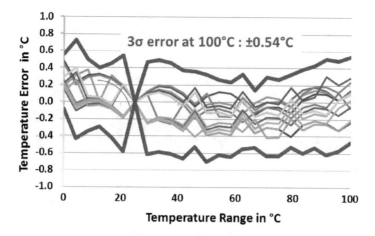

**Fig. 8.6** Statistical spread of temperature error inaccuracy for a single *vpnp*

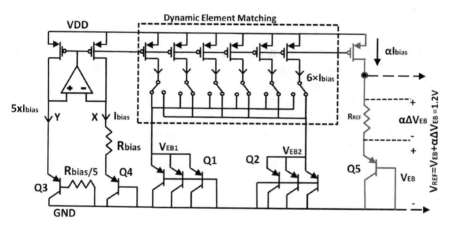

**Fig. 8.7** Bias circuit for sensor core

## 8.6.3  Sensor Core and PTAT Bias Generator

Figure 8.7 shows the implemented PTAT current generator and sensor core. The
circuitry on the left side of Fig. 8.7 generates a PTAT current, which is sourced into
the sensor core through the PMOS current mirror. Q1 and Q2 are the temperature
sensing BJTs and are comprised of three *vpnp*'s connected in parallel. The parallel
connection is intended to minimize the variations in base emitter series resistance
due to process variations, thereby, reducing variations in $V_{EB}$. PMOS current
elements are implemented using DEM, which helps suppress device mismatch
errors. Q2 carries five times more current than Q1. Hence the current ratio, *n*,

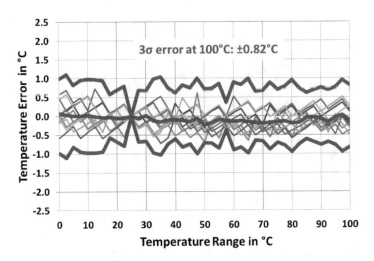

**Fig. 8.8** Error inaccuracy in the bias circuit and *vpnp*s

between the two *vpnps* is 5. The right side circuit is a typical example of BGR circuit and is shown here to illustrate that a reference circuit can easily be built if a PTAT current is available.

Figure 8.8 shows the statistical inaccuracy for the sensor core and the PTAT current generator. Comparing it with Fig. 8.6, we see that the $\pm 3\sigma$ limits have worsened to around $\pm 1\,°C$ due to mismatch and process variations in *vpnps* and mismatch in the PTAT current generator and PMOS current sources.

Next we explain how the PTAT current is generated and how it helps compensate for the error introduced due to process variations in $\beta$. Consider the simplified circuit shown in Fig. 8.7 where the voltage at node '$X$' can be written as:

$$V_X = I_{BIAS} \times R_{BIAS} + V_{EB4}. \tag{8.6}$$

Similarly, the potential at node '$Y$' is given as:

$$V_Y = I_{B3}\frac{R_{BIAS}}{n} + V_{EB3}, \tag{8.7}$$

where $I_{B3}$ is the base current of Q3 and can be written in terms of the emitter current as $I_{E3}/(\beta_3 + 1)$. We note that since $I_{E3}$ is $n \times I_{BIAS}$, we can then rewrite $V_Y$ as:

$$V_Y = I_{BIAS}\frac{R_{BIAS}}{(\beta_3 + 1)} + V_{EB3}. \tag{8.8}$$

The low input referred noise OpAmp forces nodes '$X$' and '$Y$' to be at equal potential, a necessary condition for generating a PTAT current. So $V_X = V_Y$ and we obtain:

$$I_{BIAS}\frac{R_{BIAS}}{(\beta_3 + 1)} + V_{EB3} = I_{BIAS} \times R_{BIAS} + V_{EB4} \tag{8.9}$$

which can be reduced with arithmetic manipulations to:

$$I_{BIAS} = \frac{\Delta V_{EB34}}{R_{BIAS} \times \alpha_3},$$ (8.10)

where $\Delta V_{EB34} = V_{EB3} - V_{EB4}$ and:

$$\alpha_3 = \frac{\beta_3}{1 + \beta_3}.$$ (8.11)

The current computed in Eq. (8.10) is PTAT since it is proportional to $\Delta V_{EB34}$, which is PTAT. Now if we copy this PTAT current into Q1 using the PMOS current mirror, we will end up having $I_{E1} = a \times I_{BIAS}$, where "$a$" is current mirror gain. Now $V_{EB1}$ can be determined:

$$V_{EB1} = \frac{kT}{q} \ln\left(\frac{\alpha_1 \times a}{I_{S1}} I_{BIAS}\right).$$ (8.12)

Here $\alpha$ is the multiplication factor and was described in Sect. 8.4.1.2. Plugging the value of $I_{BIAS}$ from Eq. (8.10) into Eq. (8.12) results in:

$$V_{EB1} = \frac{kT}{q} \times \ln\left(\frac{\alpha_1 \times a}{I_{S1}} \times \frac{\Delta V_{EB34}}{R_{BIAS} \times \alpha_3}\right).$$ (8.13)

Now if the two *vpnp*s, Q1 and Q3, are matched we have $\alpha_1 = \alpha_3$ and Eq. (8.13) can be reduced to:

$$V_{EB1} = \frac{kT}{q} \times \ln\left(\frac{a \times \Delta V_{EB34}}{I_{S1} \times R_{BIAS}}\right).$$ (8.14)

From Eq. (8.14) we see that the $V_{EB1}$ is independent of $\beta$ and hence the process variations in $\beta$ will have no impact on the error inaccuracy of the sensor.

### 8.6.4 12-Bit Segmented SAR ADC

The output of any ADC is essentially the ratio of its input and reference. Typically, in a BJT-based temperature sensor that is biased by a PTAT current, the PTAT voltage, $\alpha \Delta V_{EB}$, serves as the ADC input voltage while a BGR serves as the ADC reference. In [8] the final digital temperature reading is derived as:

$$\mu = \frac{\alpha \Delta V_{EB}}{V_{EB1} + \alpha \Delta V_{EB}},$$ (8.15)

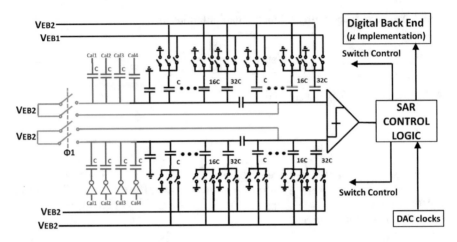

**Fig. 8.9** Block diagram of the implemented SAR ADC

where the $\alpha$ is a scaling factor and the denominator is the BGR. Also, $\Delta V_{EB} = \Delta V_{EB2} - \Delta V_{EB1}$ in Fig. 8.7. Equation (8.15) can be modified into the following form:

$$\mu = \frac{1}{1 + \dfrac{1}{\alpha \left(1 + \frac{V_{EB1}}{V_{EB2}}\right)}}. \tag{8.16}$$

The ratio $V_{EB1}/V_{EB2}$ can be measured by applying $V_{EB1}$ as an input to an ADC with $V_{EB2}$ as a reference. Figure 8.9 shows segmented 12-bit SAR ADC with such a configuration.

### 8.6.4.1 ADC Resolution Specifications

The ADC resolution should be high enough to make quantization errors insignificant in comparison to accuracy requirement of the temperature sensor. Targeting an error inaccuracy of $\pm 1\,°C$ translates to an ADC resolution requirement of $\pm 0.2\,°C$ as explained in [8]. Since $V_{EB}$ slope decreases as $2\,mV/°C$, our target value for resolution becomes 0.2mV which can typically be achieved with a 12-bit ADC.

### 8.6.4.2 Top Plate vs Bottom Plate Input Sampling

Figure 8.10 plots $V_{EB1}/V_{EB2}$ against temperature, where $V_{EB1}$ is input and $V_{EB2}$ is the reference. The figure shows the temperature dependence of $V_{EB1}/V_{EB2}$ before the ADC in blue and after being measured by the ADC in dotted red. Moreover,

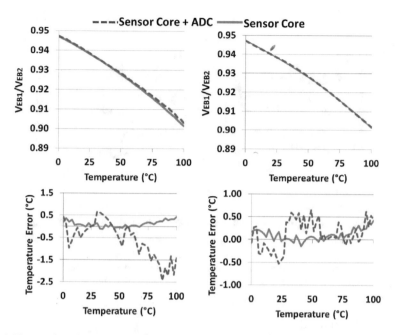

**Fig. 8.10** $V_{EB1}/V_{EB2}$ for top plate input sampling (*top left*) and bottom plate input sampling (*top right*). Temperature error for top plate sampling (*bottom left*) and bottom plate sampling (*bottom right*)

two sampling configurations are illustrated in Fig. 8.10 where top plate sampling is shown on left and bottom plate sampling is shown on the left.

In the case of top plate sampling, the voltage divider created due to presence of parasitics on DAC capacitors affect the reference voltage amplitude during each bit conversion since, unlike the input, the reference connects to the bottom plate. Hence, both input and reference are affected differently and this results in a gain error as evident from Fig. 8.10, which increases at higher temperatures. This gain error can be easily taken care of with bottom plate input sampling as in this approach both input and reference are affected by the same amount of parasitics and charge leakage. The bottom left plot in Fig. 8.10 shows the estimated temperature error for top plate input sampling and bottom right shows the error for bottom plate input sampling for a nominal run.

## 8.6.5   Statistical Simulation Results

Figure 8.11 shows the statistical simulation results of SAR ADC+PTAT current generator. After eliminating the need for a BGR circuit, we observe the inaccuracy of our sensor depends on mismatch and process variations of the BJTs and PTAT current bias and to some extent on the resolution of the interface electronics.

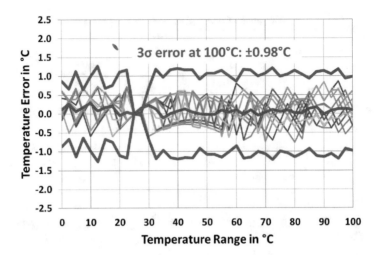

**Fig. 8.11** Statistical spread in estimated temperature error with $\pm 3\sigma$ limits for 25 trials for sensor core and reference-less SAR ADC

## 8.7 Conclusions

We discussed some of the key specifications of temperature sensors for monitoring the thermal profile in 3D integrated chips. We explained the various choices available for designing the thermal sensor core and the data converter as well as the pros and cons of each design. We presented in detail a circuit level realization of a BJT-based thermal sensor with a novel 12-bit reference-less segmented SAR ADC as the digital interface. Finally, we provided statistical simulation results to show the contributions of the sensor core and the ADC to the overall accuracy of the thermal sensor. By avoiding the need to use an explicit reference and its associated analog sources of error, we demonstrated how the accuracy of a thermal sensor can be improved.

**Acknowledgements** The authors would like to thank Mubadala Technology, Abu Dhabi, United Arab Emirates for funding this project.

## References

1. S. Borkar, 3D integration for energy efficient system design, in *IEEE Design Automation Conference*, pp. 214–219 (2011)
2. U. Kang, et al., 8Gb 3-D DDR3 DRAM using through-silicon-via technology, in *IEEE International Solid State Circuits Conference*, pp. 130–131 (2010)
3. G. Chen, et al., Circuit design advances for wireless sensing applications. Proc. IEEE **98**(11), 1808–1827 (2010)

4. J.-Q. Lu, 3-D hyperintegration and packaging technologies for micro-nano systems. Proc. IEEE **97**(1), 18–30 (2009)
5. D. Sekar, et al., A 3D-IC technology with integrated micro-channel cooling, in *IEEE International Interconnect Technology Conference*, pp. 13–15 (2008)
6. R. Jotwani et al., An x86-64 core in 32nm SOI CMOS, in *IEEE International Solid State Circuits Conference*, pp. 106–107 (2011)
7. A.P. Brokaw, A simple three-terminal IC bandgap reference. IEEE J. Solid State Circuits **9**(6), 388–393 (1974)
8. K. Souri, et al., A 0.12 $mm^2$ 7.4 $\mu$ W micropower temperature sensor with an inaccuracy of ± 0.2 ° C (3 $\sigma$) from -30 ° C to 125 ° C. IEEE J. Solid State Circuits **46**(7), 1693–1700 (2011)
9. K. Souri, et al., A precision DTMOST-based temperature sensor, in *IEEE European Solid State Circuits Conference*, pp. 279–282 (2011)
10. M. Perrott, et al., A temperature-to-digital converter for a MEMS-based programmable oscillator with better than +/- 0.5ppm frequency stability, in *IEEE International Solid State Circuits Conference*, pp. 206–208 (2012)
11. S. Mahdi Kashmiri, et al., A thermal-diffusivity-based frequency reference in standard CMOS with an absolute inaccuracy of 0.1% from 55C to 125C. IEEE J. Solid State Circuits **45**(12), 2510–2520 (2010)
12. M. van Elzakker, et al., A 1.9uW 4.4J/conversion-step 10b 1MS/s charge redistribution ADC, in *IEEE International Solid State Circuits Conference*, pp. 244–245 (2008)
13. K. Luria, et al., Miniaturized CMOS thermal sensor array for temperature gradient measurement in microprocessors, in *IEEE International Symposium on Circuits and Systems*, pp. 1855–1858 (2010)
14. U. Sonmez, et al., A 0.008$mm^2$ area-optimized thermal-diffusivity-based temperature sensor in 160nm CMOS for SoC thermal monitoring, in *IEEE European Solid State Circuits Conference*, pp. 395–398 (2014)
15. D.F. Hilbiber, A new semiconductor voltage standard, in *IEEE International Solid-State Circuits Conference*, pp. 32–33 (1964)
16. R.J. Widlar, New developments in IC voltage regulators. IEEE J. Solid State Circuits **6**(1), 1–7 (1971)
17. E.A. Vittoz, MOS transistors operated in lateral bipolar mode and their application in CMOS technology. IEEE J. Solid State Circuits **18**(3), 273–279 (1983)
18. C. van Vroonhoven, et al., Thermal diffusivity sensors for wide-range temperature sensing, in *Proceedings of IEEE Sensors*, pp. 764–767 (2008)
19. K.A.A. Makinwa, Smart temperature sensors in standard CMOS. Procedia Eng. **5**, 930–939 (2010)
20. M. Tuthill, A switched-current, switched-capacitor temperature sensor in 0.6$\mu$m CMOS. IEEE J. Solid State Circuits **33**(7), 1117–1122 (1998)
21. J. Shor, K. Luria, et al., Ratiometric BJT-based thermal sensor in 32nm and 22nm technologies, *IEEE International Solid State Circuits Conference* pp. 210–212 (2012)
22. M.K. Law, A. Bermak, et al., A sub $\mu$ W embedded CMOS temperature sensor for RFID food monitoring application. IEEE J. Solid State Circuits **45**(6), 1246–1255 (2010)
23. P. Chen et al., All-digital time-domain smart temperature sensor with an inter-batch inaccuracy of -0.7°C ±0.6° after one-point calibration. IEEE Trans. Circuits Syst. Regul. Pap. **58**(5), 913–920 (2011)
24. K.A.A. Makinwa, Temperature sensor performance survey [Online]. Available http://ei.ewi.tudelft.nl/docs/TSensor_survey.xls
25. F. Sebastiano, et al., A 1.2V 10$\mu$W NPN-based temperature sensor in 65nm CMOS with an inaccuracy of ±0.2°C (3s) from −70°C to 125°C, *IEEE International Solid State Circuits Conference*, pp. 312–313 (2010)

# Chapter 9
# EDA Environments for 3D Chip Stacks

Love Cederström

## 9.1 Introduction

The vertical stacking of integrated circuits (ICs) in 3D configurations presents new challenges for Electronic Design Automation (EDA) software [14]. An example of the current state of the art in 3D chip stacking is the Hybrid Memory Cube from Micron [8, 12] where one logic die serves as a base for up to four memory dies connected using Through-Silicon Vias (TSVs). Another commercially successful example is the high-end Xilinx Virtex 7 FPGA[1] [2, 6] which is based on a 2.5D silicon interposer. These 2.5D and 3D packaging technologies are still in their infancy in comparison with the widely used wire bonding technology. A more established technology than 2D or 3D is flip chip assembly, which despite its maturity, accounts for only 17 % of the market [9]. With TSMC planning to ramp up 3D ICs production in 2014–2016 [16], and GlobalFoundries starting to offer TSVs in their 20 nm process [4], 3D chip stacks are likely to become a viable option wherever high-volume customer electronics requires extremely small form factors.

3D integration technology is generally not available for small sample-size tape-outs (MPW[2]) [5, 16], and so design groups aiming at leveraging system performance through 3D integration should be prepared to encounter several issues in their design methodologies and tool flows. One such issue is the interplay between 3D chip heat generation and 3D physical design. This issue is treated with some depth in

---

[1] Field Programmable Gate Array.

[2] Multi project wafer.

L. Cederström (✉)
Technische Universität Dresden, Chair of Highly-Parallel VLSI-Systems and
Neuro-Microelectronics, Mommsenstr. 12 Töpler-Bau, 01069 Dresden, Germany
e-mail: Love.Cederstrom@ieee.org

© Springer International Publishing Switzerland 2016
I.M. Elfadel, G. Fettweis (eds.), *3D Stacked Chips*,
DOI 10.1007/978-3-319-20481-9_9

175

the next chapter (Chap. 10). In this chapter, we more generally consider the design processes used to implement a 3D stack and review the main differences they have with the standard system design for 2D chips. With focus on the analog, mixed–mixed signal design of high-speed transceivers, a test vehicle for a Network-on-Chip in a 3D stack is used as a case study. In the next subsections, the test vehicle is described, and in the following sections an analysis of the requirements for a rigorous 3D design methodology is provided. A notable point is that the 3D test site is that of a full system and so it requires many areas of expertise sourced out of many different working entities (e.g., research institutes, corporate divisions, companies, etc.). Although this may sound like a matter of course, it results in fact in the use of different operating systems, data management systems, languages, and software tools. Such diversity has profound implications for the EDA solutions of 3D system design.

### 9.1.1  An Evaluation Concept for Chip Stack Interconnects

One 3D integration scheme is to attach ICs to a passive silicon chip carrier known as interposer. The advantage of such approach is that it mitigates the risk of TSV formation within the ICs themselves. The interposer along with its TSVs can be developed separately while a standard chip fabrication process is used to fabricate the ICs. Another advantage is the lower cost due to the adoption of existing chip and packaging technologies for 3D integration rather than the development of entirely new technologies. Thus the risk and cost associated with including TSVs in active ICs are eliminated. The proposed 3D test vehicle with the passive interposer has three main goals:

1. Test transceivers designed and implemented in a semiconductor technology (28 nm).
2. Test TSVs developed using in-house processes in a realistic setting.
3. Analyze EDA flows and adapt them for future 3D stacking projects.

The interposer test vehicle is comprised of two identical, symmetric, bumped, flip chips that are mirrored with respect to each other and attached with C4 bumps[3] on each side of the passive interposer. TSVs connect the two chips in a face-to-face configuration. The two dies and the interposer form a 3D stack that is attached and wire-bonded to an organic package substrate. Using a standard PCB technology as package substrate in SODIMM[4] format, a low-cost solution for a 3D stack carrier is thus achieved. The SODIMM module can in turn be used with a standard low-cost socket on a system board that provides interface headers, power supply, and clock generation. In Fig. 9.1, the packaged system is designed with following components:

---

[3] Controlled collapse chip connection.

[4] Small outline dual in-line memory module.

**Fig. 9.1** The test vehicle design is comprised of two symmetric flip chip ICs in a face-to-face configuration, thus enabling a 3D chip stack that tests TSVs and 28 nm transceivers in a realistic product-like setting

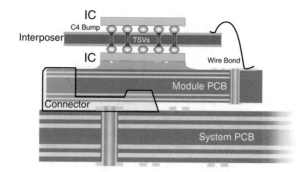

**IC:**
    Mixed-signal 28 nm dies of approximately 1.3 mm by 3 mm.
**Interposer:**
    Silicon chip carrier with connectivity for power, ground, digital inputs/outputs, and TSVs for the Z dimensional interconnect.
**Module PCB:**
    3D stack carrier for chip-on-board wire bonding. This is the organic package substrate of the 3D test vehicle.
**System PCB:**
    Circuit board to connect the test vehicle to power supply, clocks, measurement pin headers, and an extension connector for an FPGA.

## 9.1.2  Design Environments

When designing and implementing a full system, the design and implementation tasks are typically performed by several groups with each having its physical (e.g., geographical locations) and virtual (e.g., chip or package design, system testing, etc.) contexts. In such multi-groups projects, concurrent engineering is typically chosen as the best methodology toward successful product development [13]. This generally means that the design process is integrated and managed in an organized matrix of scheduled tasks with clear, well-defined contracts between the different engineering groups across all their physical and virtual contexts. In the semiconductor industry, such management implies a concurrent approach to design environments and software tools, an example of which is the use of uniform computer systems with a consistent design tools and a database accessible by all groups involved in the system design project. While such uniformity may be achieved for traditional IC chip design, it is still not achievable when it comes to chip, package, and system co-design. The generic context in the latter case is one of non-uniformity with many heterogeneous environments having to interface with each other during the product design cycle. In the specific context of a 3D stack system design, the development of die, interposer, package substrate, and system PCB, one can identify several tasks and subtasks, each with its own physical and virtual environment. They include:

1. 3D chip stack design and implementation: (a) develop TSV technology; (b) model electrical TSVs; (c) design interposer; (d) manufacture silicon interposer with TSVs; (e) design IC; and (f) manufacture IC in a 28 nm process.
2. Package design and implementation: (a) design a module PCB for carrying the 3D stack; (b) manufacture and assemble module PCB.
3. System design and implementation: (a) design the system including power supply and clock generation and (b) manufacture and assemble system PCB.

These numerous tasks obviously fall outside the expertise of any one person or even one whole group of people be they in an academic department or a research institute. They must therefore be distributed. For our 3D test vehicle, the various tasks are addressed by the following laboratories and institutes all attached to the Technical University of Dresden: Communication Laboratory—RF Engineering (IfN-HF); Institute of Circuits and Systems—Highly-Parallel VLSI-Systems and Neuromorphic Circuits (HPSN); and Institute of Semiconductors and Microsystems—Department For Semiconductor Technology (IHM). Figure 9.2 shows the various tasks and how they are distributed across the different research laboratories and institutes. It is clear that a system design procedure valid across the board will need to accommodate disparate computer systems and a multitude of software tools for various design tasks such as physical design, physical verification, circuit and systems simulation, and component modeling. In other words, the design environment is not only *distributed* but also *heterogeneous*. Combining

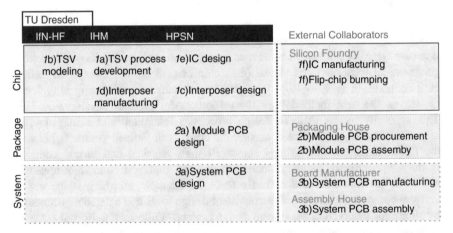

**Fig. 9.2** Implementing the 3D test site requires collaboration across several institutes, operating on the chip, package and system level. The components are a 3D stack at the chip level, a PCB for chip-on-board assembly at the package level, and a PCB with test interface connectors, power supply, and clock generation at the system level

design and manufacturing in a design-for-manufacturing (DFM) flow will even complicate matters further due to the drastically different contexts of design and fabrication. In the next section, each of the individual design tasks will be described with subsequent sections addressing how these tasks are linked together in order to achieve a high degree of concurrent engineering in a highly distributed heterogeneous environment.

## 9.2   Contemporary Electronic Designing Automation

It goes with saying that EDA software tools play a crucial role in chip and system design and verification. The EDA industry started in the 70s with Calma among others who sold desk size workstations bundled with design software [7]. Today personal computers are a commodity and a plethora of EDA packages from several vendors are available, with the Internet used for instant distribution of licenses and executables. Software tool vendors addressing the various steps the full design process, from semiconductor device engineering (TCAD) to wiring harness design and everything in between, belong to what has become a multi-billion dollar industry [10]. As described in Sect. 9.1.2, the net result of having many expert collaborators in a system design project is a distributed, heterogeneous design environment. For example, the PCB design typically runs on Windows whereas IC design typically runs on Linux, with package design using one or the other depending on the tool and the vendor. In this section, details on the methods used to design each individual physical component of the 3D test vehicle are given, starting from the IC itself and moving up the hierarchy to the system board.

### 9.2.1   Integrated Circuits

The overall 3D test vehicle is meant to operate as a digital system, yet the nature of the high-speed TSV transceivers requires custom analog and mixed-signal (AMS) design methods. Designing a mixed-signal IC is a rather complex task, and at the 28 nm process node, it requires an even more involved set of verification steps. The flow can be roughly divided into three main steps: (1) digital register transfer level (RTL) design; (2) custom digital/analog design; and (3) physical design with Place and Route (PnR). At each of these steps, technology information of the silicon foundry is required. Such information is provided by process design kit (PDK), which is applied to both the compilation of digital design blocks using standard cells and macros (e.g., on-chip memory) and the implementation of the custom analog/mixed-signal cells. Aside from providing physical information to enable cell placement and interconnect routing, the PDK also contains circuit simulation models such as the transistor level BSIM[5] compact model.

---

[5] Berkeley short-channel insulated gate field effect transistor model.

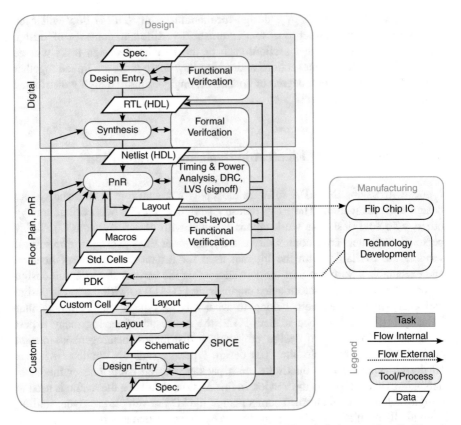

**Fig. 9.3** Designing a mixed-signal IC is a complex process: To each design step (*small oblongs*) corresponds one or several simulation and verification steps (*large oblongs*), that are iteratively applied to the produced design data (*rhombi*). One common approach for system-on-chip designs is to combine a SPICE-based analog flow (*bottom*), with RTL-based digital flow (*top*) with the physical design executed simultaneously during place-and-route and final verification (*middle*)

A typical AMS design flow is described in Fig. 9.3. To each design step corresponds one or several verification steps to ensure that the design step leads to an implementation that satisfies the design specifications. For custom implementations, design entry and layout followed by SPICE[6] simulations is used iteratively to achieve the targeted performance. On the other hand, digital design is based on describing the design with a hardware description language (HDL) such as Verilog or VHDL. This can be carried out within a logic synthesis environment, where the HDL description can be verified for functional correctness against a set of specifications and then technologically mapped onto a target IC library with timing constraints taken into account. Finally, the chip layout is generated based standard

---

[6] Simulation Program with Integrated Circuit Emphasis.

and custom cell geometry information combined with the technology information provided by the PDK. The last step is to run Design and Electrical Rule Checking (DERC) to make sure that physical (e.g., spacings, CD, run lengths, overlaps, etc.) and electrical (e.g., timing, noise, power) design rules are satisfied.

For the specific 3D test case, the Cadence® Virtuoso® suite together with the Spectre® simulator were employed for custom design, the Synopsys® Design Compiler® for digital implementation and Cadence® Encounter PnR for integrating the compiled digital blocks made of library standard cells with the custom cells. Design signoff was achieved with Mentor® Calibre® for layout versus schematic (LVS) and design rules check (DRC); Ansys® Redhawk or Cadence Voltus for power analysis (IR); and static timing analysis (STA) is performed using Synopsys® PrimeTime. It should be noted that this design flow is composed of a diverse set of EDA software tools from several vendors, including the Big Three: Synopsys®, Cadence®, and Mentor® [15]. This hybrid approach is very common as it combines tool-selection flexibility with best-in-class methodologies at each individual step of the design flow. The drawback is of course that seamless tool integration is required so that the output of one tool (e.g., Synthesis) can be fed as input to another (e.g., PnR). Very often, the effort required for tool integration is far from trivial. Standardization trends such as Verilog (IEEE standard 1364) and UPF[7] (IEEE standard 1801-2009) are to be welcomed as they greatly facilitate multi-vendor, tool interoperability, and enable the setup of seamless, concurrent system design environments.

## 9.2.2  Interposers

An interposer refers to a device connecting two different parts of a system. An example of an interposer is the Printed Circuit Board (PCB) or the Multi-Chip Module (MCM). In modern packaging it denotes a chip carrier containing at least one conductive redistribution layer (RDL) on silicon that can connect several chips, either on both sides in a 3D integration scheme or side by side in a 2.5D integration scheme. The interposer is normally a passive carrier that provides mechanical support and electrical interconnections. When the silicon carrier contains active devices, the interposer is said to be active and can be considered an IC in it self.

The specifications of a passive interposer destined to be a chip carrier with TSVs and pads for C4 bumps and wire bonding are derived mainly from the physical connectivity requirements. For error-free interposer design, die pin locations and net names have first to be propagated to the package design environment. Next, interposer design rules, physical outline requirements, and allowed IO locations for the chip stack are captured. To preserve design content (e.g., IC pin locations) and maintain connectivity, the die image is required. Since the PCB/package design is not on the same system as the IC design, there is a need for simple file-based

---

[7] Unified Power Format.

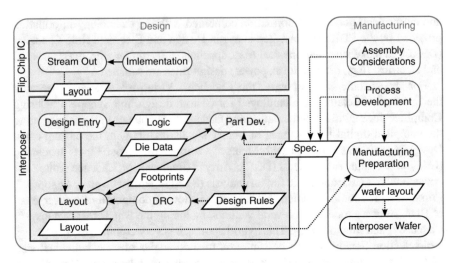

**Fig. 9.4** As the interposer is a passive device the design flow is comparatively simple, mainly the concern is to ensure that connectivity is correct and that the design will allow PCBs to be routable. This can be achieved by importing a die graphical format (e.g., GDSII) into a package/PCB design environment (Stream-Out → Layout) and here extract logical and physical die data to develop a custom part (Part Dev.). The die is then logically (Design Entry) and physically (Layout) connected to produce an interposer pinout that accommodate assembly considerations

interface, to pass information from the IC domain to the PCB/package domain. This interface should be lightweight, i.e., of small size and easy to use. Once the bond area and IC are available as design objects with physical footprints and logical connectivity information, a netlist can be entered for connecting the ICs to each other and to the most suitable bond pads. The latter data can be used as specifications in order to combine the IC objects with an interposer IO object (BGA-balls/C4 or wire bond pads) and define a full 3D layout at the interposer level.

Figure 9.4 depicts a complete flow for implementing a passive interposer, from obtaining die information to exporting a layout for manufacturing. One of the main challenges in such flow is the crossing between the IC design domain and the packaging domain, especially that there are multiple IC and package design software solutions with incompatible data formats. In some foundry processes an optical shrink is applied to the design at tape-out, which effectively excludes any common database, LEF/DEF[8] or other *design* data format solutions for transferring design contents between the package and IC design domains. On the one hand, the GDSII stream-out data format does contain geometries of the die as manufactured, and so post-fabrication details such as die shrink can be captured. Using this GDSII approach, one single industry standard data format can then be used for design data exchange between IC and packaging. With such GDSII standard, it is possible not only to use two identical dies as in our 3D test vehicle, but also

---

[8] Library Exchange Format, Design Exchange Format.

any number of difference dies, which can come from any number of design teams or IP sources, regardless of their internal IC design methodologies, databases, or data formats. An advantage of the GDSII approach is that in streaming out only critical design information such as bump locations and TSV keep-out zones, IC intellectual property can be protected. Furthermore, using a standard *production* data format improves the likelihood of first-time system design success, especially when several IC design teams are collaborating. The major drawback of the GDSII format is that it contains no logical design intent, and so it is not possible to feed it into the *logical* design hierarchy of the full system. However, the complex task of deriving connectivity can be greatly alleviated by an interface that can automatically extract information from the GDSII text objects to create *design* objects native to the packaging EDA environment.

The Cadence® Allegro® System-in-Package (SiP) solution provides an interface that identifies text objects whose origins are within bump pad geometries and interpret them as pin names, thus making it possible to create "smart" design objects from "dumb" GDSII graphics. Once an object containing both physical pin location and corresponding logic data (e.g., net name or pin name) is created, it can be exported as both a physical footprint and a logical part, and included in the design library, thus making it possible to instantiate several copies of the same part. Disregarding design rules such as line spacings and widths, the physical specifications for an interposer layout mainly consist of bond pad size, location, and outline. This can be captured as a footprint together with a symbol that maps physical pins to logical net names. Using the Cadence® Allegro® Part Developer together with Allegro® Designer is one way of conveniently specifying the design boundaries within which the top and bottom dies have to reside. The Allegro® System Connectivity Manager (SCM) has a tabular interface as contrasted with the more commonly used schematic design entry interfaces. For systems with parts that have many IOs (larger ICs and packages), this presents an effective way of keeping track of logical connectivity. When a design has been entered logically and is ready for layout, physical and electrical design rules associated with the interposer need to be taken into account. In an R & D phase or when using in-house technology, there is usually no available PDK containing physical design rules for TSVs and RDL, which means that such design constraints need to be entered manually. In contrast to an IC design, a passive interposer requires relatively few lithography masks for manufacturing, which makes it feasible to manually derive design rules and "floorplan" the interposer (physical outline with bond wire pads) directly from the specifications. With a constraints manager, design rules can be entered, modified, and incorporated in the interposer layout and DRC steps. Once the interposer layout is completed, individual interposers are instantiated to fill up an entire wafer for the fabrication technology at hand. As a passive interposer resides in the silicon technology domain, the GDSII format is the de-facto standard for mask creation, and so any interposer layout tool should support the streaming out of such format.

### 9.2.3  Printed Circuit Boards

PCB design is based on the concept of assembling active and passive components. A PCB typically includes complex integrated circuits, single transistors, diodes, capacitors, inductors, and resistors. These are then normally logically entered into an comprehensive schematic design representation that integrates both active and passive components. Designing a PCB differs vastly from designing an IC. For one thing, the PCB "library" is made of a limited number of available and well-defined parts which results in a "smaller" design search space. Another major difference is the cost of making a design error. While in the IC case, the design error will result in another fab run to fix the design, the error in the PCB case is likely to be repairable by one of the most available tools in Electrical Engineering, namely, hand soldering! The need for simulation-based tools for full PCB design verification has not been pressing at all.

The relatively low complexity of PCB design allows the DRC and LVS steps to be executed repeatedly during the design process. This means that for low-performance PCB design, once the layout is 100 % connected and DRC-clean, the PCB is basically done. For high-performance PCBs, where high-speed interconnections are paramount, the main challenges are those of signal integrity (SI) and power integrity (PI), which have emerged as sign-off requirements. Even for relatively standard PCBs with multi-gigabit interconnect, SI analysis now requires the use of 3D electromagnetic field solvers, whose usage in the past was limited to microwave and RF designs. Furthermore, thermal simulations integrated with the PCB electrical design flow have now become common in contrast with the traditional approach of leaving thermal analysis to mechanical system and packaging engineers. Such electromagnetic and thermal tools are now part of the Chips-Package-System (CPS) methodologies which cover signal integrity, power integrity, and thermal analysis. Such methodologies are now required for the design and analysis of high-performance systems.

Figure 9.5 describes a PCB design flow that is similar to that of the interposer in Fig. 9.4. The PCB flow is however based on available parts, having well-characterized packages, and on the use of the Extended Gerber format (RS-274X) for PCB physical information transfer. For the 3D test vehicle, the very same PCB flow can be used for both Module and System PCBs although the point tools used in each of these PCBs will be different. For System PCB, a graphical design entry like the Allegro® DEHLD[9] schematic editor and a standard PCB layout tool are viable options. In Module PCB, the presence of chip-on-board bonding requires a richer feature set in the design tool than standard PCB design. This is because one needs to handle bondwire profiles, among other things. Since the Module PCB acts as a package with only the die stack and decoupling capacitors, it is advantageous to use a tabular design entry like the Allegro® SCM for managing the many IOs and

---

[9] Design entry HDL.

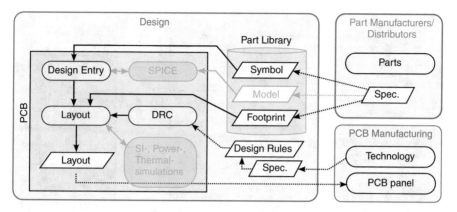

**Fig. 9.5** PCB implementations start with design entry, normally using standard parts represented by schematic symbols. Footprints containing physical information are used at the connectivity-driven layout stage. As layout verification, DRC and LVS are performed iteratively to satisfy physical and electrical constraints, including spacing and runlength matching for high-speed signal lines

their connectivity. By staying in the same design environment the footprint created for the interposer can be used unaltered with connectivity maintained. The Allegro® SiP layout with its support for wire bonding in combination with the SCM turned out to be valid tool options for our Module PCB (Fig. 9.1).

## 9.3   Connecting the Dots for a 3D Stack

In the previous sections, the design of the different components of a chip stack was described as if they were separate consecutive steps: (1) IC design → (2) Interposer design → (3) Module PCB design → (4) System PCB design. Not having any feedback from the top node in the design hierarchy is clearly risky and often results in non-working or under-performing designs, especially in the presence of cross-level interdependencies in the design hierarchy. This is typically the case for analog interfaces such as high-speed input/outputs (IOs).

In Fig. 9.6, the complete design flow of the 3D test vehicle is depicted. This flows combines all the previously treated design tasks (IC, interposer, module PCB, and systems PCB) but adds the TSV modeling required for 3D chip stack integration. Since TSVs are proximity sensitive (e.g., coupling effects), the TSV geometries of interest are exported only when a proposed layout and a pin assignment scheme are available. This in turn forces an iterative process from the IC physical design stage (PnR) to analog design stage of the transceivers through interposer/package design and TSV modeling. It is worth noting that this iterative process spans three design domains, namely, package, chip, and interposer TSV.

The TSV modeling (as detailed in Chap. 3) is based on using a 3D electromagnetic (EM) field solver tool to extract the frequency behavior (S/Z-parameters) of the

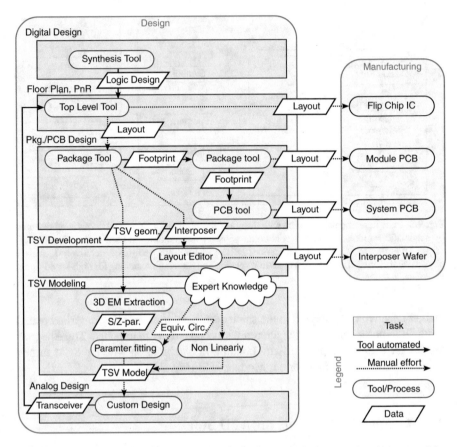

**Fig. 9.6** To implement a 3D stack, different design tasks have to be integrated so that connectivity intent is consistent from the IC level up to the system PCB. Additionally the TSVs for the high-speed chip stack interconnections have to be modeled to account for coupling effects, which would require input from the interposer design level, this creating a large iterative loop (i.e., spanning many hierarchical levels and design partners)

TSV, which can then be used to fit an equivalent circuit to the geometries at hand. For 3D EM extraction, ANSYS HFSS is used, and for parameter fitting Agilent ADS is used to construct a VerilogA circuit model. Nonlinearities such as depletion regions around the TSVs are not included in the EM parameter extraction and have to be modeled by hand using Matlab. By applying the described flow, each of the individual components of the 3D interposer system was successfully manufactured (Fig. 9.7) and assembled in a single die configuration enabling validation of several sub-systems such as PLLs and ADCs. It has also enabled sanity checks in the form of low speed signaling via interposer loopback structures for TSV-transceivers. At the time of print, the 3D stacking is still pending.

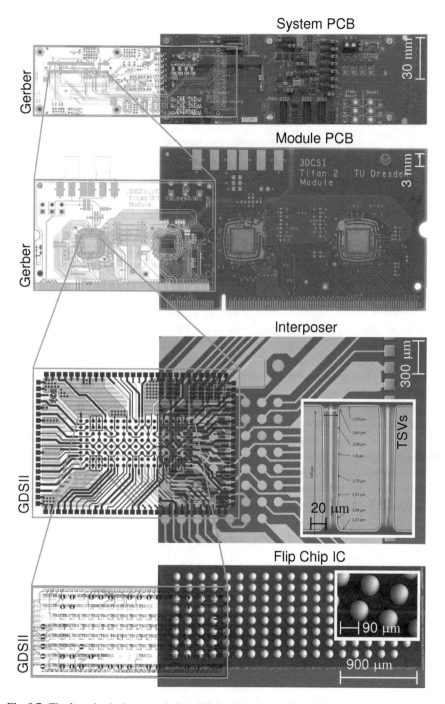

**Fig. 9.7** The four physical parts as designed (*left side*) and manufactured (*right side*) showing how they fit together in the final system, where of course two instances of the flip chip IC are required for a 3D stack assembly. The interposer micrograph shows a sample without TSVs, *inset* from trial production run

## 9.3.1  Analyzing the 3D Flow

The tripartite representation, or Y-chart, of integrated circuit design proposed in 1983 by Gajski and Kuhn [3] serves as the basis of our analysis. The Y-chart was introduced in an era where the digital flow, now standard practice, was still emerging. The three-axis partitioning it proposes for the design process is in fact generic and valid for the 3D interposer design flow [1, 11]. According to Gajski and Kuhn, the three axes of digital IC design are:

- Physical representation, from mask geometries up to full chip layout planning.
- Functional representation, from boolean expressions up to algorithms and systems
- Structural representation, from circuits up to registers and switches for memories and processors.

In Fig. 9.8, the 3D design flow is dissected along these three axes from the device level (TSV, transistor) at the center, up to the system level at the periphery. The concentric circles indicate the various correspondences between design representations, with arrows indicating EDA processes, tools, and data flow (e.g., OA[10] data base). Dotted arrows indicate processes that are not fully automated where manual effort is still considerable. The structural representation would be in the form of HDL (digital), schematics (custom analog IC), and connectivity tables (interposer).

It is important to observe that the TSV placed at the very center of the hierarchy is *physically* located on the interposer which *structurally* is above the individual IC. That the Y-chart is organized in this manner stems from the fact that the TSV is an intrinsic functional part of the capacitive transceiver. In other words, the transceiver is *functionally* on one level but *physically* on two. In fact, one of the conundrums of modern IO design lies with problems like this very one: the crossing of the hierarchy levels. This crossing is even more pronounced with 3D stacks that use analog mixed-signal components. The ambiguity of the structural/physical versus functional placement causes the large iterative loop in the design cycle. This loop is rendered with the arrows crossing the innermost concentric circle. The solution we present relies on the graphics data format GDSII, for transferring information across the interfaces between the IC and the interposer, and from the interposer to TSV modeling.

The crossing from the IC design domain into the packaging and PCB domains while maintaining design consistency can be achieved by propagating tool-specific data such as footprints and symbols. This of course has the drawback locking the design in EDA-system chosen for IC design. The design context of our 3D interposer system with two dies, one interposer, two PCBs (Fig. 9.1), involves three design and development partners and four external contractors (Fig. 9.2) each with their own physical representation (Fig. 9.7) of the design, which means that it is impossible to lock the system design in one EDA system. It also means that there in no full

---

[10] Open access.

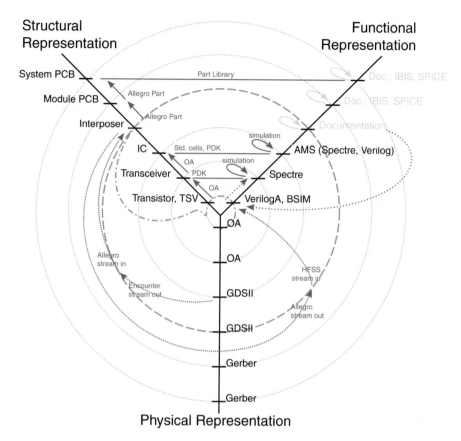

**Structural Representation**

**Functional Representation**

System PCB — Part Library — Doc., IBIS, SPICE

Module PCB — Allegro Part — Doc., IBIS, SPICE

Interposer — Allegro Part — Documentation

IC — Std. cells, PDK — simulation — AMS (Spectre, Verilog)

OA — simulation

Transceiver — PDK — Spectre

OA

Transistor, TSV — VerilogA, BSIM

OA

OA

Allegro stream in

Encounter stream out — GDSII — Allegro stream out

HFSS stream in

GDSII

Gerber

Gerber

**Physical Representation**

**Fig. 9.8** A Y-chart displaying the development cycle of a 3D chip stack: the *dashed circle* at the interposer level and the *dash-dotted curve* illustrate the conundrum of chip stack interconnect design. The design loop is closed by using GDSII stream-out to stream-in, capturing the connectivity as it is manufactured. The interposer design hierarchy boundary can thus be confidently crossed

logical representation of the system design. What is missing is a tool that can cross the boundaries between the different domains of the physical representation, giving a complete system view in a *distributed, heterogeneous* environment [8]. For obvious reasons, the problem is hard to tackle. Indeed, who could predict all possible custom IC design flows, data management methods, and computer systems? But the system must be designed, and it is left up to design methodology team to tackle this daunting task with ad hoc approaches. In the work presented here, it is shown that a "correct by design" method is possible based on the use of standard manufacturing data formats for design representation and for the transfer of data across physical domains. The proposed flow is safe and flexible in terms of usability in a distributed context but is cumbersome due to the large iterative loop as illustrated in Fig. 9.8

## 9.3.2  Merging Design Tasks Through Migration

The unidirectional transfer of data from the IC domain to the packaging domain is a serious impediment for productivity and a severe limitation on its usefulness, especially for a 3D system like ours. In the case of the capacitive transceivers, this problem was not of great concern since the small form factor (small pitch) and the functionality (symmetrically placed transceivers) have basically restricted the design to a tightly defined pinout and TSV configuration. However, in a chip stack with finer pitch TSVs and arbitrary positioning, a reduced design/modeling cycle for the TSVs would be beneficial or even mandatory. Putting packaging in charge of the physical implementation of the top metal and pin locations shortens the iteration loop described in Fig. 9.6. It enables crossing directly from packaging to modeling and then back to the custom analog transceiver before a complete top level IC implementation is ready, thus speeding up this development loop significantly. By creating such an interface and assigning the top level power distribution to the packaging engineer, the top level of the IC is effectively no longer solely owned by the place-and-rout stage of the design. It actually splits up and becomes a joint effort of the packaging engineer and the IC design engineer who are forced to cooperate closely, as described in Fig. 9.9. The benefits of this split become apparent, especially when the system has strict requirements on power/thermal management or, as in our transceiver case, on achievable data rates.

Since it is impossible to enforce the same IT infrastructure (hardware, OS, tools, licenses, etc.) across various design partners, design methodology teams have to propose solid solutions that can overcome incompatibilities and ensure consistency

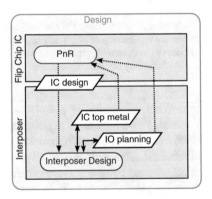
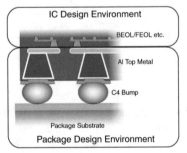

**Fig. 9.9** *Left*: The described interface can in general be used to bridge the gap between IC and package design in *both* directions in a distributed manner (i.e., without a common database and software license setup). *Right*: Migrating the responsibility for IC top metal and bump location to the packaging engineer

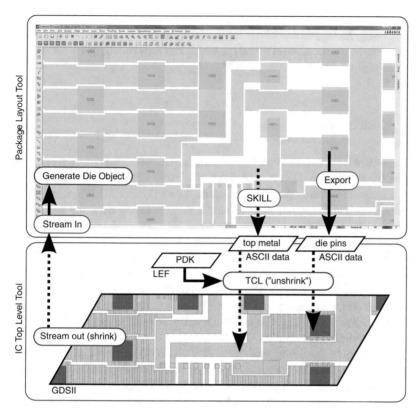

**Fig. 9.10** By using the API of the package design environment, an interface can be created that produces ASCII data for the top metal geometries and die pins for export, which can be interpreted and scaled with TCL scripts in the IC design environment

and interoperability. EDA software APIs[11] can be, and are often, utilized toward such goal. APIs effectively stitch tools of different vendors on different systems together through custom interfaces or design data export capabilities. In Fig. 9.10, a flow for migrating the chip IO planning top level metal to a packaging environment is proposed. For the analyzed design flow, this can been implemented using the API of Cadence® Allegro® (SKILL) to access the design database and create ASCII files of the top level metal routing for the IOs and the power mesh. This can then be easily read through the Encounter® scripting interface using TCL.[12] Similarly, an already available option to create an ACSII text data with pin information (locations, name, pad stack, number, etc.) can be used in combination with an import script. The developed Allegro® SKILL and TCL scripts in Encounter® let

---

[11] Application programming interfaces.

[12] Tool command language.

the package/substrate/board designer not only drive signal assignment to bumps but also drive top metal pad and bump locations. Through this, the packaging engineer would own the top metal level of the IC, which enforces collaboration for successful chip-package co-design.

## 9.4 Conclusion

EDA flow implementations have been likened to cars where most design tasks would benefit from having a Honda Accord, while many companies and methodology groups insist on building their own Formula-1 cars [7]. For 3D ICs, there is a need for such Formula-1 car, but there is also a more pressing need for an EDA flow that is as easy to drive as the Accord. What 3D chip stacks with TSVs bring to the table is distributing the task of producing a part, which historically has been relatively centralized, so that packaging has functional implications at the IC level. In a sense, this is the case for all IO design, which is not new in terms of SI, PI, and thermal management. What is new is the use of Si, PI, and thermal specification at the sign-off stages of the design. What this means is that system integration would be migrated down from the PCB level, and at the same time up from SoC[13] level to the package level.

The analysis of the design flow for a TSV-based 3D stack reveals that there is a profound impact on tool compatibility. Such stacks are expected to be developed and implemented in a highly distributed context, and design environments need to accommodate such context. Mandatory in a chip stack are data channels and interfaces between different ICs. The TSV is at the center of functionality for such interfaces, and its modeling and parasitic extraction cuts across design domains, causing large iterative design loops that must be unraveled to prevent obstructively long turn-around times. In the presented case study, the design flow was held together using the decades old workhorse of GDSII. For design flows to become production-ready, vertical tool integration along all three axes is required: from process engineers at the Fab up to system designer; creating structural, functional, and physical consistency. The distribution of design partners drives an increased demand on interface flexibility, which will further aggravate integration problems between different parts of EDA. With EDA being a highly competitive and fast moving technology market with different suppliers having different points of strength, early adopters of 3D integration will be expected to use different state-of-the-art design methods and point tools from different vendors to an achieve design success. The take-away messages that conclude this chapter are:

- Be prepared to script your own design interfaces as needed to create consistent and reliable domain crossing.

---

[13] System-on-Chip.

- Make sure that no individual tool creates a lock-in to any specific supplier, in other words demand detailed information on APIs.
- Use physical *manufacturing* data formats for flexibility. Such a complex task as designing real application 3D stack-based systems will be a teamwork spanning technologies, domains, companies, and most probably continents.

**Acknowledgements** The work and analysis presented in this chapter would not have been possible without the combined efforts of the mentioned research departments of the Technische Universität Dresden, especially Michael Haas (IfN-HF), Sebastian Killge (IHM), Johannes Görner (HPSN), Dennis Walter (HPSN), and Sebastian Höppner (HPSN) have contributed greatly (as described in other chapters). The author would also like to acknowledge Stefan Scholze for detailed discussions on IC design flow; Stefan Schiefer and Stephan Hartmann for PCB development support; all of the HPSN group. Furthermore gratitude has to be expressed to RacyICs GmbH, channel partner of GLOBALFOUNDRIES; Frank Dresig of GLOBALFOUNDRIES; Werner Schneider of Microelectronics Packaging Dresden GmbH; and last but not least to Professor Rene Schüffny (HPSN) for continued support.

# References

1. G. Dost, G. Herrman, Entwurf und Technologie von Mikroprozessoren, in *Taschenbuch Mikroprozessortechnik* (Fachbuchverlag Leipzig, München/Wien, 1999), pp. 357–412
2. C. Erdmann, D. Lowney, A. Lynam, A. Keady, J. McGrath, E. Cullen, D. Breathnach, D. Keane, P. Lynch, M. De La Torre, R. De La Torre, P. Lim, A. Collins, B. Farley, L. Madden, A heterogeneous 3D-IC consisting of two 28nm FPGA die and 32 reconfigurable high-performance data converters, in *IEEE International Solid-State Circuits Conference Digest of Technical Papers (ISSCC)* (2014), pp. 120–121
3. D. Gajski, R. Kuhn, Guest editors' introduction: new VLSI tools. Computer **16**(12), 11–14 (1983)
4. Globalfoundries, GLOBALFOUNDRIES demonstrates 3D TSV capabilities on 20nm technology (2013), http://www.globalfoundries.com/newsroom/press-releases/2013/12/28/globalfoundries-demonstrates-3d-tsv-capabilities-on-20nm-technology. Accessed 10 Dec 2014
5. Globalfoundries, GlobalShuttle MPW program schedule (2014), http://www.globalfoundries.com/services/globalshuttle. Accessed 10 Dec 2014
6. L. Madden, E. Wu, N. Kim, B. Banijamali, K. Abugharbieh, S. Ramalingam, X. Wu, Advancing high performance heterogeneous integration through die stacking, in *Proceedings of the European Solid-State Device Research Conference (ESSDERC)* (2012), pp. 18–24
7. P. McLellan, *EDA Graffiti* (Paul McLellan, 2010) https://www.createspace.com/en/community/message/109020
8. P. McLellan, SemiWiki - GSA 3DIC (2014), https://www.semiwiki.com/forum/content/3346-gsa-3dic.html. Accessed 16 Dec 2014
9. P. McLellan, Will 3DIC ever be cheap enough for high volume products (2014), https://www.semiwiki.com/forum/content/4100-will-3dic-ever-cheap-enough-high-volume-products.html. Accessed 16 Dec 2014
10. D. Nenni, P. McLellan, *Fabless: The Transformation of the Semiconductor Industry*. SemiWiki.com LLC (2013)
11. D. Payne, A big boost for equivalency checking (2013), https://www.semiwiki.com/forum/content/2355-big-boost-equivalency-checking.html. Accessed 14 Jan 2015

12. S. Pugsley, J. Jestes, H. Zhang, R. Balasubramonian, V. Srinivasan, A. Buyuktosunoglu, A. Davis, F. Li, NDC: analyzing the impact of 3D-stacked memory+logic devices on MapReduce workloads, in *IEEE International Symposium on Performance Analysis of Systems and Software (ISPASS)* (2014), pp. 190–200
13. S.C. Shalak, H.-P. Kemser, N. Ter-Minassian, Defining a product development methodology with concurrent engineering for small manufacturing companies. J. Eng. Des. **8**(4), 305–328 (1997)
14. A. Sheibanyrad, F. Petrot, A. Jantsch, *3D Integration for NoC-Based SoC Architectures*, 1st edn. (Springer, New York Dordrecht Heidelberg London, 2010)
15. E. Sperling, What are EDA's big three thinking? (2014), http://semiengineering.com/what-are-edas-big-three-thinking/. Accessed 13 Jan 2015
16. TSMC, TSMC future R&D plans (2014), http://www.tsmc.com/english/dedicatedFoundry/technology/future_rd.htm. Accessed 10 Dec 2014

# Chapter 10
# Integrating 3D Floorplanning and Optimization of Thermal Through-Silicon Vias

Puskar Budhathoki, Johann Knechtel, Andreas Henschel, and Ibrahim (Abe) M. Elfadel

## 10.1 Introduction

Continuous scaling of CMOS technology into deep sub-micron ranges causes larger interconnect delay, higher device density, and higher power density. The complexity of physical design also increases with the increase of device density and circuit complexity. The implementation of 3D circuits based on 3D integration technology [1] is a promising More-than-Moore technology to boost performance [6] and reduce wirelength [13], both while maintaining compact packing densities.

In 3D ICs, multiple dies containing active devices are stacked on top of each other using vertical interconnections to form one single integrated circuit. A distinct advantage of 3D IC technology is that it helps assemble disparate technologies deployed in different dies, thus supporting heterogeneous systems-on-chip integration. Figure 10.1 illustrates a 3D IC with three dies stacked and integrated using vertical Through-Silicon Vias (TSVs). In the finished package, the top substrate layer is typically covered with a heat spreader and a heat sink that are not shown. Note that some TSVs are aligned for bottom-to-top connectivity in the stack.

P. Budhathoki • A. Henschel
Department of Electrical Engineering and Computer Science, Institute Center for Smart Infrastructure (iSmart), Masdar Institute of Science and Technology, Abu Dhabi, United Arab Emirates
e-mail: pbudhathoki@masdar.ac.ae; ahenschel@masdar.ac.ae

J. Knechtel • I.M. Elfadel (✉)
Department of Electrical Engineering and Computer Science, Institute Center for Microsystems (iMicro), Masdar Institute of Science and Technology, Abu Dhabi, United Arab Emirates
e-mail: jknechtel@masdar.ac.ae; ielfadel@masdar.ac.ae

© Springer International Publishing Switzerland 2016
I.M. Elfadel, G. Fettweis (eds.), *3D Stacked Chips*,
DOI 10.1007/978-3-319-20481-9_10

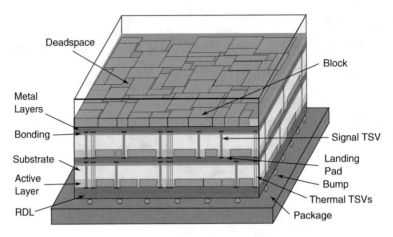

**Fig. 10.1** 3D IC with three dies, stacked using Face-to-Back bonding technology

3D IC technology has been successfully applied for memory integration, e.g., Hybrid Memory Cube (HMC)[12], and for memory-logic integration, e.g., 3D-MAPS [17]. Both applications demonstrate increased performance, higher bandwidth, and lower power consumption as compared with existing 2D technology. However, the compact packing density, the reduced thermal spreading due to thinned dies and the inevitable use of (relatively poor heat conductive) adhesives for vertical wafer bonding are the main contributors to a significant increase of temperature in 3D ICs. A case in point is depicted in Fig. 10.2a, where under certain assumptions/circumstances, the maximum temperature of the critical die increases rapidly as the number of stacked layers increases while the temperature of the die close to the heat sink remains almost constant. This is shown in Fig. 10.2b.

High temperatures compromise circuit reliability, increase leakage power, and decrease circuit performance. The widespread acceptance of 3D IC integration requires the tackling of this thermal challenge head-on. Low-cost, effective thermal management guidelines are needed to maintain the operating temperature within acceptable levels in the 3D stack.

### 10.1.1 Challenges for 3D Integration

Aside from the thermal challenge, other key barriers hampering the widespread acceptance of 3D ICs include: complex design and manufacturing processes for TSV formation, wafer thinning, and wafer bonding; 3D IC testing; business model for 3D IC product development; and distributed EDA environments. For the latter challenge the reader is referred to Chap. 9 of this book where two-sided silicon interposer test vehicle is presented along with the EDA methodology flow needed to verify the system design. For the specific case of multi-layer chap stacking, we

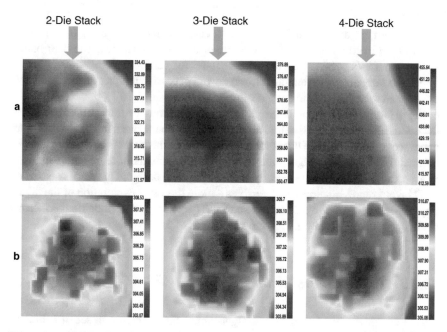

**Fig. 10.2** (**a**) Thermal profile of the critical (*bottom*) die without thermal-TSV placement. (**b**) Thermal profile of top die (adjacent to the heatsink) without thermal-TSV placement for the n200 benchmark

also have the additional CAD challenge of deploying efficient 3D data structures that can support the complexity of 3D systems, limit redundancy, and cover the full search for best physical design solutions. In particular, such 3D data structures should support the vertical dependencies between functional modules.

The large number of devices per unit area in 3D ICs will notably increase the temperature of the dies that are not adjacent to the heat sink which, in turn, degrades the performance of the whole system. Additionally, the integration of heterogeneous components with different stacking technologies (Wafer-to-Wafer, Die-to-Wafer, and Die-to-Die) introduce overhead in testing methodologies. Stacking options (Face-to-Face, Face-to-Back, Back-to-Back) can impact physical parameters such as routing regions, wire length distributions, packing density, TSV geometries, and chip footprint.

## 10.1.2  Motivation and Goal

In recent years, considerable efforts in research have been undertaken to address thermal issues at the early stage of 3D physical design. Goplen et al. [8] indicate that thermal TSVs constitute an important path for vertical heat flow. Thermal TSVs and regular inter-die connections (i.e., signal TSVs) can be leveraged to increase the vertical heat transfer and reduce die temperatures [4, 5, 15]. Lee et al. [16] study the arrangement of thermal TSVs in multi-chip modules and found that the heat removal is directly proportional to the size of thermal via islands. The distribution of thermal TSVs, which are the major channel for vertical heat flow, has a major impact on temperature reduction. However, and excessive number of thermal TSVs would lead to serious routing congestion in 3D ICs. Additionally, thermal TSVs are costly to fabricate. Thermal-TSV optimization algorithms are thus needed to address the trade-off between location and number of TSVs on the one hand and on the other hand, reduction of on-chip temperatures. The goal of this work is to develop and test thermal-TSV placement algorithms to minimize the maximum-temperature gradient to acceptable levels during the early stage of physical design. Our work is based on a 3D layout representation which enables a more thorough exploration of the 3D IC physical design space, which will result in 3D IC design that mitigate the thermal challenge without compromising the advantages of 3D IC integration.

### 10.1.3  Problem Formulation

The input for the 3D floorplanning algorithm is the following: a set of $n$ blocks, $\{B_1, B_2, B_3, \ldots, B_n\}$, and a maximum number of dies. The connectivity between the blocks is specified by a netlist, where each block $B_i$ is described by a pair $(h_i; w_i)$ in which $h_i$ and $w_i$ are, respectively, the height and the width of the $i$th block. Moreover, the maximum number of dies is represented by $k$. A 3D floorplan is a lower-left-corner coordinate assignment $(x_i, y_i, l_i)$ for each block $B_i$ such that no two blocks overlap and $1 \leq l_i \leq k$.

We formulate the 3D floorplanning task with thermal-driven TSV placement as a weighted multi-objective optimization problem, seeking to minimize the occupied area of the floorplan ($A$), total wirelength ($w$), and number of signal TSVs ($n_{\text{vias}}$):

$$\alpha \cdot A + \beta \cdot w + \gamma \cdot n_{\text{vias}} \tag{10.1}$$

where $\alpha$, $\beta$, and $\gamma$ represent priorities for the three objectives. We assume that the lowermost die $d_0$ is connected to the package board. For each net $m$, wirelength is calculated using half-perimeter wirelength $w(m, d_i)$ on each related die $d_i$. Wirelength costs are defined as

$$C_{\text{WL}} = \sum_m \left( l_{\text{TSV}} \times \text{TSVs}(m) + \sum_{d_i \in m} w(m, d_i) \right) \tag{10.2}$$

where $\text{TSVs}(m)$ denotes the required signal TSV count for net $m$ and $l_{\text{TSV}}$ the "cost" of the TSV. The total number of inserted signal TSVs needs to be minimized due to limited routing resources and fabrication cost. We propose to address the thermal-TSV placement problem with a post-placement optimization stage in order to further improve the vertical heat flow on thermal-driven floorplanning results. The thermal-TSV placement problem can be formulated as minimizing the required number of thermal TSVs, for any given 3D floorplan, to satisfy a maximal-temperature threshold constraint.

As for the setup of thermal modeling, a heat sink is attached to the silicon substrate of the top die. The ambient temperature for the whole chip stack is considered constant at room temperature, $27\,°C$.

## 10.2  Methodology for Integrating 3D Floorplanning and Thermal Management with TSV Placement

We propose a design flow that integrates thermal management with area- and wirelength-driven floorplanning and subsequent TSV placement. Floorplanning is based on an extension of the classical 2D corner block list (CBL) representation which encodes 3D ICs using an ordered sequence $\{CBL_1, CBL_2, \ldots, CBL_n\}$ of CBL tuples and one global sequence $\{a_1, a_2, \ldots, a_n\}$ to implement interconnect structure. This extension was first presented in [14].

Figure 10.3 illustrates the design flow of our two-stage approach [3]. The first stage has been tailored for optimization of area and wirelength, along with signal TSVs optimization. The second stage employs thermal-constraint assessment. In this second stage, an iterative method is applied in which the thermal conductivity of design regions exhibiting locally maximal temperatures are modified by deploying a vertical path (with additional TSVs) to transfer heat away towards the heat sink. The goal of this stage is to satisfy the thermal constraints using as few additional TSVs as possible. The main contributions of our work are summarized in the gray boxes highlighted in Fig. 10.3. In the following subsection, we describe the individual modules employed by our 3D floorplanning flow and motivate their choice over alternatives.

### 10.2.1  3D Floorplanning

As mentioned above, we have selected a 3D extension of the corner block list (CBL) layout representation [10, 14, 18] as the 3D data structure used in our 3D floorplanning. We deemed CBL suitable for our purposes for several reasons. For one thing, the runtime complexity ($O(n)$) is favorable in comparison with other 3D layout representations as shown by [7]. For another, the solution space ($O(n!3^{n-1}2^{4n-4})$) is comparably small, and the third reason CBL can represent both slicing and non-slicing (such as "wheels") floorplans.

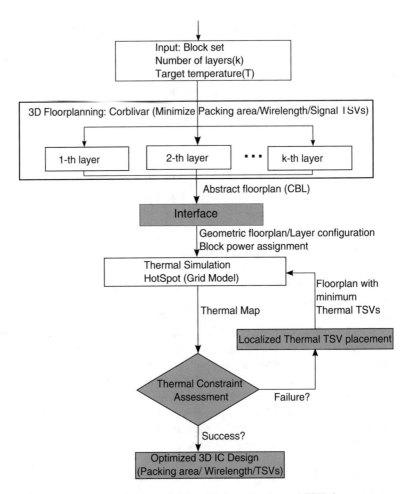

**Fig. 10.3** Methodology flow for thermal-driven 3D floorplanning and TSV placement

As floorplanning module, we employ Corblivar [14], a very recent CBL-based floorplanning tool. This extension was first presented in [14], and is publicly available at http://www.ifte.de/english/research/3d-design/. Corblivar is tailored for multi-objective floorplanning with emphasis on planning massively parallel interconnect structures in 3D ICs. Massively parallel interconnect (e.g., TSV stacks in case of TSV-based 3D integration) are typically used to limit power-supply noise and to improve the vertical heat flow towards the heat sink [2, 8]. In addition, area, wirelength, and thermal management objectives can be included in Corblivar's optimization formulation. Furthermore, Corblivar is able to restrict the number of dies as well as their outlines, which is often required for manufacturing purposes.

## 10.2.2   Floorplanning/Thermal Management Interface

Given a 3D floorplan, this interface module provides an output format readable by the thermal-simulation program. This module also generates layer configurations containing geometric and material characteristics properties of each stacked layer and die along with the floorplan and power trace files needed for thermal simulation.

## 10.2.3   Thermal Simulation

HotSpot is a thermal-simulation tool which provides temperature estimation of a functional module and the whole chip by employing the principle of thermal–electrical duality [11]. HotSpot has been successfully used for many floorplanning tasks, predominantly in the realm of 2D thermal management. Temperatures at the center of functional modules are calculated by solving the RC network of thermal capacitances and resistances. Recently, various extensions have been proposed to enable multi-layer simulations, assuming that the floorplan for each die or layer is given and intra- and inter-die material properties are provided. The HotSpot extension presented in [19] allows modeling of TSVs with related thermal resistivity and specific heat-capacity values in specific delimited regions. The grid model of HotSpot provides internal grid temperatures without aggregating them into per-block temperatures, thus ensuring accurate estimation. However, accurate thermal simulations for larger-scale benchmarks and realistic applications are computation-ally expensive in this grid model. By modifying the grid-model resolution during simulation, we trade off accuracy with speed. It is important to note that this HotSpot extension does not account for a secondary heat path, i.e., conduction through bumps towards the package and PCB. In practice, this results the lowermost die farthest away from the heatsink experiencing the highest temperatures, which makes it thermally critical die. When a secondary heat path is considered, the thermally critical die is more likely to be found in the middle of the 3D IC stack.

## 10.2.4   Thermal-Constraint Assessment and TSV Placement

The layered thermal maps from the HotSpot extension discussed above are parsed and scrutinized for temperature-constraint violations. An iterative approach is applied until the maximum temperature of each grid cell falls below the threshold temperature. The thermal-TSV placement is initialized by setting the thermal conductivity and specific heat capacity of each grid cell to the default values of the materials used, i.e., copper for TSVs and silicon for the chip.

During every cycle of alternating thermal simulation and thermal-TSV placement, the conductivity $k_{new}$ is updated for the grid cell $c_{max}$ exhibiting maximum temperature, according to the formula [23]:

$$k_{new} = k_{old} \frac{T_{curr}}{T_{threshold}} \qquad (10.3)$$

where $k_{old}$ is the previous conductivity of the grid cell, $T_{curr}$ is the current temperature, and $T_{threshold}$ is the threshold temperature for the whole chip. The required TSV density in the grid cell $c_{max}$ is then calculated by the formula [23]:

$$v = \min\left(v_{max}, C \cdot \frac{k_{new} - k_{old}}{k_{via} - k_{old}}\right) \qquad (10.4)$$

where $v_{max}$ is the maximum allowable TSV density in each grid cell, $C$ is a user-specified constant, and $k_{via}$ is the thermal conductivity of the TSV.

The minimum number of thermal TSVs in each grid cell needed to maintain the threshold temperature is derived from the required TSV density in that delimited region, i.e., cell $c_{max}$. Here, the TSV density impacts the TSV area fraction used in total resistance calculation which, in turn, impacts the thermal behavior in the thermal RC-network model. Naturally, the smaller the grid cells, the more accurate the thermal approximation would be. Next, we calculate the compound thermal resistivity, based on the parallel resistance using the material characteristics of silicon and copper thermal TSVs:

$$R_{total} = \frac{1}{(A_{TSV} * \frac{1}{R_{Cu}}) + (A_{\overline{TSV}} * \frac{1}{R_{Layer}})} \qquad (10.5)$$

where $A_{TSV}$ represents the area fraction of the die with TSVs while $A_{\overline{TSV}}$ represents the fraction without TSVs. $R_{Cu}$ and $R_{Layer}$ encode the resistivity of copper and inter-die materials (e.g., bonding layers), respectively. The HotSpot extension assumes that specific heat capacity has a lower impact than thermal resistivity and does not consider its effect for temperature estimation. However, compound specific heat capacity based on mass weighted mean, as proposed by Tonpheng et al. [22], is considered during TSV placement in our model.

Finally, temperature profiles of 3D chips are updated and the process is repeated. The algorithm terminates when the maximum temperature of each grid cell is below the threshold temperature. Figure 10.4 illustrates the algorithm for thermal-TSV placement within the grid cells.

**Fig. 10.4** Localized
thermal-TSV placement
algorithm

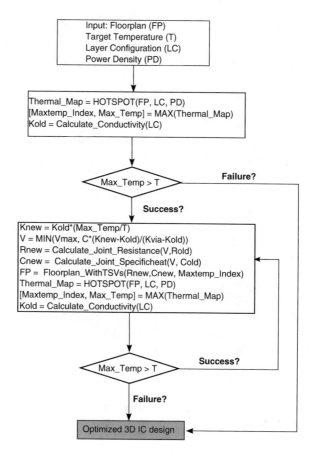

## 10.3   Experimental Setup

The experiments are performed on the Gigascale System Research Center (*GSRC*)
benchmark set [9] with varying number of stacked dies. The design flow is flexible in
that users can specify the number of stacked dies as well as the maximum threshold
temperature of the whole chip. Dies are considered as $100\,\mu$m thick where blocks
reside in the active silicon layer. Die outlines are flexible and range from 5 mm ×
5 mm to 15 mm × 15 mm. The specification of the 3D chip along with material
properties and dimensions [20, 21] are listed in Table 10.1.

The chip temperature is verified by the HotSpot extension presented in [19].
Thermal models differ for varying 3D-integration technologies. We mainly focus
on Face-to-Back stacking, Die-to-Die bonding, and via-in-the-middle TSV manufacturing. The thermal-TSV islands are modeled in the passive silicon substrate
as well as the bonding layers between every two adjacent dies. The dimensions of
thermal TSVs are $5\,\mu$m × $5\,\mu$m and the pitch is $10\,\mu$m. The algorithm for our design
flow as described in Figs. 10.3 and 10.4 is implemented in C++.

**Table 10.1** Dimensions and material properties of the chip model

| Part (material) | Area (mm²) | Thickness (μm) | Thermal conductivity (W/m-K) | Specific heat capacity (J/kg-K) |
|---|---|---|---|---|
| Active Si layer | 5×5–15×15 | 2 | 117.5 | 700 |
| Passive Si layer | 5×5–15×15 | 98 | 117.5 | 700 |
| BEOL | 5×5–15×15 | 12 | 2.25 | 517 |
| Bonding layer | 5×5–15×15 | 20 | 0.2 | 2187 |
| Heat spreader | 30×30 | 1000 | 400 | 397 |
| Heat sink | 60×60 | 6900 | 400 | 397 |
| TSV (Cu) | Dimensions: 5 μm × 5 μm, | | Pitch: 10 μm | |

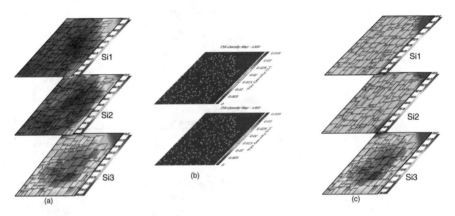

**Fig. 10.5** Results for a three-die stack of GSRC benchmark $n300$. (**a**) Thermal profiles of dies without thermal TSVs. (**b**) Required TSV densities to maintain threshold temperature. (**c**) Thermal profiles of dies after placement of thermal TSVs

## 10.4  Experimental Results and Discussions

Figure 10.5a shows the thermal profiles for the $n300$ GSRC benchmark when blocks are arranged into three stacked dies. The farther away the die is from the heat sink (which is stacked on top of the chip), the larger its temperature increase. The maximum temperature for the critical die (i.e., the die farthest away from the heat sink) is approximately 389 K while it is 318 K for the die nearest to the heat sink. Note that the target threshold temperature of the whole chip is set to 360 K. The grid-model resolution is set to 64 × 64. Thereafter, thermal-TSV placement is conducted by enforcing the required TSV density in the grid cells that have temperatures greater than the threshold temperature. This is also shown in Fig. 10.5b. The percentage of thermal-TSV density for each cell is represented by its color: a red square represents a grid cell with maximal density and a blue square represents a grid cell with minimal TSV density. After thermal-TSV placement, the magnitude of the temperature is greatly reduced since thermal TSVs efficiently transfer heat between different dies as shown in Fig. 10.5c. However, Fig. 10.5c

**Table 10.2** Comparisons of different metrics for the GSRC benchmarks

| Metric | Our, 4 dies | | | Wong and Lim [23], 4 dies | | |
|---|---|---|---|---|---|---|
| | n100 | n200 | n300 | n100 | n200 | n300 |
| Chip area ($\mu m^2 \times 10^5$) | 0.50 | 0.50 | 0.80 | 0.83 | 0.74 | 1.36 |
| Wirelength ($\mu m \times 10^5$) | 1.52 | 2.71 | 3.88 | 1.55 | 3.10 | 4.68 |
| Threshold temperature (°C) | 87.7 | 75.8 | 56.4 | 87.7 | 75.8 | 56.4 |
| Average TSV density | 0.006 | 0.010 | 0.028 | 0.007 | 0.029 | 0.027 |

shows slight temperature increase in the die near the heat sink after placement of thermal TSVs. This is due to an increase in heat-sink base temperature when thermal TSVs conduct significant amount of heat from multiple dies towards the heat sink.

Table 10.2 shows a comparison between our approach and the Integrated Thermal Via Floorplanning (IVF) algorithm [23] in terms of wirelength, chip area, and average TSV density. We observe that our approach shows promising results for all three objectives (wirelength, chip area, and thermal-TSV density) and is able to maintain threshold temperatures for different *GSRC* benchmarks. Note that in [2], regular thermal-TSV placement is used whereby each die contains regularly but sparsely placed thermal TSVs that are vertically aligned so that they can serve as heat-pipes running through the whole 3D IC. The work proposed in this chapter reduces temperature by an additional 10–15 % for exemplary benchmarks in comparison with the regular thermal-TSV placement proposed in [2].

Table 10.3 provides the overall statistics and rankings for different designs of a three-die stack for the GSRC n300 benchmark. The z-score of a metric $X$ is defined as $z = \frac{X-\mu}{\sigma}$ where $\mu$ and $\sigma$ are the mean and standard deviation of $X$ over all experiments. The same n300 benchmark was considered several times (with fixed-die outlines of 10 mm × 10 mm) for our two-stage approach. The rank without TSVs is determined with $\alpha = 1$ for the wirelength metric in Eq. (10.2). Similarly, the rank with TSVs is determined with equal distribution of cost for wirelength, signal TSVs ($n_{vias}$) and thermal TSVs ($T_{vias}$). Our thermal-driven, two-stage approach is able to find the optimal 3D IC design for all objective functions.

Figure 10.6 shows several thermal effects as a function of thermal-TSV density for different GSRC benchmarks [3]. First, it shows a rapid temperature increase for critical dies as the number of stacked dies is increased. Second, this temperature increase can be reduced by approximately 100 K just with 0.5 % of average thermal-TSV density in hotspots for the considered four-die design. Due to stacking of blocks and compact layout packing, the thermal-TSV demand increases with die count as expected. Our two-stage approach simultaneously decreases wirelength, chip area, maximal temperature of the chip, and number of thermal TSVs required to achieve the pre-determined threshold temperature of the whole 3D IC. We also observe that reduction of maximum temperature does not scale in proportion with the number of thermal TSVs: for more than 1 % average thermal-TSV density, the temperature plateaus at a value dependent on the number of dies and the design. In summary, the observations suggest that thermal-design quality can be maintained

**Table 10.3** Statistics and selection of optimized 3D IC design with and without TSVs for n300 benchmark

| IC design | $\omega$ ($\mu$m $\times 10^5$) | $\omega$ (z-score) | $n_{vias}$ | $n_{vias}$ (z-score) | $T_{vias}$ | $T_{vias}$ (z-score) | Avg. z-score | Rank w/o TSVs | Rank with TSVs |
|---|---|---|---|---|---|---|---|---|---|
| 1 | 3.72 | 0.62 | 2040 | −0.493 | 11,515 | −0.1017 | 0.0084 | 4 | 4 |
| 2 | 3.58 | −1.15 | 2051 | 0.8089 | 11,251 | −0.7124 | −0.352 | 1 | 2 |
| 3 | 3.64 | −0.426 | 2051 | 0.8089 | 11,675 | 0.2683 | 0.217 | 3 | 5 |
| 4 | 3.75 | 0.915 | 2035 | −1.085 | 11,220 | −0.7841 | −0.318 | 5 | 3 |
| 5 | 3.76 | 1.09 | 2053 | 1.0456 | 12,368 | 1.8713 | 1.33 | 6 | 6 |
| 6 | 3.59 | −1.04 | 2035 | −1.085 | 11,325 | −0.5412 | −0.889 | 2 | 1 |
| Average | 3.67 | | 2044.2 | | 11,559 | | | | |
| Std. dev. | 0.77 | | 8.4479 | | 432.299 | | | | |

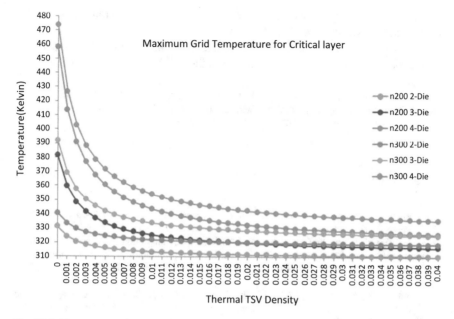

**Fig. 10.6** Temperature drop vs. thermal-TSV density for the critical dies of several *GSRC* benchmarks

by limiting the die count. Our proposed *localized TSV placement* approach does achieve the objective of minimizing the number of thermal TSVs and determining their exact locations. According to our experiments, both number and location have strong impact on reducing the hotspot temperature.

## 10.5   Conclusions

We have developed a flow for thermal-aware 3D IC physical design based on efficient 3D representations of 3D chip floorplans. The framework was applied to the multi-objective 3D floorplanning problem. Thermal-TSV insertion, with full thermal simulation, was conducted in an iterative loop with the goal of fining the minimal thermal-TSV density needed to reduce hotspot temperatures.

Our work addresses the challenge of integrating 3D floorplanning with thermal management. Therefore, multi-objective floorplanning is followed by post-placement optimization of thermal TSVs. Not only does such optimization improve the vertical heat flow on thermal-driven floorplanning results but also it helps in mitigating routing congestion by a sparser deployment of thermal TSVs. The results show that our novel approach can optimize interconnect wire-length, chip area, number of TSV, and hotspot temperatures simultaneously. Our proposed two-stage approach, namely, thermal-driven floorplanning followed by *localized TSV placement*, effectively limits on-chip temperature of 3D ICs within a reasonable level. Finally, we showed that the benefits of thermal-TSV insertion strongly increase with the number of dies. A possible extension of our work is to make the number of dies in the 3D stack a parameter in the multi-objective optimization.

**Acknowledgements** The authors would like to thank Mubadala Technology, Abu Dhabi, United Arab Emirates for funding support of the TwinLab collaboration between Masdar Institute, UAE, and TU Dresden, Germany, Ref. 372/002/6754/102d/146/64947.

# References

1. E. Beyne, The rise of the 3rd dimension for system integration, in *2006 International Interconnect Technology Conference* (IEEE, Burlingame, 2006), pp. 1–5
2. P. Budhathoki, J. Knechtel, A. Henschel, I.A.M. Elfadel, Integration of thermal management and floorplanning based on three-dimensional layout representations, in *Proceedings of the International Conference on Electronics, Circuits, and Systems* (IEEE, Abu Dhabi, 2013), pp. 962–965
3. P. Budhathoki, A. Henschel, I.A.M. Elfadel, Thermal-driven 3D floorplanning using localized TSV placement, in *2014 IEEE International Conference on IC Design & Technology (ICICDT)* (IEEE, Austin, 2014), pp. 1–4
4. J. Cong, J. Wei, Y. Zhang, A thermal-driven floorplanning algorithm for 3D ICs, in *IEEE/ACM International Conference on Computer Aided Design, 2004. ICCAD-2004* (2004), pp. 306–313
5. J. Cong, G. Luo, Y. Shi, Thermal-aware cell and through-silicon-via co-placement for 3D ICs, in *2011 48th ACM/EDAC/IEEE Design Automation Conference (DAC)* (2011), pp. 670–675
6. W.R. Davis, J. Wilson, S. Mick, J. Xu, H. Hua, C. Mineo, A.M. Sule, M. Steer, P.D. Franzon, Demystifying 3D ICs: the pros and cons of going vertical. IEEE Des. Test Comput. **22**(6), 498–510 (2005)
7. R. Fischbach, J. Lienig, J. Knechtel, Investigating modern layout representations for improved 3D design automation, in *Proceedings of the 21st Edition of the Great Lakes Symposium on Great Lakes Symposium on VLSI, GLSVLSI '11* (ACM, New York, 2011), pp. 337–342
8. B. Goplen, S. Sapatnekar, Thermal via placement in 3D ICs, in *Proceedings of the 2005 International Symposium on Physical Design, ISPD '05* (ACM, New York, 2005), pp. 167–174
9. GSRC Benchmarks, http://vlsicad.eecs.umich.deu/BK/GSRCbench/ (2000)
10. X. Hong, G. Huang, Y. Cai, J. Gu, S. Dong, C.-K. Cheng, J. Gu, Corner block list: an effective and efficient topological representation of non-slicing floorplan, in *IEEE/ACM International Conference on Computer Aided Design, 2000. ICCAD-2000* (IEEE, San Jose, 2000), pp. 8–12
11. W. Huang, S. Ghosh, S. Velusamy, K. Sankaranarayanan, K. Skadron, M.R. Stan, HotSpot: a compact thermal modeling methodology for early-stage VLSI design. IEEE Trans. Very Large Scale Integr. Syst. **14**(5), 501–513 (2006)
12. J. Jeddeloh, B. Keeth, Hybrid memory cube new dram architecture increases density and performance, in *Symposium on VLSI Technology (VLSIT), 2012* (IEEE, Honolulu, 2012), pp. 87–88
13. J.W. Joyner, R. Venkatesan, P. Zarkesh-Ha, J.A. Davis, J.D. Meindl, Impact of three-dimensional architectures on interconnects in gigascale integration. IEEE Trans. Very Large Scale Integr. Syst. **9**(6), 922–928 (2001)
14. J. Knechtel, E.F. Young, J. Lienig, Structural planning of 3D-IC interconnects by block alignment, in *ASP-DAC* (2014), pp. 53–60
15. J. Knechtel, E.F. Young, J. Lienig, Planning massive interconnects in 3D chips. IEEE Trans. Comput. Aided Des. Integr. Circuits Syst. **34**, 1808–1821 (2015)

16. S. Lee, T.F. Lemczyk, M. Yovanovich, Analysis of thermal vias in high density interconnect technology, in *Eighth Annual IEEE Semiconductor Thermal Measurement and Management Symposium, 1992. SEMI-THERM VIII* (1992), pp. 55–61
17. S.K. Lim, 3D-MAPS: 3D massively parallel processor with stacked memory, in *Design for High Performance, Low Power, and Reliable 3D Integrated Circuits* (Springer, New York, 2013), pp. 537–560
18. Y. Ma, X. Hong, S. Dong, C. Cheng, 3D CBL: an efficient algorithm for general 3D packing problems, in *48th Midwest Symposium on Circuits and Systems, 2005* (IEEE, Covington, 2005), pp. 1079–1082
19. J. Meng, K. Kawakami, A.K. Coskun, Optimizing energy efficiency of 3-D multicore systems with stacked DRAM under power and thermal constraints, in *Proceedings of the 49th Annual Design Automation Conference, DAC '12* (ACM, New York, 2012), pp. 648–655
20. J.-H. Park, A. Shakouri, S.-M. Kang, Fast thermal analysis of vertically integrated circuits (3D ICs) using power blurring method, in *InterPACK*, vol. 9 (2009), pp. 19–23
21. A. Sridhar, A. Vincenzi, M. Ruggiero, T. Brunschwiler, D. Atienza, 3D-ICE: fast compact transient thermal modeling for 3D ICs with inter-tier liquid cooling, in *Proceedings of the International Conference on Computer-Aided Design, ICCAD '10* (IEEE, Piscataway, 2010), pp. 463–470
22. B. Tonpheng, J. Yu, O. Andersson, Thermal conductivity, heat capacity, and cross-linking of polyisoprene/single-wall carbon nanotube composites under high pressure. Macromolecules **42**(23), 9295–9301 (2009)
23. E. Wong, S.K. Lim, 3D floorplanning with thermal vias, in *Proceedings of Design, Automation and Test in Europe, 2006. DATE'06* (IEEE, Munich, 2006), pp. 1–6

# Part II
# Photonic and Opto-Electronic 3D Integration

*The time is short, and much remains to be done before you are fit to proclaim the Gospel of Three Dimensions to your blind benighted countrymen in Flatland.*

*Edwin A. Abott, Flatland*

# Chapter 11
# Introduction to Optical Inter- and Intraconnects

Niels Neumann, Ronny Henker, and Marcus S. Dahlem

## 11.1 Historical Overview

Starting with the invention of the semiconductor laser and the development of low-loss optical fibers in the 1960s, there has developed an unbroken trend to replace electrical communication systems with optical equivalents. Most information processing is (still) performed in the electronic domain, and despite the oft-mentioned drawback of conversion between the optical and electrical domains, optical communication systems still present significant advantages over the electrical counterparts. Optical waveguides offer tremendous bandwidths (several terahertz) and low-loss propagation, allowing for long transmission distances without the need for signal regeneration. Additionally, optical systems are not susceptible to electromagnetic crosstalk or interference, and enable waveguide crossings which ease the bus routing.

Optical systems have been used in two major areas: long-haul data transmission and data center connectivity. Long-haul systems work in the second and third optical windows, around 1300 nm and 1550 nm, respectively. To optimize performance,

N. Neumann (✉)
Technische Universität Dresden, Chair for RF Engineering, 01062 Dresden, Germany
e-mail: niels.neumann@tu-dresden.de

R. Henker
Technische Universität Dresden, Chair for Circuit Design and Network Theory, 01062 Dresden, Germany
e-mail: ronny.henker@tu-dresden.de

M.S. Dahlem
Department of Electrical Engineering and Computer Science, Institute Center for Microsystems (iMicro), Masdar Institute of Science and Technology, Abu Dhabi, United Arab Emirates
e-mail: mdahlem@masdar.ac.ae

© Springer International Publishing Switzerland 2016
I.M. Elfadel, G. Fettweis (eds.), *3D Stacked Chips*,
DOI 10.1007/978-3-319-20481-9_11

single-mode silica fibers are used, avoiding distortions induced by modal dispersion. However, chromatic dispersion, polarization mode dispersion, absorption, and nonlinearities can still cause impairments on the transmitted signals. The second optical window has been chosen to take advantage of the chromatic dispersion minimum of silica fiber, while the third optical window has minimum attenuation. The single-mode operation also allows for the use of devices based on mode coupling, such as gratings which can be employed to build sharp filters for Dense Wavelength-Division Multiplexing (DWDM).

However, in data centers, transmission distances usually do not exceed 1–10 km. Therefore, requirements on signal distortions are more relaxed, but connectivity is extremely important. Due to larger fiber diameters, multi-mode fiber systems are much simpler to implement, since the mechanical precision needed is much lower than for single-mode fibers. This type of systems usually works in the first optical window, around 850 nm, where silicon-based photodiodes and economical lasers (e.g., vertical-cavity surface-emitting lasers—VCSELs) can be built. These systems are optimized for cost, and usually have a lower accumulated data rate per fiber due to lower line rates, less spectrally efficient modulation formats and far less WDM channels, all of which contribute to lowering costs.

More recently, optical systems have evolved to cover smaller and smaller transmission distances. Board-to-board and on-board systems were driven by the data center paradigm, i.e., multi-mode waveguides and comparably low spectral efficiency schemes. However, when further reducing the connectivity distances to enable on-chip and chip-to-chip communication (e.g., on a 3D integrated chip stack), additional constraints must be taken into account. The omnipresent material is silicon, which can be used to build low-loss waveguides for wavelengths above 1.1 μm. Furthermore, the on-chip feature sizes naturally match the compact dimensions of silicon single-mode waveguides. This path, where silicon is used as the main optical material, normally for wavelengths around 1550 nm, leads to the silicon photonics approach. In this scenario, however, VCSELs are not suitable to be used as sources anymore. Rather hybrid laser sources built with III–V materials are typically used.

There is yet another scheme where VCSELs are used as laser sources. Given that they are a mature technology and widely available, VCSELs can be integrated in a 3D chip stack. Their use implies an operating wavelength around 850 nm and multi-mode operation. The central element of a chip stack is the interposer, where light-carrying Through-Silicon Vias (TSVs) can be implemented.

In the following chapters, the concepts and devices needed for chip-to-chip and on-chip optical communications are discussed. Chapter 12 explores various types of optical TSVs. The underlying manufacturing processes are introduced and compared to the equivalent ones for electrical TSVs. The transmission channel formed by optical TSVs is modeled based on geometrical optics. Finally, a demonstration system consisting of a VCSEL and a photodiode connected through an optical TSV is experimentally investigated. Chapter 13 covers some of the optical components needed in a typical optical interconnect system. Lasers act as light sources and can be used for electro-optical conversion by applying direct modulation. Alternatively, external modulators can be used. Inversely, photodetectors convert optical signals back to the electrical domain. Characterization and measurement procedures (both optical and electrical) of all these integrated optical devices are also reviewed. Chapter 14 focuses on the design and analysis of an optomechanical tuning mechanism for a silicon microring resonator, using a silicon cantilever. This low-power tuning mechanism can be used for implementing on-chip WDM schemes. Both mechanical and optical numerical computations are presented. Chapter 15 tackles the important issue of the influence of thermal loads in on-chip optical communication links. Different approaches that minimize the thermal dependence of the silicon structures are discussed, and an athermal Mach–Zehnder interferometer is experimentally demonstrated. Chapter 16 deals with IC design issues which are essential to enable integrated optical transceivers. In particular, laser drivers have a great potential for optimization: data rates can be increased by supporting wider bandwidths and higher order modulation formats. Both approaches are explored. Vertical inductors pave the way for such advanced laser driving ICs. Accordingly, this class of IC elements is investigated as well. Finally, in Chap. 17, the design of a highly efficient interdigitated back contacted solar cell is explored and benchmarked against other relevant work in the field. A complete review is presented and the main results obtained by the Masdar Institute's group are highlighted. The inclusion of this topic is motivated by the fact that the integration of power sources (in this case, a solar cell) in the 3D stack is possible and would open opportunities for new 3D integrated chips.

In order to underline the interaction between the optical and electrical components, the following two sections discuss system aspects, requirements for the system architecture, different options for components, and the idea of an electronic–photonic integrated circuit.

## 11.2  System Aspects

The optical intraconnect system core (Fig. 11.1) includes an electro-optical conversion element, the transmission medium, and an opto-electrical conversion element.

As can be seen, even though this is a link for optical signal transmission, the major part of the transceiver consists of electrical integrated circuits (ICs).

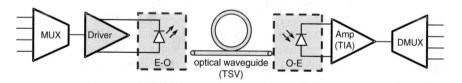

**Fig. 11.1** Optical intraconnect system

At the transmitter part, several parallel electrical channels with small bit rates are multiplexed (MUX) to a high bit rate signal. This signal will then be driven and converted from the electrical into the optical domain by an electro-optical modulation, e.g., via a directly modulated laser-like VCSELs. The modulated optical signal is launched into and transmitted over an optical waveguide. This can be carried out in one or multiple channels using multiple wavelengths (WDM system). One or more photodiodes (PD) detect the optical signal and convert it back to the electrical domain. There, the photocurrent will be transformed into an amplified voltage by a transimpedance (TIA) and limiting amplifier (LA). Finally, a clock-data-recovery (CDR) extracts the signal clock and regenerates the data signal. Afterwards the electrical signal is demultiplexed (DMUX) to the electrical sub data rates again.

In order to accommodate electrical links, the ICs in optical links have to fulfill several requirements, including high bandwidth and low power consumption resulting in high energy efficiency. Furthermore, as higher integration comes into play with 3D packaging, compact design and new integration methodologies as well as circuit structures are required. To achieve more energy-efficient optical intraconnects, several degrees of freedom for the optimization can be applied:

1. The power consumption of the components needs to be reduced.
2. High bandwidths have to be provided for the inherently high optical data rates in the order of tens of Gbit/s. This requires improvement of optical components and ICs.
3. Lower optical losses on the transmission medium and for the coupling are required, which entail improved packaging, assembling, and alignment techniques as well as an improvement of the passive optics. With lower losses, the requirements for the electro-optical devices can be relaxed in terms of optical powers and therefore, and as a result the IC requirements on signal integrity can be relaxed as well.
4. The ICs have to be improved with regard to higher bandwidths, lower power consumption, and lower noise.

Along with these improvements, 3D integration also brings along some inherent advantages for the application of optical intraconnects. Since the transmission distances are very short, transmission losses can be neglected as will be described in Chap. 12. Thus, transmission power can be reduced to a minimum which in turn reduces the requirements for the ICs in terms of providing enough modulation swings and bias for the optical components. On the other hand, super low-power and efficient electro-optical converters are necessary to benefit from this improvement.

Power dissipation and heat production are of high importance in chips and also chip stacks where a large amount of processing power is concentrated in small space. For optical transmission systems, data rate dependent ($xR_{bit}$) and data rate independent ($P_0$) contributions to the power dissipation $P_{diss}$ exist:

$$P_{diss} = P_0 + xR_{bit} .$$

(11.1)

The laser source has a threshold current and voltage that defines a bias point with minimum power $P_{0,las}$ where it can be operated. Obviously, the lower the threshold voltage and current are, the lower the data rate independent power dissipated by the laser

$$P_{0,las} = V_{bias,las}I_{bias,las} \qquad (11.2)$$

Unlike electrical waveguides, the loss of optical waveguides does not increase with frequency. The optical loss contributes only to the static power dissipation. The energy spent on modulation is data rate dependent. For the high-data rates targeted in this class of systems, these contributions are dominant.

Today's optical components (especially modulators and lasers) are optimized for off-chip and off-board communications. Consequently they provide and demand significant amounts of power which have to delivered by the driving circuitry. As a rule of thumb, circuits driving optical components are more power hungry than their the ones driving electronic components. This is the main reason the distance where optical communication is more energy-efficient than electrical connections is in the range of centimeters. It is also the reason why optical on-chip systems have not yet seen wide usage other than in optical board-to-board communication. Therefore, optimizing optical intraconnects for communication over distances up to a few centimeters can be successful only if the performance of both optical and electronic components is improved.

For an optimized IC design, different semiconductor technologies can be used. While standard CMOS provides low-cost volume fabrication and highly energy-efficient circuits, it needs additional methods to achieve high bandwidth. For instance bandwidth peaking techniques using inductors can be used for this purpose. However, planar inductors demand large chip area, which is not suitable for highly integrated 3D packages. BiCMOS offers higher bandwidths but at the cost of higher power consumption. Similarly, III/V technologies provide even higher bandwidth at even higher power consumption. Therefore, the technology choice is always a trade-off between achievable bandwidth, energy efficiency, and chip area. However, it has been shown that BiCMOS ICs can achieve a well-balanced performance to meet all those requirements.

Another important system aspect is the choice of a modulation scheme. A direct modulation of the laser current at its operation point only allows for amplitude modulation formats. Furthermore, direct modulation of VCSELs is currently possible up to bandwidths of up to 20–25 GHz. The use of several techniques in combination with the ICs can enhance this bandwidth as will be explained later in Sects. 16.3 and 16.4. This limits the spectral efficiency but enables a simple system design without modulator and with just a photo diode as receiver.

Complex (amplitude and phase) optical modulation formats require an optical IQ modulator at the transmitter side and either delay line interferometers (for PSK formats) or a coherent receiver. A noteworthy third option is using an electrical subcarrier modulated with any advanced modulation format (e.g., QAM or OFDM) which is amplitude modulated (direct modulation or external modulation) on the

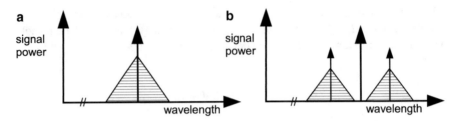

**Fig. 11.2** (**a**) Optical baseband transmission, (**b**) use of electrical subcarrier for optical transmission

optical carrier. Figure 11.2 exhibits example spectra of this setup inspired by Radio-over-Fiber systems. In scenarios with direct modulation (e.g., VCSEL-based systems), this scheme could be interesting in configurations that also have a wireless part to distribute signals to the antenna elements. Consequently, an IF signal of the RF wireless signal could be used. However, the optical amplitude modulation produces two sidebands which lowers the spectral efficiency.

For silicon photonics, off-chip lasers that may be shared among different links and on-chip modulators are preferred. They can be designed as IQ modulators (e.g., in a dual parallel Mach–Zehnder modulator configuration). This paves the way for advanced modulation formats at the cost of higher system complexity. Additional elements introduce loss and consume power which has to be considered in an overall optimization.

The prospect of the superior features of optical communications is a strong driver to enhance the properties of all parts of an on-chip optical communication system: active and passive optical devices including waveguides such as TSVs as well as the co-designed electronic circuits. In the following chapters, several options for tackling these challenges are discussed in detail. Space limitations do not allows us to exhaustively cover *all* possible options.

## 11.3   Electronic–Photonic Integrated Circuits

Most modern communication infrastructures are based on optical networks, which connect nodes in different continents, countries, and cities, all the way to the end user with fiber-to-the-home. Optical fiber cables are also used to interconnect servers, computers, and other diverse electronic equipment (e.g., audio and video). A step further is to bring optical data links into these electronic devices, connecting different internal blocks (e.g., optical data bus for connecting motherboard, hard drives, and graphics cards). Further down the way, those optical connects can be used on the same electronic board to connect different integrated circuits and other elements. Ultimately, optical links can be implemented to interconnect multiple cores or other blocks within the same microprocessor chip. At this scale and integration level, most of the optical blocks are realized in silicon.

The number of cores in a single microprocessor has been increasing over the years, with hundreds or thousands on the same chip in the near future. This predictable growth will make the traditional on-chip interconnects (which are done electrically) unable to handle bandwidth, power, noise, and delay requirements, which are currently major constraints when designing integrated circuits. The performance improvement required by the datacom industry can be enabled by on-chip optical interconnects in a way similar to the one with which low-loss optical fibers revolutionized the telecom industry. In particular, the increasing on-chip bandwidth requirements can leverage the wavelength-division multiplexing (WDM) capability of optical systems.

The field of Silicon Photonics has seen significant growth over the last two decades. In particular, silicon-on-insulator (SOI) platforms are suitable for fabricating high-quality planar waveguide optical devices. With an electronic band gap around $1.1\,eV$ at room temperature, silicon is transparent for wavelengths above $1.1\,\mu m$. Long-haul data transmission systems traditionally operate in the second and third optical windows, where silicon is transparent. In fact, substantial work has been done at the conventional C-band telecom window (1530–1565 nm). The low propagation loss at telecom wavelengths makes silicon a suitable material for designing devices compatible with existing optical fiber networks and systems. Additionally, silicon's high index of refraction, combined with silica (in SOI platforms), results in high-index-contrast structures that can be scaled down to hundreds of nanometers, with micron-sized bending radii. Silicon is also compatible with many of the traditional complementary metal-oxide semiconductor (CMOS) industry fabrication processes, making it both time- and cost-effective to mass produce Photonic Integrated Circuits (PICs).

The possibility of integrating electronics and photonics on the same material platform leads to the concept of Electronic–Photonic Integrated Circuits (EPICs), illustrated in Fig. 11.3. In this platform, several optical building blocks (e.g., waveguides and WDM filters) can be combined with electronic building blocks (e.g., microprocessors and memory elements). Additional material layers can be added to form a hybrid platform, which can add important active optical functions such as emission and amplification. The main optical building blocks include integrated optical sources, efficient fiber-to-chip vertical and horizontal couplers, low-loss waveguides, photonic crystals, optical switches, electro-optic and all-optical modulators, WDM systems (optical add-drop multiplexers), optical delay lines and memory elements, and photodetectors. For a growing number of applications, on-chip optical sensors (e.g., temperature, pressure, optical intensity, refractive index, chemical, and biological) are also desired, either for monitoring and controlling functions during chip operation, or for building compact lab-on-a-chip systems. Many of the components mentioned above (modulators, WDM filters, delay lines, sensors, etc.) can be built in configurations based on optical microcavities, particularly using microring resonators.

**Fig. 11.3** Electronic–Photonic Integrated Circuit concept, where optical and electronic building blocks are combined on the same platform: *SW* optical switch, *MO* optical modulator, *ADC* analog-to-digital converter, CPU—microprocessor

# Chapter 12
# Optical Through-Silicon Vias

Sebastian Killge, Niels Neumann, Dirk Plettemeier, and Johann W. Bartha

## 12.1   Introduction

The traditional approach solutions for electrical connections are copper-based. The major performance indicator for copper-based electrical connection is signal transmission speed. For increasing bandwidth, an optical connection in an application-specific integrated circuit (ASIC), even over short distances becomes more an more interesting. Optical connection lines have the potential to outperform copper-based connections in terms of bandwidth. On the other hand, they are more complex due to electro-optical and opto-electrical conversion needed in the system as mentioned in Chap. 11. Another side benefit is that the transmission distances especially for chip-to-chip or chip-stack-to-chip communication in optical domain are far less critical than in electrical domain. For such optical chip-to-chip or chip-stack-to-chip communication, optical TSVs are mandatory.

Principally, there are two possibilities for optical transmission in a TSV. The first uses an air-filled TSV, i.e., free-space transmission. Hereby, a direct communication is managed through a TSV etched through silicon with no other additional filling. The second is to create an optical waveguide within the TSV. To guide the optical field inside a TSV, its filling has to have a higher refractive index than the adjacent materials. Having a comparably high refractive index, silicon has to be shielded from the TSV filling using a lower refractive index material.

S. Killge (✉) • J.W. Bartha
Technische Universität Dresden, Institute of Semiconductors and Microsystems - IHM, 01062 Dresden, Germany
e-mail: sebastian.killge@tu-dresden.de; Johann.Bartha@tu-dresden.de

N. Neumann • D. Plettemeier
Technische Universität Dresden, Chair for RF Engineering, 01062 Dresden, Germany
e-mail: niels.neumann@tu-dresden.de; Dirk.Plettemeier@tu-dresden.de

© Springer International Publishing Switzerland 2016
I.M. Elfadel, G. Fettweis (eds.), *3D Stacked Chips*,
DOI 10.1007/978-3-319-20481-9_12

The properties of the optical transmission strongly depend on the wavelength of the source and corresponding waveguide material used. The intended scenario is to use a 850 nm source.

## 12.2 Modeling

The transmission behavior inside TSVs can be categorized in two major classes: waveguiding and non-waveguiding. From the manufacturing point of view which is discussed in Sect. 12.3 non-waveguiding TSVs are the most suitable approach. In that case, the optical via is simply a hole in the silicon interposer. The initial beam width $w_0$ of the source (wavelength $\lambda_0 = 850$ nm for VCSEL-based solutions) can be modeled as a Gaussian beam for free-space propagation (refractive index $n = 1$). It widens with transmission distance $z$ by

$$w(z) = w_0 \sqrt{1 + \left(\frac{\lambda_0 z}{n \pi w_0^2}\right)^2}. \tag{12.1}$$

The far-field approximation with the beam angle ($\Phi_0 = 12°$ for typical VCSELs)

$$w(z) = \frac{\lambda_0 z}{n \pi w_0} = \Phi_0 z \tag{12.2}$$

can be used for common TSV geometries. If the beam widens more than the TSV diameter $d_{TSV}$, parts of the beam with incident angle $\varphi_1$ at the interface between substrate and air are lost in the substrate ($n_{sub}$) according to Snellius' Law

$$\varphi_2 = \arccos\left(\frac{n_{air}}{n_{sub}} \cos \varphi_1\right). \tag{12.3}$$

However, due to the high refractive index contrast between air and silicon ($n_{sub} = n_{Si} = 3.66$ at 850 nm), most of the light is reflected back into the TSV as shown in Fig. 12.1. This is not a waveguiding mechanism but it lowers the losses dramatically (especially over short distances in the range of a few hundred microns which are typical for that kind of TSVs). The number of reflections can be determined taking into account the TSV geometry (diameter $d_{TSV}$ and length $l_{TSV}$) and the beam angle $\Phi_0$. For each number of reflections $k \in \mathbf{N}$, the maximum angle

$$\varphi_{kR} = \arctan\left(\frac{\frac{d_{TSV}}{2}}{\frac{l_{TSV}}{2k-1}}\right) \tag{12.4}$$

can be computed. The maximum number of reflections in the given geometry is limited by the beam angle $\varphi_{kR} \leq \Phi_0$. Different reflection factors for the TE and TM polarized parts of the wave

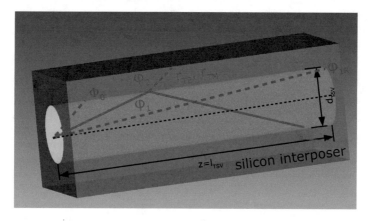

**Fig. 12.1** Freespace TSV model

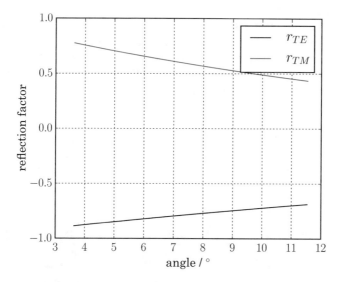

**Fig. 12.2** TE and TM reflection factors for silicon–air interface depending on the incident angle $\varphi_1$

$$r_{\text{TE}} = \frac{n_{\text{air}} \sin \varphi_1 - n_{\text{sub}} \sin \varphi_2}{n_{\text{air}} \sin \varphi_1 + n_{\text{sub}} \sin \varphi_2} \qquad (12.5)$$

$$r_{\text{TM}} = \frac{n_{\text{air}} \sin \varphi_2 - n_{\text{sub}} \sin \varphi_1}{n_{\text{air}} \sin \varphi_2 + n_{\text{sub}} \sin \varphi_1} \qquad (12.6)$$

have to be taken into account. The smaller the incident angle $\varphi_1$ is the higher the reflection factor (see Fig. 12.2) is.

When assuming a homogeneous power distribution (i.e., for multimode signals), integrating over the angular spectrum for all occurring number of reflections $k$ yields the total loss due to transmission into silicon

$$P_{\text{loss,TE|TM}} = \sum_k \int_{\varphi_{kR}}^{\min(\varphi_{(k+1)R}, \Phi_0)} |r_{\text{TE|TM}}|^{2(k-1)}(1 - |r_{\text{TE|TM}}|^2)\mathrm{d}\varphi . \qquad (12.7)$$

TE and TM contribute equally to the total loss of the TSVs due to its cylindrical geometry

$$P_{\text{loss}} = \frac{P_{\text{loss,TE}} + P_{\text{loss,TM}}}{2} . \qquad (12.8)$$

For example, a typical silicon TSV with a length $l_{\text{TSV}} = 370\,\mu\text{m}$ and a diameter $d_{\text{TSV}} = 48\,\mu\text{m}$ has $P_{\text{loss}} = 2.3\,\text{dB}$ loss.

To lower the loss, the TSV can be filled by a transparent material to create a waveguide similar to multimode fibers where only the material loss is of interest (see Fig. 12.3). For SU-8-based rib waveguides, attenuations as low as 0.36 dB/cm could be achieved [1]—a negligible value for TSVs with a length of a few hundred microns that is negligible. However, to achieve guiding behavior, the filling has to have a higher refractive index than its adjacent material. One option would be using a material with a higher refractive index than silicon. Such a kind of material is rarely available, so the silicon has to be shielded from the waveguide material. Silicon dioxide ($n_{\text{cl}} = 1.4525$ at 850 nm) is a good candidate because it can be manufactured as a very conformal layer by a well-known and controllable oxidation

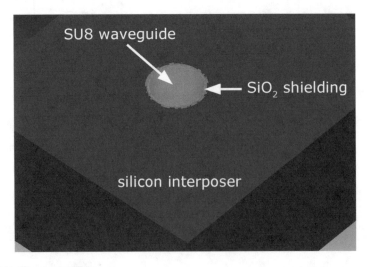

**Fig. 12.3** Waveguide TSV

process. With $SiO_2$ as the adjacent layer, SU-8 can be used as waveguide material ($n_c = 1.56$). Numerical simulations show that the thickness of the shielding layer should be at least half of the wavelength of the signal to ensure that the evanescent field will not penetrate into silicon. The maximum numerical aperture being handled with that refractive index scheme is

$$A_N = \sqrt{n_c^2 - n_{cl}^2} = 0.569 . \tag{12.9}$$

This corresponds to a beam angle of about $35°$ being more than sufficient for VCSELs and other lasers.

In multimode waveguides mode dispersion dominates. Assuming mode equilibrium (the worst case probably not achieved over short distances as in TSVs), the available bandwidth for length $L$ is determined by the difference of the propagation time between the fastest ($t_{g,min} \approx \frac{L}{c_0} n_c$) and slowest ($t_{g,max}$) mode

$$t_{g,max} - t_{g,min} = t_{g,min} \left( 1 - \frac{n_{cl}}{n_c} \right) . \tag{12.10}$$

The available bandwidth depends on the pulse shape. For rectangular pulses (implying a pulse broadening $\sigma = \frac{t_{g,max} - t_{g,min}}{\sqrt{12}}$), a baseband bandwidth

$$f_B = \frac{0.2}{\sigma} = \frac{\sqrt{0.48}}{t_{g,max} - t_{g,min}} \tag{12.11}$$

can be computed. For the waveguide mentioned above, a bandwidth of 1.8 THz for 1 mm transmission length can be achieved. Please note that the actual bandwidth is expected to be at least one order of magnitude higher. The numerical aperture of the source is lower than the one supported by the waveguide and mode equilibrium is unlikely to be achieved after millimeter distances. Thus, much less power is in the slow modes—so there will be less interference.

## 12.3 Fabrication of Optical TSVs

The TSV waveguide concept requires a $SiO_2$ surface as waveguide cladding and additionally a waveguide material with a higher refractive index in the core (e.g., SU-8). To achieve the aforementioned features, several important process steps had to be optimized: Deep Reactive Ion Etching (DRIE), thermal oxidation of $SiO_2$, and filling with waveguide material. Two general strategies have been considered to realize TSVs without the application of special wafer support techniques: (a) Using wafers with standard thickness, e.g., 525 μm for 4 in., or (b) using thinned wafers of a thickness that can be handled without temporarily bonding of a support substrate, e.g., 180–200 μm. In both cases different challenges arise: In (a),

cylindrical holes with high aspect ratios (AR) and smooth inner surfaces have to
be realized, followed by a conformal and time-efficient deposition of a thin $SiO_2$
cladding layer and the waveguide material. In (b), significant problems arise due
to the lower resistance of the thinned wafer to layer stress induced by deposited
films resulting in bowing or even breakage. Mainly, photolithographic, etching, or
deposition processes are affected. Handling of thin wafers by operators and facilities
might also be problematic. However, one of the key features is the manufacturing
of TSVs having the required high aspect ratio (AR > 10). The manufacturing
process (b) for the optical TSVs is similar to copper-based TSVs as discussed in
Chap. 2 regarding silicon via etching and thermal oxidation. As discussed below,
the advantages using thinned wafers prevail.

Figure 12.4 depicts a scheme of the technology concept for waveguide TSV
for optical chip-stack inter-/intraconnects. The etch-stop on the p-doped silicon
wafer backside is realized by depositing a 50 nm aluminum layer via physical
vapor deposition (PVD). Then, front and back sides of the wafers are coated with
photoresist (Fig. 12.4a), followed by an anisotropic deep silicon etch (SOI) of the
40 μm TSV-holes (Fig. 12.4b). The TSVs are produced by the Bosch process, which
is also known as DRIE. The etching is performed via cycling between a deposition
($C_4F_8$) and an etching ($SF_6$) step. The DRIE process has to be optimized to attain
cylindrical contact holes with aspect ratios in the range of 4–10 with etching rates
from 2 μm min up to 4 μm min depending on the actual dimensions. During the
deposition step a sidewall passivation film of $C_4F_8$ is created while in the following

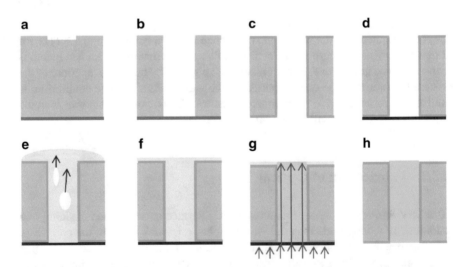

**Fig. 12.4** Technology scheme for waveguide TSV for optical chip-stack inter-/intraconnects.
(**a**) Photolithographic masking; (**b**) etching; (**c**) thermal oxidation; (**d**) temporary backside bonded
PDMS-membrane; (**e**) spin-coating of SU-8 and vacuum TSV filling; (**f**) spin-off: reduce SU-8
residual layer; (**g**) SU-8 backside exposure and development; (**h**) TSV waveguide

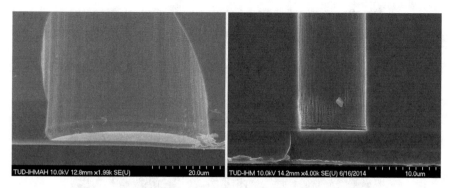

**Fig. 12.5** *Left*: SEM of the bottom of an optical TSV (40 μm diameter, 180 μm length) with backside stop layer after DRIE and on the *right*: SEM of the bottom of an optical TSV (20 μm diameter, 200 μm length) after DRIE, detailing minimal notching on 50 nm Al stop layer

etching step the trench bottom is etched isotropically by $SF_6$. Each step depends on a number of process parameters like gas flows, the power of the inductively coupled plasma, the platen power or on the platen time, etc. The optimized etch profiles with positively tapered angles are generated by carefully balancing between the two steps. The etching process is stopped on the back side of the wafer side by a backside stop layer (Fig. 12.5).

After stripping the photoresist and selective etching of aluminum, fluoropolymer residues remaining on the TSV walls are successfully removed by dry plasma enhanced or wet chemical methods (the latter used 1-methyl-pyrrolidone, NMP, at 55 °C). The developed optimum parameters of the subsequent smoothing process guarantee a TSV sidewall surface roughness below 20 nm. The previously rough inner surface of the vias (<100 nm) is smoothed by a plasma-enhanced step using oxygen ($O_2$), argon (Ar), and nitrogen trifluoride ($NF_3$) (Fig. 12.6). As mentioned before, the process also has to be optimized regarding the bottom of the holes to achieve a notching free shape (Fig. 12.5). Inherent to the Bosch process are undercuts which arise at the top and the bottom of the holes. To remove them and to create tapered profiles, a further etching step is required to widen the holes at the wafer top and bottom. Thus, cylindrical holes with high aspect ratios from 5:1 to 20:1 are realized (Fig. 12.7). Afterwards, a 2 μm $SiO_2$ cladding of the waveguide with an optical refractive index $n = 1,4525$ (wavelength $\lambda_0 = 850$ nm) is produced by thermal oxidation (Fig. 12.4c). Being independent of the aspect ratio, the oxidation step results are independent of the aspect ratio, it results in a conformal layer all over the wafer scale (Fig. 12.8). This furnace process is temperature- and time-dependent and allows the manufacturing of oxide thicknesses up to about 2 μm.

Next, the TSV backside is covered with a temporary bonded polydimethylsiloxane (PDMS)—membrane or a photoresist (Fig. 12.4d). The SU-8 solution acting as waveguide material is deposited on the wafer front side as a thin film via spin-coating. Homogeneous filling of the TSVs is achieved by applying vacuum and

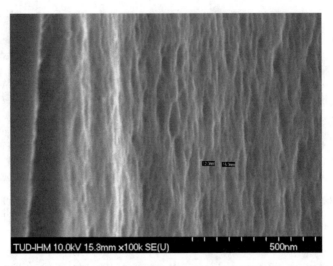

**Fig. 12.6** SEM after plasma-enhanced smoothening, using oxygen ($O_2$), argon (Ar), and nitrogen trifluoride ($NF_3$) (max. 20 nm surface roughness at TSV sidewall, 20 μm diameter, 200 μm length)

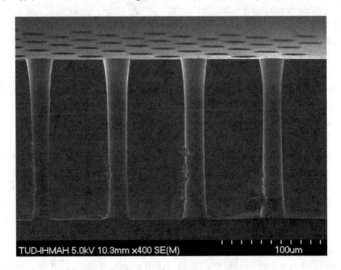

**Fig. 12.7** SEM after silicon via etching onto a stop layer releasable afterwards and additional enlargement of TSV diameter on wafer front- and backside (20 μm diameter, 200 μm TSV length)

tuning temperature (Fig. 12.4e). Reduction of the high solvent concentration during the following development of SU-8 can be achieved by vacuum pre-treatment, drying, additional pre- and soft-bake on different levels of vacuum and temperature. This diminishes gaps, voids and blisters in the material as well. A defect-free filling up to AR 20:1 was demonstrated. With an additional spin-off the top-layer thickness is reduced (Figs. 12.4f and 12.9).

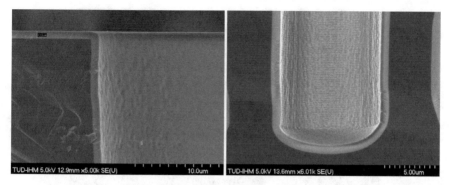

**Fig. 12.8** SEM of a uniform thermal-$SiO_2$ layer with a thickness about 900 nm as cladding of the TSV waveguide; details—*Left*: TSV side wall top (40 μm diameter, 200 μm TSV length); *right*: TSV side wall bottom (10 μm diameter, 200 μm TSV length)

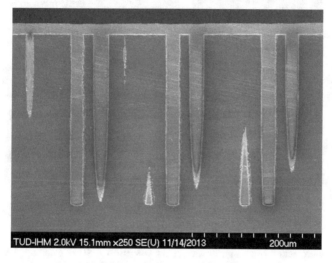

**Fig. 12.9** SEM of a uniform thermal-$SiO_2$ layer with a thickness about 900 nm as cladding of the TSV waveguide; details: TSV side wall bottom (10 μm diameter, 200 μm TSV length)

In the following step, an exposure of the SU-8 layer to UV light from wafer backside is performed in order to achieve a cross-linked SU-8 inside the TSV (Fig. 12.4g). The exposed SU-8 residual resist layer on the top surface is removed in the development step, it remains only in the cross-linked TSV. The SU-8 residues on top of the SU-8 are related to the residue thickness. Finally, an additional resist stripping or PDMS-membrane removal uncovers the wafer backside. This results in a uniform-filled waveguide TSV (Fig. 12.4h; Figs. 12.10 and 12.11) for optical chip-stack inter-/intraconnects.

**Fig. 12.10** SEM of an optical waveguide TSV (filling: SU-8; cladding: 2000 nm SiO$_2$; 40 μm TSV diameter, 380 μm TSV length)

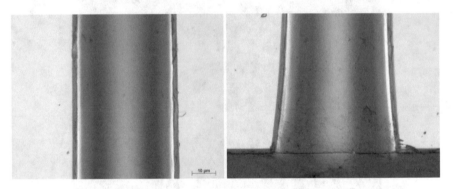

**Fig. 12.11** Microsection of an optical waveguide TSV (filling: SU-8; cladding: 2000 nm SiO$_2$; 40 μm TSV diameter, 380 μm TSV length). *Left*: central part of TSV, *right*: TSV bottom

## 12.4  TSV Characterization

The TSV system measurements have been carried out using VCSELs and photodiodes operating at 850 nm [2]. The bandwidth of the VCSELs is limited to around 20 GHz. Therefore, the VCSEL is expected to limit the data rate—the photodiodes have a slightly higher bandwidth. The VCSELs and photodiodes are fiber-coupled to 50/125 μm multimode fiber (OM3). This fiber has a numerical aperture $A_N =$ 0.2 which yields a bandwidth of around 15 THz for 1 mm transmission distance according to Eqs. (12.9)–(12.11). Hence, no effects due to mode dispersion are

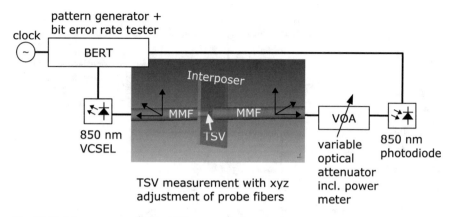

**Fig. 12.12** Measurement setup for TSV system characterization

expected even if the laser sources are connected with short pigtails to the TSV. When placing the fibers at each side of the TSV, coupling loss is introduced. The setup is shown in Fig. 12.12.

For system characterization, the dependency of bit error rate (BER) and received optical power (ROP) is good since loss in the measurement setup is compensated. The optical attenuator in the setup (Fig. 12.12) is swept to set the different optical powers. In a back-to-back measurement, the data rate of 18 Gbit/s was found to operate error free. Therefore, it has been used for the BER vs. ROP measurement as depicted in Fig. 12.13. No error floor can be seen and there is no penalty between the back-to-back measurement and the measurements with the air-filled and the polymer-filled TSVs. That means, the bandwidth of the TSVs is sufficient. Any dispersive effects would be visible in the eye diagrams (Fig. 12.14). However, the measurements shown for the back-to-back case, the polymer-filled TSV, and the air-filled TSV do not exhibit any signs of dispersion.

The attenuation measured with the setup shown in Fig. 12.12 can be divided into the contributions from the TSV itself and the coupling loss due to the input and output fibers. Depending on the TSV technology, absorption inside the waveguiding material (for waveguide TSVs) or power lost into the substrate (for free-space TSVs) is the main reason within the TSVs. The coupling losses of the input and output fibers strongly depends on the TSV diameter.

For the air-filled free-space TSVs, the refractive index of the substrate material $n_{sub}$ influences the reflection factors. The higher the contrast between air and substrate the higher the reflection factors are, and the lower is the lost of power in the substrate. The loss measurements have been carried out at 650 nm on air-filled TSVs with diameters between 20 and 50 μm. The setup has been calibrated by coupling the probe fibers directly without any TSV in between. Consequently, the measured values include the coupling loss between the fibers and the TSV which was minimized by choosing suitable fibers with appropriate diameters. Losses down

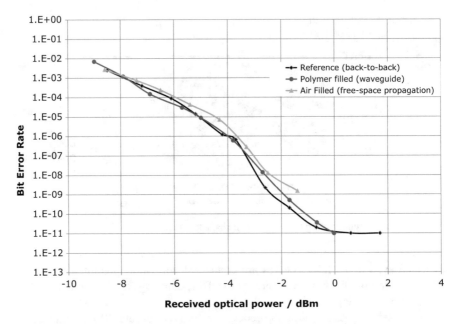

**Fig. 12.13** Bit error rate vs. received optical power (18 Gbit/s NRZ) for back-to-back measurement, with polymer-filled TSV (i.e., waveguide) and with air-filled TSV (i.e., free-space)

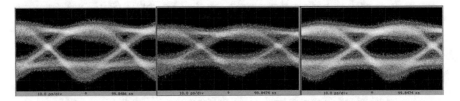

**Fig. 12.14** Eye diagrams for the TSV measurements at 18 Gbit/s: back-to-back (*left*), polymer-filled TSV (*middle*), air-filled waveguide (*right*)

to 1.3 dB could be achieved. For 850 nm, similar values are expected and could accomplished with the data transmission setup using the VCSELs.

Additionally, smaller TSVs (15 μm diameter) with copper-coated sidewalls have been measured using single mode fiber as input. Taking into account the refractive index of copper being smaller than 1 (air), this scheme acts as a waveguide. The losses strongly depend on surface roughness leading to absorption in copper. For straight connections and short distances (as in TSVs), comparably low attenuations are expected. In the measurements, 3.5 dB loss could be achieved. However, the smaller TSV diameter renders the alignment difficult because single mode fibers had to be used. Yet, this is an encouraging which offers the use of electrical, lined TSV also for optical transmission.

**Fig. 12.15** Optical microscope images of air-filled TSV (*left*), back-illuminated air-filled TSV with copper cladding (*middle*) and polymer-filled TSV (*right*)

The classical waveguide setup for TSVs with SU-8 as core and silicon dioxide as cladding (see Fig. 12.3) has also been demonstrated. In experiments, losses of around 3 dB could be shown. The authors expect attenuation levels below that of free-space TSVs by fine-tuning the manufacturing process. Special attention has to be paid on surface roughness of the silicon dioxide cladding is pivotal for waveguide properties. Nanometer-scale roughness has been achieved. Analyzing the TSVs shows that air bubbles in the polymer and gaps between the SU-8 core and the cladding correlate with increasing loss (up to 10 dB).

Figure 12.15 shows microscope images of the three discussed optical TSV types: air-filled (left) with free-space propagation, air-filled with copper cladding (middle), and SU-8 polymer-filled with silicon dioxide cladding.

## 12.5  Conclusion

In this chapter, three options to realize optical TSVs have been introduced: air-filled TSVs with silicon walls, with copper walls, and waveguide TSVs with SU-8 core and silicon dioxide cladding. The manufacturing steps were detailed in detail and important parameters to tune the process were discussed. Models for the transmission behavior of these three TSV types were developed, which predicted bandwidth of more than 1 THz and low attenuation are predicted. The measurement results matched the values predicted by the models very well. At a data rate of 18 Gbit/s (limited by the available electro-optical conversion), no impairments caused by TSVs could be noticed.

All three approaches are promising: The air-filled TSVs are simple to manufacture and show a sufficient performance for short distances. SU-8-filled waveguide TSVs have the potential for lowest loss but require a well-adjusted manufacturing process. Finally, TSVs with copper-walls can be co-used as electrical and optical TSVs boosting the area efficiency.

**Acknowledgements** The authors would like to acknowledge A. Hiess, W. Haas, M.Junige, U. Merkel, K. Richter, A. Jahn, S. Waurenschk, F. Winkler, C. Wenzel, V. Neumann from the Institute of Semiconductors and Microsystems, TU Dresden for their assistance and hard work, without whom research in this project would have been impossible.

# References

1. J.-S. Kim, J.-W. Kang, J.-J. Kim, Simple and low cost fabrication of thermally stable polymeric multimode waveguides using a UV-curable epoxy. Jpn. J. Appl. Phys. **42**, 1277 (2003)
2. J.A. Lott, A.S. Payusov, S.A. Blokhin, P. Moser, N.N. Ledentsov, D. Bimberg, Arrays of 850 nm photodiodes and vertical cavity surface emitting lasers for 25 to 40 Gbit/s optical interconnects. Phys. Status Solidi C **9**(2), 290–293 (2012)

# Chapter 13
# Integrated Optical Devices for 3D Photonic Transceivers

**Seyedreza Hosseini, Michael Haas, Dirk Plettemeier, and Kambiz Jamshidi**

## 13.1 Semiconductor Lasers

A light amplification by stimulated emission of radiation or LASER is composed of an active material which has the optical gain and optical resonator. Semiconductor lasers are lasers based on semiconductor gain media, where optical gain is created by stimulated emission under conditions of a high carrier density. The carriers are pumped using electrical current injection or optical pumping methods to higher energy states and will come back to lower states with stimulated photon radiation. Commercial laser diodes emit at various wavelengths. For the optical resonators, different structures such as Fabry–Perot cavities, ring cavities, distributed feedback (DFB) cavities, and distributed Bragg reflectors (DBRs) are used. Common materials for semiconductor lasers and other optoelectronic devices are III/V semiconductors such as:

S. Hosseini (✉)
Technische Universität Dresden, Integrated Photonic Devices Lab, Dresden University of Technology, 01062 Dresden, Germany
e-mail: seyedreza.hosseini@tu-dresden.de

K. Jamshidi
Junior Professor, Integrated Photonic Devices, Dresden University of Technology, 01062 Dresden, Germany
e-mail: Kambiz.Jamshidi@tu-dresden.de

M. Haas
Technische Universität Dresden, Chair for RF Engineering, 01062 Dresden, Germany
e-mail: michael.haas@tu-dresden.de

D. Plettemeier
Technische Universität Dresden, Chair for RF Engineering, 01062 Dresden, Germany
e-mail: Dirk.Plettemeier@tu-dresden.de

© Springer International Publishing Switzerland 2016
I.M. Elfadel, G. Fettweis (eds.), *3D Stacked Chips*,
DOI 10.1007/978-3-319-20481-9_13

235

- GaAs (gallium arsenide)
- AlGaAs (aluminum gallium arsenide)
- GaP (gallium phosphide)
- InGaP (indium gallium phosphide)
- InGaAs (indium gallium arsenide)
- InGaAsP (indium gallium arsenide phosphide)

These are all direct bandgap semiconductors. Since the photon energy of a laser diode is close to the bandgap energy, it is necessary to have compound material systems for having different emission wavelengths. For the ternary and quaternary semiconductor compounds, the bandgap energy can be continuously varied in some substantial range. In AlGaAs $= Al_xGa_{1-x}As$, for example, an increased aluminum content (increased x) causes an increase in the bandgap energy.

### 13.1.1  III–V Lasers

#### 13.1.1.1  Edge-Emitting Laser

A quantum well semiconductor laser consists of a thin layer of a semiconductor medium which is active region, embedded between other semiconductor layers of wider band gap. The narrow width of active region causes the quantum confinement effect. The wavelength of the light emitted by a quantum well laser is determined by the width of the active region rather than just by the band gap of the material from which it is constructed [1]. It means shorter wavelength can be obtained through using these kinds of lasers. The efficiency of a quantum well laser is also higher than a conventional laser diode due to the stepwise form of its density of states function. The current threshold is less than the conventional lasers. As an example of these types of lasers, we can mention a GaAs quantum well embedded in AlGaAs, or InGaAs in GaAs. The thickness of such a quantum well is typically $\approx$5–20 nm (Fig. 13.1).

If a large amount of optical gain or absorption is required, multiple quantum wells (MQWs) can be used, with a spacing typically chosen large enough to avoid overlap of the corresponding wave functions.

#### 13.1.1.2  Vertical-Cavity-Surface Emitting Laser

In a vertical-cavity-surface emitting laser (VCSEL), the reflector and the output coupling mirror are parallel to the quantum film and the laser light propagates perpendicular to the semiconductor wafer surface (surface emitting). The VCSEL utilizes a Fabry–Perot optical cavity, which is composed of two parallel mirrors on the top and bottom layers. These mirrors can be made from a sequence of two layers with varying refractive index which are called DBR mirrors. For having light emission from the surface, the mirrors reflectivity should be near unity.

The VCSEL has several advantages over edge-emitting diodes. The VCSEL is cheaper to manufacture in high quantity, is easier to test, and is more efficient. Furthermore, VCSELs have wavelength versatility, high beam quality, low threshold currents, and potential for cheap mass production with hundreds or thousands of them grown on a single wafer.

## 13.1.2   Lasers on Silicon

### 13.1.2.1   Hybrid Lasers

Indirect band gap semiconductors such as silicon do not show strong light emission. Despite this, the development of a silicon laser is so important because silicon is the major material option for integrated circuits in electronic and photonic components.

In recent years, many research groups have been working on lasing from silicon. In this regard, some new solutions such as hybrid silicon lasers, Raman laser, which takes advantage of Raman scattering, and Germanium-on-Silicon laser have been introduced and utilized.

A hybrid silicon laser is a type of laser which consists of a silicon waveguide on a Silicon-On-Insulator (SOI) wafer and a III–V direct semiconductor material such as Indium(III) phosphide and Gallium(III) arsenide as top layers on a silicon waveguide. The III–V epitaxial wafer is designed with different layers such that the active layer can emit light when it is excited either by photons or by electrons. Due to the close vicinity of active layers and silicon waveguide, the emitted light can be easily coupled to the silicon waveguide.

The ability to generate light is a property of indium phosphide and other known direct band gap semiconductors. The design of the individual silicon waveguides is critical for specifying the hybrid silicon laser performance. For example, the confinement factor can be dramatically impacted by varying the waveguide width. This is due to the tendency of mode sinking more in the silicon as the waveguide width is increased [2].

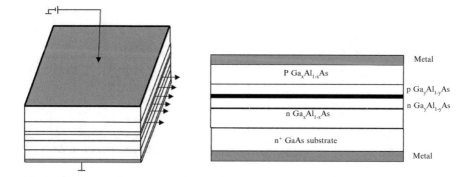

**Fig. 13.1**  GaAs quantum well laser schematic

One of the hybrid silicon laser fabrication techniques is called plasma assisted wafer bonding. Wafer bonding is widely used to make the SOI wafers used in most silicon photonic devices. By using this technique many hybrid silicon lasers can be fabricated simultaneously on a silicon wafer. These light sources could be used for optical communications when it is integrated with silicon photonics and overcome one of the major hurdles in the silicon photonic industry.

### 13.1.2.2   Ge-on-Si Laser

Another recent development is related to Ge-on-Si (or SiGe-on-Si) epitaxial growth. Pure Ge has a significant mismatch with Si in terms of its lattice constant and thermal expansion coefficient. Ge is an indirect semiconductor in which the energy gap from the top of the valence band (0.8 eV) to the $\Gamma$ valley Fig. 13.2 is close to the actual band gap (0.66 eV). Because of this pseudo-direct gap behavior, a new approach for obtaining optical gain around 1550 nm has been proposed. Tensile strain can lead germanium to emit light efficiently at lower energies and higher wavelengths than used traditionally in telecom applications. It is shown that lasing is possible at wavelengths around 1600 nm by using tensile strain and by filling the empty states with n-type doping [3]. Schematic band structures of bulk, tensile strained intrinsic, and tensile strained n+ germanium are shown in Fig. 13.2 [4].

It can be seen that germanium has two direct and indirect gaps. Since indirect gap has lower energy-of 136 meV, germanium is an indirect material. Therefore, its efficiency for light emission is low. However, the difference between the direct and indirect gaps can be decreased by introducing tensile strain. Tensile strain can be introduced using the difference in the temperature expansion coefficients between silicon and germanium [5]. It should be mentioned that both direct and indirect gaps

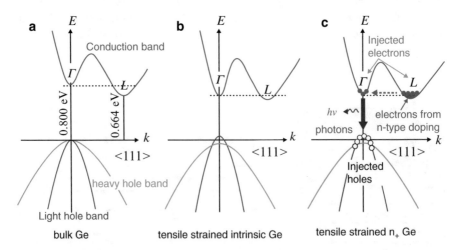

**Fig. 13.2** Schematic band structures of bulk, tensile strained intrinsic, and tensile strained *n*+ Germanium (from [4])

will be changed by inducing tensile strain [6]. For example, a tensile strain of 0.25 % is needed to have a direct gap of 0.76 eV (corresponding to 1630 nm). In this strain, indirect gap is still smaller than direct gap; the rest of the difference between direct and indirect gaps can be compensated by filling electrons into the indirect gap via n-type doping. Since the energy states below the direct valley in the conduction band are fully occupied by extrinsic electrons from n-type doping, injected electrons are forced into the direct valley and recombine with holes, resulting in an efficient direct gap light emission. Recently, a CMOS-compatible approach has been developed at innovations in high-performance (IHP) Microelectronics to induce tensile strain across germanium by depositing a silicon nitride layer acting as stressor [7].

Theoretically, Ge becomes a direct gap material at 2 % tensile strain according to the deformation potential theory [8]. A thermally induced tensile strain of 0.24 % with a phosphorous doping level of $1019 \, \text{cm}^3$, caused first room-temperature Ge-on-Si laser, was implemented by Liu et al. [9]. The presented laser is an optically pumped edge-emitting multimode Ge-on-Si laser which has a gain spectrum of 1590–1610 nm.

### 13.1.2.3  Erbium-Doped Laser

Rare earth (erbium and ytterbium) doped lasers using silicon compatible methods have been proposed. $Al_2O_3 : RE^{3+}$ above concentric silicon nitride rings consist of micro laser structures which have been reported in [10]. A SiNx bus wave guide has been used to pump injection and laser output. All of these material depositions were made on Si substrate which helps in the integration of this type of laser. Er-doped micro ring laser with 300 nm gap between bus wave guide and ring and pumping laser of 976 nm, the output is a single mode laser at 1559.82 nm with suppression of >30 dB. Also, with Yb-doped micro ring laser with 400 nm gap and pumping of 970.96 nm, the structure has been shown a single mode laser at 1042.74 nm with side mode suppression of >40 dB.

## 13.2  Optical Modulators

An optical modulator is a device which modulates the electrical signals along the optical signal path. There are various techniques and structures that can be used for this purpose. It is well-known that changing the electric field can cause changes in refractive index real and imaginary parts, which are called electro-refraction and electro-absorption, respectively. Pockels effect, the Kerr effect, and the Franz–Keldysh effect are the most important manifestations of electro-refraction and electro-absorption phenomena. However, it has been shown that these effects are weak in pure silicon at the communications wavelengths of 1300 nm and 1550 nm [11]. The most common method in silicon device modulation is free carriers plasma dispersion effect.

### 13.2.1 Modulators Based on the Plasma Dispersion Effect

The number of free electrons and holes in p-n or p-i-n semiconductor junctions can be changed by applying an electrical voltage. The free carrier variation can result in changes of the material refractive index. This effect can be explained using Drude relations based on semi-classical methods [12]

$$\Delta n = -\frac{e^2 \lambda^2}{8\pi^2 c^2 \epsilon_0 n} \left( \frac{\Delta N_e}{m_e} + \frac{\Delta N_h}{m_h} \right) \tag{13.1}$$

$$\Delta \alpha = -\frac{e^3 \lambda^2}{4\pi^2 c^3 \epsilon_0 n} \left( \frac{\Delta N_e}{m_e^2 \mu_e} + \frac{\Delta N_h}{m_h^2 \mu_h} \right), \tag{13.2}$$

where $\Delta n$ is the refractive index change, $\Delta \alpha$ is the absorption coefficient change, $e$ is the electronic charge, $\lambda$ is the wavelength, $c$ is the speed of light in a vacuum, $\epsilon_0$ is the permittivity of free space, $n$ is the refractive index of unperturbed silicon, $m_e$ and $m_h$ are, respectively, the effective masses of electrons and holes, $\mu_e$ and $\mu_h$ are, respectively, the mobilities of electrons and holes, $\Delta N_e$ and $\Delta N_h$, are respectively the variations in concentration of electrons and holes.

The $\Delta n$ and $\Delta \alpha$ variations can also be captured using Soref's empirical equations [12]. For a wavelength equal to 1.55 $\mu$m we have:

$$\Delta n = -8.8 \times 10^{-22} \Delta N_e - 8.5 \times 10^{-18} \Delta N_h^{0.8} \tag{13.3}$$

$$\Delta \alpha = 8.5 \times 10^{-18} \Delta N_e + 6 \times 10^{-18} \Delta N_h. \tag{13.4}$$

CMOS-compatible silicon based modulators have been proposed recently to modulate the output light of laser in optical communications, which support large bandwidths and consume low energy per bit [13]. Either Mach–Zehnder or ring resonator structures have been used to modulate the intensity of light based on carrier plasma dispersion effect in which the refractive index of the silicon waveguide is changed by varying the number of carriers in it. Number of carriers in the waveguide can be varied using three techniques, carrier accumulation (two sides of waveguide as capacitor plates), carrier injection (forward biased p-i-n diode, intrinsic layer as waveguide), or carrier depletion (reverse biased p+-p-n-n+ diode) techniques.

Ring resonators have been used in several applications like optical filters, microwave photonics, delay generation, dispersion compensation, and optical modulators [14–16]. Due to the high index contrast between Si and SiO2, the bending radius of rings in CMOS compatible SOI platforms can be very small (several micrometers). This property has been used to fabricate micro ring resonators [17]. The resonance frequency of the ring can be varied by changing the number of carriers inside the ring waveguide. The intensity of the resonator output light can be modulated by alternating its operation between resonance and out of resonance states [18].

### 13.2.1.1   Ring Resonator

The normalized output power is

$$P_t = \frac{\alpha^2 + |t|^2 - 2\alpha |t| \cos(\theta + \phi_t)}{1 + \alpha^2 |t|^2 - 2\alpha |t| \cos(\theta + \phi_t)},\qquad (13.5)$$

where $\alpha$ is the loss coefficient of the ring, $t = |t|\, e^{i\phi_t}$ is the coupling coefficient, $\theta = 4\pi^2 n_{\text{eff}} \frac{R}{\lambda}$ is phase difference caused by ring, and $R$ is the ring radius (Fig. 13.3).

The typical output spectrum is shown in Fig. 13.4.

As is well known from free carrier plasma dispersion effect, the voltage changes the effective refractive index. In other words, the voltage or modulation signal causes small changes in optical transmission behavior, and this is the optical signal modulation.

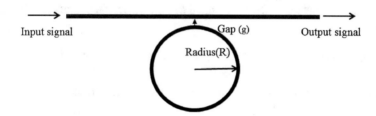

**Fig. 13.3**  The ring resonator schematic

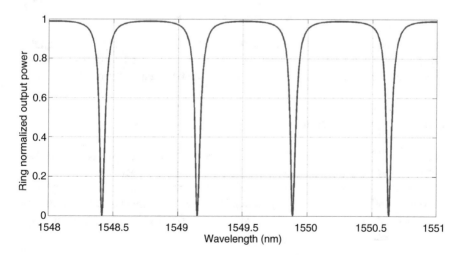

**Fig. 13.4**  Normalized output power of ring modulator

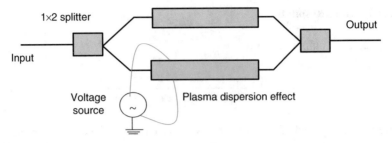

**Fig. 13.5** Mach–Zehnder modulator schematic

### 13.2.1.2   Mach–Zehnder

Mach–Zehnder modulator is based on the Mach–Zehnder interferometer whose schematic is shown in Fig. 13.5.

If we assume the input electrical signal $E_i$ symmetric, the output electrical signal will be:

$$E_o = E_i e^{j\phi_u} + E_i e^{j\phi_d} = E_i e^{j\frac{2\pi}{\lambda} n_0 L}\left(1 + e^{j\Delta\phi}\right) = E_i e^{j\frac{2\pi}{\lambda} n_0 L}\left(1 + e^{j\frac{2\pi}{\lambda}\Delta n L}\right), \quad (13.6)$$

where $\phi_u$ and $\phi_d$ are the phase shift of upper and lower arms, respectively. The signal intensity at the output of the modulator is:

$$\frac{I_o}{I_i} = 4\cos^2\left(\frac{\pi}{\lambda}\Delta n L\right). \quad (13.7)$$

This is classic form of amplitude modulation (AM) and the modulation signal is hidden inside the refractive index change. As per the free carrier plasma dispersion effect, the refractive index of the junction area can be changed using voltage.

### 13.2.1.3   Figures of Merit

The bandwidth of modulation signal specifies how much data (in bit per second) can be transferred via the modulator. High modulation speeds are necessary for interconnect applications, while for some other applications (e.g., sensing) only relatively moderate modulation speeds are needed. The extinction ratio is the ratio of output optical power when the light source is on ($P_1$) to output optical power when the light source is off ($P_0$) in dB:

$$ER = 10\log_{10}\left(\frac{P_1}{P_0}\right). \quad (13.8)$$

The ratio of input power ($P_{in}$) to output optical power in case of logical one state (OOK) ($P_1$) is called the insertion loss:

$$\text{IL} = 10 \log_{10} \left( \frac{P_{in}}{P_1} \right).$$

(13.9)

The signal which goes through a material with an effective index change equal to $\Delta n_{eff}$ would experience phase shifting as

$$\Delta \Phi = \frac{2\pi}{\lambda_0} \Delta n_{eff} L,$$

(13.10)

where $\lambda_0$ is the free space wavelength, and $L$ is material length. As it was stated before, the refractive index change is a function of the applied voltage. The voltage which creates a phase shift equals to $\pi$ is called $V_\pi$. The product $V_\pi L$ is known as the modulation efficiency and is an important feature in each optical modulator.

### 13.2.1.4  State of the Art

There have been great efforts for designing and fabricating optical modulators with high data rates. Several optical modulators with data bit rate up to 60 Gbit/s and modulation efficiency ($V_\pi L$) down to 0.056 Vcm reported. High bit rate ring modulators of 60 Gbit/s have been proposed using higher voltage levels of 6 V [19]. 50 Gbit/s modulation from a 1 mm long phase shifter MZM arm based on carrier depletion with 3.1 dB extinction ratio has been reported in [20]. 40 Gbit/s modulation with 2.7 Vcm modulation efficiency has been reported for 3.5 mm long of MZM arm in [21]. Hybrid silicon electro-absorption modulator with 9.4 dB extinction ration and 50 Gbit/s has been demonstrated for 1.3 μm transmission [22]. A traveling-wave eletro-absorption modulator based on the hybrid silicon platform with 50 Gbit/s transmission rate, driving voltage swing of 2 V and extinction ratio of 9.8 dB has been reported in [23]. Optical modulators based on photonic crystal silicon with 10 Gbit/s data rate and a bias voltage of 0 V have been shown to have low modulation efficiency of less than 0.056 Vcm [24]. A silicon/organic hybrid modulator with slow light effect and low $V_\pi L$ 0.56 Vmm has been reported in [25]. A silicon p+-i-n+ diode Mach–Zehnder with impressive $V_\pi L$ figure of merit of 0.36 Vmm and data rate of 10 Gbit/s has been reported in [26]. These modulator examples illustrate the point that the best features do not appear together as there are so many parameters on which the figures of merit depend. This means that there is a trade-off between data rate and modulation efficiency in order to find the optimum design for a given set of specifications. Ring modulators (in comparison with Mach–Zehnder ones) seem to be the best candidate in the realization of transmitters for optical interconnects [16]. To further enhance the bandwidth density of the ring modulators, advanced modulation formats have been proposed by Bell Labs [27].

## 13.2.2   Ge Modulators

Another candidate for the realization of low-power modulators is to use electro-absorption germanium modulators. Electro-absorption modulators work on the basis of Franz–Keldysh effect in bulk germanium, which is much stronger than the carrier plasma dispersion effect of silicon [4]. These modulators need low voltage levels and they have wide bandwidth and can support high bandwidth sources. Recently, a germanium modulator, based on this method, has been reported which consumes around 100 fJ/bit at 25 Gbit/s using voltage levels of 4 Vpp [28].

## 13.2.3   Pockel Effect Based Modulators

Silicon does not show Pockel's effect since it has a symmetric crystalline, but recently several groups have shown that this effect exists in strained silicon [29, 30]. Silicon nitride has been deposited on silicon using different thermal coefficients, a proper strain is produced which changes the symmetry of silicon and causes a second order nonlinear effect which can be used for electro-optical modulation. Second order nonlinear coefficients of more than 100 pm/V have been achieved, which results in producing a modulator with modulation efficiency of around 100 Vcm. This value is still too high; therefore, several techniques need to be used to improve the modulation efficiency in order to bring it to the levels that can be provided by an integrated circuit.

## 13.3   Photodetectors

The main idea of light (photon) detection comes back to the photoelectric effect explained by Einstein in 1905. Quantum mechanics tells us that if a photon with a specified energy (equal to energy difference of two energy states) collides with matter, an electron can absorb the energy and be transferred from a lower energy state to a higher energy state and produce an electric current.

Photodetectors are devices used for the detection of light and optical powers in most cases. There are many types of photodetectors that may be appropriate for particular applications. Photodiodes are semiconductor devices with a p-n junction or p-i-n structure where light is absorbed in a depletion region and generates a photocurrent. The responsivity $R$ is defined as the ratio of the detector output current to its input optical power and unit typically is A/W. Another important feature in photodetectors is dark current which is how much electric current would flow when there is no light or no photons entering the device. Another important performance parameter for photodetectors is their 3 dB bandwidth that shows the frequency range over which the photodetector can be used without losing the signal quality at output.

## 13.3.1 Germanium-on-Silicon Photodetectors

Different approaches for getting high quality Ge epitaxy on Si have been discussed recently in [31]. Germanium is an inherently indirect bandgap semiconductor. Strain has a significant effect on the band structure and the optoelectronic properties of the semiconductor epitaxial layers. So, it is important to know how the strain can affect optoelectronic properties. The effect of tensile strain on the band structure of Ge at 300 K is shown in Fig. 13.2. Ge has an indirect gap of 0.664 eV at the $L$ valley and is a direct gap of 0.800 eV at the $\Gamma$ valley ($k = 0$). When Ge is exposed to tensile stress, both the direct and indirect gaps are compressed but this compression would be stronger for the indirect gap. A dark current density of 0.15 mA cm$^2$ at a reverse bias of $-1$ V is among the lowest report for Ge-on-Si devices [32]. Ge-on-Si photodetector state of the art is shown in Table 13.1. Note that the reverse bias voltage is always 1 V unless otherwise mentioned.

## 13.3.2 Sub-Bandgap Photodiodes

Other devices include sub-bandgap photo detectors based on various physical phenomena as introduced in [33]. Some of these phenomena are:

- Mid bandgap absorption based on implantation into a SOI wafer
- Surface state absorption based on surface state detection

**Table 13.1** Comparison between various germanium photodetector designs [31]

| Responsivity (A/W) @ 1550 nm | 3 dB bandwidth (GHz) | Dark current density (mA/cm$^2$) | Dark current (μA) | Diode design | Year | References |
|---|---|---|---|---|---|---|
| *Normal incidence design* | | | | | | |
| 0.13 @ 1.3 μm,0 V | 2.3 @ 3 V | 0.2 | 0.2 | p-i-n | 1998 | [32] |
| 0.75 | 2.5 | 15 | 0.14 | p-i-n | 2002 | [34] |
| 0.035 | 38.9 @ 2 V | 100 | 0.31 | p-i-n | 2005 | [35] |
| 0.56 | 8.5 | 110 | 0.79 | p-i-n | 2005 | [36] |
| – | 36.5 @ 2 V | $1 \times 10^6$ | $4 \times 103$ | MSM | 2005 | [37] |
| – | 39 @ 2 V | 375 | 0.075 | p-i-n | 2006 | [38] |
| 0.28 | 17 @ 10 V | 180 | 0.57 | p-i-n | 2006 | [39] |
| 0.037 | 15 | 27 | 0.035 | p-i-n | 2007 | [40] |
| *Waveguide-coupled design* | | | | | | |
| 1.08 | 7.2 | $1.3 \times 10^3$ | 1 | Top, p-i-n | 2007 | [41] |
| 1 | 25 @ 6 V | $6.5 \times 10^5$ | 130 | Butt, MSM | 2007 | [42] |
| 0.85 | 26 | – | 3 | Bottom, p-i-n | 2008 | [43] |
| 1.1 | 32 | $1.6 \times 104$ | 1.3 | Butt, p-i-n | 2009 | [44] |
| >1.1 <1540 nm | 50 @ −5 V | 8900 @ −5 V | – | Butt, MSM | 2009 | [45] |

- Internal photoemission absorption based on optical excitation of electrons in a metal over the Schottky barrier
- Two photon absorption based on absorbing two photons at exactly the same time

## 13.4   Measurement Procedures and Results

Device characterization is an important step for validating the design and manufacturing process as well as providing valuable feedback to improve the performance. For the performance assessment of the fabricated optical modulators, different approaches with specific measurement equipment and methods are available. Naturally, each of the concepts has its individual advantages and disadvantages. One can distinguish between optical, optoelectronic, and electrical measurement methods.

1. The optical measurement methods are used to characterize manufactured ring modulators in relation to the modulation spectrum, the extinction ratio (ER), the free spectral range (FSR), and the frequency stability. Furthermore, the insertion loss (IL) due to coupling losses and waveguide attenuation, the chromatic dispersion, and the polarization behavior can be measured.
2. Electrical measurement methods are useful for testing the matching of the RF-port or -ports ($S_{11}$).
3. Optoelectronic measurements are used for characterizing the effectiveness of the modulation and the supported bandwidth of the optical modulator.

The effectiveness can be characterized by applying a voltage on the RF-port and measuring the frequency shift of the optical spectrum. The needed voltage swing to adequately modulate the signal is determined. The bandwidth can be determined by the 3-dB decline of the $S_{21}$ parameter of the modulator.

As for the modulation efficiency and the frequency bandwidth of the optical modulator, some general design rules can be obtained. For instance, it is well known that by increasing the doping level, the operating bandwidth will increase and the modulation efficiency will decrease. Both of these outcomes are desirable. The limiting factor is the insertion loss, which forces the choice of a high doping level. Another limiting parameter is the bias voltage. Increasing the bias voltage could result in higher bandwidth but would cause higher modulation efficiency as well. The latter outcome is undesirable. Based on these general design rules, a ring resonator has been designed with the following parameters: n-type and p-type doping of $5 \times 10^{17}$ cm$^{-3}$, bias voltage of 3 V, ring radius of 150 $\mu$m and a gap between waveguide and ring of 210 nm. Next, the three different measurement methods are discussed in much more detail.

### 13.4.1 Optical Characterization

The optical characterization can be carried out by using a tunable laser source and an optical power meter. By tuning the laser wavelength and measuring the received optical power the optical spectrum of the device can be sampled. The insertion loss and extinction ratio can be determined. This method offers the advantage of achieving a high dynamic range limited only by the dynamic range of the power meter and a relatively high spectral resolution, and the capability to measure over a broad wavelength range also depending on the tunable laser. The dynamic range can be as high as 80–100 dB, the frequency resolution in the range of 0.1 pm (12.5 MHz) and a wavelength range of approximately 130–160 nm. One disadvantage is that phenomena like chromatic dispersion cannot be measured due to the lack of phase information and that the measurement equipment is relatively expensive, especially the tunable laser source with a high spectral resolution. Furthermore, the setup is sensitive to Fabry–Perot effects due to the high coherence length of the laser.

A second option is to replace the expensive tunable laser by a broadband light source such as the gain spectrum of an erbium-doped fiber amplifier (EDFA). Without optical input signal, the amplifier works as an optical amplified spontaneous emission (ASE) noise source. The power is then detected by an tunable optical filter and a power meter or an optical spectrum analyzer with sufficient frequency resolution. This makes the measurement system a little bit cheaper but reduces the dynamic range to 60–80 dB due to the limited output power of the ASE source, the spectral resolution to approximately 10 pm (1.25 GHz), and the measurement wavelength range to 45 nm in the case of an EDFA gain spectrum. Our modulator has been measured using the broadband light source option. The results are shown in Fig. 13.6. An FSR of approximately 6.4 nm and an ER of 16 dB at around 1555 nm could be measured by using a forward bias of 3 V and a current of 30 mA. The optical spectrum may serve as fast feedback signal to adjust the optical probes for insertion loss optimization.

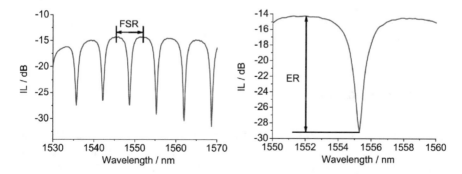

**Fig. 13.6** Measured transmission spectrum of the ring modulator with a forward bias of 3 V using an EDFA as noise source

Another measurement option would be to employ an external modulator biased for suppressed carrier modulation using just the first-order side frequencies and tuning the other side frequencies with a sinusoidal modulation signal. This measurement setup has a dynamic range in the order of the tunable laser and the spectral resolution can be very high depending on the frequency steps of the RF-modulation signal generator. However, the frequency range is limited depending on the bandwidth of the external modulator (typical below 30 GHz) used for the frequency scanning.

To measure further propagation impairments besides the optical amplitude (like chromatic dispersion) the optical phase is needed. The optical phase can be measured by using coherent optical systems with an optical vector analyzer which, e.g., uses the phase detection feature of an optical Mach–Zehnder interferometer [46]. With a polarization diversity receiver also polarization effects like polarization mode dispersion and polarization dependent loss can be measured [46]. For all measurements a bias voltage should be applied. This can be achieved by contacting the RF-ports of the modulator via an RF-probe. Usually the RF-ports are accessible by ground-signal (GS) or ground-signal-ground (GSG) pads.

### 13.4.2    Electrical Characterization

$S_{11}$, which is the ratio between reflected power and input power on a port, can be measured with RF-probes using a probe station, a network analyzer and a voltage source providing the bias voltage. The modulator can be operated in forward- and backward biasing direction by applying a positive or negative voltage respectively to the internal bias port of the network analyzer. For designing high-speed modulators, the modulator should be operated in the backward biasing direction, due to the shorter carrier lifetime [47].

The micrograph image of the ring modulator can be seen in Fig. 13.7. The lower three pads are the GSG RF-ports of the modulator, the upper two pads can be used to apply an additional DC bias to stabilize the frequency response of the modulator and the structures on the left and the right site are the optical input and output ports realized by optical grating on integrated optical waveguides.

### 13.4.3    Optoelectronic Characterization

The optoelectronic measurement is used to determine the $S_{21}$ of the modulator which is the ratio between the modulation power at the RF-port to the received electrical modulated power. From its frequency characteristic the 3 dB-bandwidth of the modulator can be determined. The complete measurement setup as schematic can be seen in Fig. 13.8. The RF-probe and the optical probes over the calibration substrate for the RF-probe are shown in Fig. 13.9.

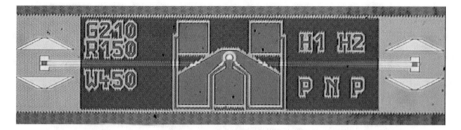

**Fig. 13.7** Micrography of the modulator

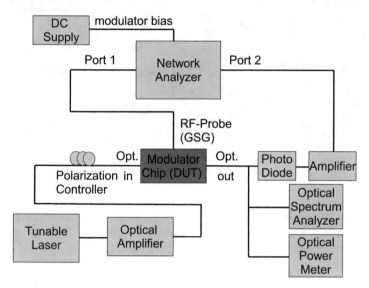

**Fig. 13.8** $S_{21}$ measurement setup

An optical signal is amplified and fed into the modulator via optical probes. A polarization controller is used to adjust the input polarization of the modulator. The optical signal is modulated over the RF-port by applying a RF-signal and a bias voltage. The RF-signal is frequency swept by the network analyzer. The modulated output light is received by a broadband photo diode, amplified, and detected with port 2 of the network analyzer. If the frequency response of the photo diode, the amplifier, and the cable is flat no calibration is needed if only the $S_{21}$ bandwidth shall be determined and not the losses itself. This can be the case depending on the equipment up to some GHz. By measuring the frequency response of the photo diode, the amplifier, and the cable a correction of the measured values can be performed. To determine the needed RF-power to fully drive out the modulator the frequency shift of the optical output spectrum can be measured by applying a DC bias sweep on the RF-port. For this purpose, the tunable laser is removed and the optical amplifier is used as an ASE source and the RF-power on port 1 of the network analyzer is switched off. The optical spectrum is then measured with an

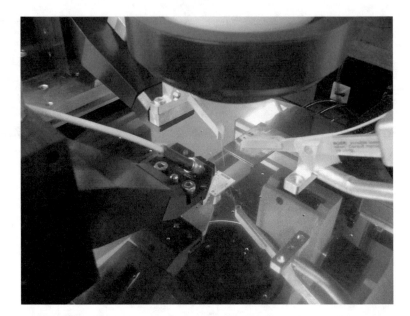

**Fig. 13.9** Laboratory measurement setup (electrical and optical probes over a calibration substrate for the RF-probe)

optical spectrum analyzer. By using other optical characterization methods such as a tunable laser and a power meter or optical vector analyzer, it is also possible to measure the frequency shift introduced into the optical spectrum.

## 13.5 Conclusion

In this chapter, several integrated photonic devices used for the realization of optical interconnects have been reviewed. On the transmitter side, semiconductor lasers to produce continuous wave optical signal and various optical modulators to map electrical data into optical signals have been described. On the receiver side, waveguide photodiodes to map optical signals to electrical signals for telecom wavelengths have been presented. Finally, measurement setups and procedures for the characterization of integrated photonic devices have been briefly reviewed.

## References

1. P.S. Zory (ed.), *Quantum Well Lasers* (Academic, San Diego, 1993)
2. D. Liang, J.E. Bowers, Recent progress in lasers on silicon. Nat. Photonics **4**(8), 511–517 (2010)

3. R. Camacho-Aguilera, L.C. Kimerling, J. Michel, J. Liu, X. Sun, Ge-on-Si laser operating at room temperature. Opt. Lett. **35**(5), 679–681 (2010)
4. X. Sun, X.X. Wang, Y. Cai, L.C. Kimerling, J. Michel, J.F. Liu, R. Camacho-Aguilera, Ge-on-Si optoelectronics. Thin Solid Films **520**(8), 3354–3360 (2012)
5. L.C. Kimmerling, J. Michel, J. Liu, High performance Ge-on-Si photodetectors. Nat. Photonics **4**(8), 527–535 (2010)
6. L.C. Kimerling, J. Michel, X. Sun, J. Liu, Toward a germanium laser for integrated silicon photonics. IEEE J. Sel. Top. Quantum Electron. **16**(1), 124–131 (2010)
7. Y. Yamamoto, M. Lisker, T. Schroeder, A. Ghrib, M. de Kersauson, M. El Kudi, P. Boucaud, B. Tillack, G. Capellini, G. Kozlowski, Tensile strained Ge layers obtained via a Si-CMOS compatible approach, in *International Silicon-Germanium Technology and Device Meeting* (2012)
8. C.G. Van de Walle, Band lineups and deformation potentials in model-solid theory. Phys. Rev. B **39**, 1871–1883 (1989)
9. J. Liu, X. Sun, R. Camacho-Aguilera, L.C. Kimerling, J. Michel, Ge-on-Si laser operating at room temperature. Opt. Lett. **35**(5), 679–681 (2010)
10. J.D. Bradley, E.S. Hosseini, Purnawirman, Z. Su, T.N. Adam, G. Leake, D. Coolbaugh, M.R. Watts, Monolithic erbium- and ytterbium-doped microring lasers on silicon chips. Opt. Express **22**(10), 12226–12237 (2014)
11. G.T. Reed, A.P. Knights, *Silicon Photonics: An Introduction* (Wiley, New York, 2004)
12. R.A. Soref, B.R. Bennett, Electrooptical effects in silicon. IEEE J. Quantum Electron. **QE-23**, 123–129 (1987)
13. G.T. Reed, G. Mashanovich, F.Y. Gardes, D.J. Thomson, Silicon optical modulators. Nat. Photonics **4**(8), 518–526 (2010)
14. J.K.S. Poon, J. Scheuer, Y. Xu, A. Yariv, Designing coupled-resonator optical waveguide delay lines. J. Opt. Soc. Am. B **21**(9), 1665–1673 (2004)
15. B. Franke, O. Dyatlova, A. Alsaadi, U. Woggon, H.J. Eichler, K. Jamshidi, S. Meister, T. Schneider, Compact electrically tunable delay generator on silicon, in *Quantum Electronics and Laser Science (CLEO)* (2012)
16. H. Yu, M. Pantouvaki, J. Van Campenhout, K. Komorowska, P. Dumon, P. Verheyen, G. Lepage, P. Absil, D. Korn, D. Hillerkuss, J. Leuthold, R. Baets, W. Bogaerts, Silicon carrier-depletion-based Mach–Zehnder and ring modulators with different doping patterns for telecommunication and optical interconnect, in *14th International Conference on Transparent Optical Networks (ICTON), 2012* , pp. 1–5 (2012)
17. A. Yariv, Critical coupling and its control in optical waveguide-ring resonator systems. IEEE Photon. Technol. Lett. **14**(4), 483–485 (2002)
18. Q. Xu, B. Schmidt, S. Pradhan, M. Lipson, Micrometre-scale silicon electro-optic modulator. Nature **435**(7040), 325–327 (2005)
19. X. Li, Z. Li, T. Chu, J. Yu, X. Xiao, H. Xu, Y. Yu, 60 Gbit/s silicon modulators with enhanced electro-optical efficiency, in *Optical Fiber Communication* (2013)
20. J.-M. Fedeli, S. Zlatanovic, Y. Hu, B. Ping, P. Kuo, E. Myslivets, N. Alic, S. Radic, G.Z. Mashanovich, D.J. Thomson, F.Y. Gardes, G.T. Reed, 50-Gb/s silicon optical modulator. IEEE Photon. Technol. Lett. **24**(4), 234–236 (2012)
21. Y. Hu, G. Mashanovich, M. Fournier, P. Grosse, J.-M. Fedeli, D.J. Thomson, F.Y. Gardes, G.T. Reed, High contrast 40Gbit/s optical modulation in silicon. Opt. Express **19**, 11507–11516 (2011)
22. J.D. Peters, Y. Tang, J.E. Bowers, $1.3\mu$m hybrid silicon electroabsorption modulator with bandwidth beyond 67 GHz, in *Optical Fiber Communication Conference and Exposition (OFC/NFOEC)* (2012)
23. S. Jain, J.D. Peters, U. Westergren, Y. Tang, H.-W. Chen, J.E. Bowers, 50 Gb/s hybrid silicon traveling-wave electroabsorption modulator. Opt. Express **19**(7), 5811–5816 (2011)
24. M. Shinkawa, N. Ishikura, H.C. Nguyen, Y. Sakai, T. Baba, 10 Gb/s operation of photonic crystal silicon optical modulators. Opt. Express **19**, 13000–13007 (2011)

25. S. Chakravarty, B.S. Lee, W. Lai, J. Luo, A.K.Y. Jen, C.-Y. Lin, X. Wang, R.T. Chen, Electro-optic polymer infiltrated silicon photonic crystal slot waveguide modulator with 23 dB slow light enhancement. Appl. Phys. Lett. **97**, 093304–093303 (2010)
26. L. Sekaric, A. Yurii, W.M. Green, M.J. Rooks, A. Vlasov, Ultracompact low RF power 10 Gb/s silicon Mach–Zehnder modulator. Opt. Express **15**, 17106–17113 (2007)
27. L.L. Buhl, Y.K. Chen, P. Dong, C. Xie, Silicon microring modulators for advanced modulation formats, in *Optical Fiber Communication Conference* (2013)
28. N.-N. Feng, D. Feng, S. Liao, X. Wang, P. Dong, H. Liang, C.-C. Kung, W. Qian, J. Fong, R. Shafiiha, Y. Luo, J. Cunningham, A.V. Krishnamoorthy, M. Asghari, 30GHz Ge electro-absorption modulator integrated with $3\mu$m silicon-on-insulator waveguide. Opt. Express **19**(8), 7062–7067 (2011)
29. B. Chmielak, M. Waldow, C. Matheisen, C. Ripperda, J. Bolten, T. Wahlbrink, M. Nagel, F. Merget, H. Kurz, Pockels effect based fully integrated, strained silicon electro-optic modulator. Opt. Express **19**(18), 17212–17219 (2011)
30. R.S. Jacobsen, K.N. Andersen, P.I. Borel, J. Fage-Pedersen, L.H. Frandsen, O. Hansen, M. Kristensen, A.V. Lavrinenko, G. Moulin, H. Ou, C. Peucheret, B. Zsigri, A. Bjarklev, Strained silicon as a new electro-optic material. Nature **441**(7090), 199–202 (2006)
31. J. Michel, J. Liu, L.C. Kimerling, High-performance Ge-on-Si photodetectors. Nat. Photonics **4**(8), 527–534 (2010)
32. S.B. Samavedam, M.T. Currie, T.A. Langdo, E.A. Fitzgerald, High-quality germanium photodiodes integrated on silicon substrates using optimized relaxed graded buffers. Appl. Phys. Lett. **73**(15), 2125–2127 (1998)
33. A.P. Knights, J.K. Doylend, P.E. Jessop, Silicon photonic resonator-enhanced defect-mediated photodiode for sub-bandgap detection. Opt. Express **18**(14), 14671–14678 (2010)
34. S. Fama, L. Colace, G. Masini, G. Assanto, H.-C. Luan, High performance germanium-on-silicon detectors for optical communications. Appl. Phys. Lett. **81**(4), 586–588 (2002)
35. M. Jutzi, M. Berroth, G. Wohl, M. Oehme, E. Kasper, Ge-on-Si vertical incidence photodiodes with 39-GHz bandwidth. IEEE Photon. Technol. Lett. **17**(7), 1510–1512 (2005)
36. J. Liu, J. Michel, W. Giziewicz, D. Pan, K. Wada, D.D. Cannon, S. Jongthammanurak, D.T. Danielson, L.C. Kimerling, J. Chen, F.Ö. Ilday, F.X. Kaertner, J. Yasaitis, High-performance, tensile-strained Ge p-i-n photodetectors on a Si platform. Appl. Phys. Lett. **87**(10), 103501 (2005)
37. M. Rouviere, L. Vivien, X. Le Roux, J. Mangeney, P. Crozat, C. Hoarau, E. Cassan, D. Pascal, S. Laval, J.-M. Fideli, J.-F. Damlencourt, J.M. Hartmann, S. Kolev, Ultrahigh speed germanium-on-silicon-on-insulator photodetectors for 1.31 and 1.55$\mu$m operation. Appl. Phys. Lett. **87**(23), 231109 (2005)
38. M. Oehme, J. Werner, E. Kasper, M. Jutzi, M. Berroth, High bandwidth Ge p-i-n photodetector integrated on Si. Appl. Phys. Lett. **89**(7), 071117 (2006)
39. Z. Huang, N. Kong, X. Guo, M. Liu, N. Duan, A.L. Beck, S.K. Banerjee, J.C. Campbell, 21-GHz-bandwidth germanium-on-silicon photodiode using thin SiGe buffer layers. IEEE J. Sel. Top. Quantum Electron. **12**(6), 1450–1454 (2006)
40. T.H. Loh, H.S. Nguyen, R. Murthy, M.B. Yu, W.Y. Loh, G.Q. Lo, N. Balasubramanian, D.L. Kwong, J. Wang, S.J. Lee, Selective epitaxial germanium on silicon-on-insulator high speed photodetectors using low-temperature ultrathin Si0.8Ge0.2 buffer. Appl. Phys. Lett. **91**(7), 073503-1–073503-3 (2007)
41. D. Ahn, C.Y. Hong, J. Liu, W. Giziewicz, M. Beals, L.C. Kimerling, J. Michel, J. Chen, F.X. Kärtner, High performance, waveguide integrated Ge photodetectors. Opt. Express **15**(7), 3916–3921 (2007)
42. L. Vivien, M. Rouvière, J.-M. Fédéli, D. Marris-Morini, J.F. Damlencourt, J. Mangeney, P. Crozat, L. El Melhaoui, E. Cassan, X. Le Roux, D. Pascal, S. Laval, High speed and high responsivity germanium photodetector integrated in a silicon-on-insulator microwaveguide. Opt. Express **15**(15), 9843–9848 (2007)
43. G. Masini, S. Sahni, G. Capellini, J. Witzens, C. Gunn, High-speed near infrared optical receivers based on Ge waveguide photodetectors integrated in a CMOS process. Adv. Opt. Technol. **2008**, 5 (2008)

44. D. Feng, S. Liao, P. Dong, N.-N. Feng, H. Liang, D. Zheng, C.-C. Kung, J. Fong, R. Shafiiha, J. Cunningham, A.V. Krishnamoorthy, M. Asghari, High-speed Ge photodetector monolithically integrated with large cross-section silicon-on-insulator waveguide. Appl. Phys. Lett. **95**(26), 261105 (2009)
45. L. Chen, M. Lipson,  Ultra-low capacitance and high speed germanium photodetectors on silicon. Opt. Express **17**(10), 7901–7906 (2009)
46. D.K. Gifford, B.J. Soller, M.S. Wolfe, M.E. Froggatt,  Optical vector network analyzer for single-scan measurements of loss, group delay, and polarization mode dispersion. Appl. Opt. **44**(34), 7282–7286 (2005)
47. A. Liu, L. Liao, D. Rubin, J. Basak, Y. Chetrit, H. Nguyen, R. Cohen, N. Izhaky, M. Paniccia, Recent development in a high-speed silicon optical modulator based on reverse-biased pn diode in a silicon waveguide. Semicond. Sci. Technol. **23**(6), 064001 (2008)

# Chapter 14
# Cantilever Design for Tunable WDM Filters Based on Silicon Microring Resonators

Hossam Shoman and Marcus S. Dahlem

## 14.1 Introduction

Research into optical circuits for telecommunications began in the 1970s [1]. In order to achieve high data rates for faster transmission, bandwidth should be considered and maximized when designing communication devices. One way to achieve higher bandwidth is through replacing the electronic switches by appropriate optical equivalent devices [2, 3]. In the electronics industry, silicon has been the dominant material. The interest in silicon as a platform for photonics has also been growing in the last decades. For most applications, the ability to control the index of refraction is crucial. Materials that lack inversion symmetry, such as lithium niobate ($LiNbO_3$), gallium arsenide (GaAs), and indium phosphide (InP), exhibit excellent Pockels electro-optic effect (proportional to the applied electric field), while silicon does not due to its inversion symmetry. However, similar results can be obtained in silicon through alternative methods such as the injection of free carriers. Additionally, choosing silicon as the platform for designing optical components would enable monolithic integration of optics and electronics on a single chip. Most advances in the field of integrated photonics have been directed towards the telecommunications industry at 1.55 $\mu$m [3]. Microring resonators are essential building blocks for enabling electronic–photonic integrated circuits, in particular for designing filters that can be used in optical communications, utilizing wavelength-division multiplexing (WDM) schemes [4, 5]. In WDM systems, light is routed

H. Shoman
Masdar Institute of Science and Technology, Abu Dhabi, United Arab Emirates
e-mail: hshoman@masdar.ac.ae

M.S. Dahlem (✉)
Department of Electrical Engineering and Computer Science, Institute Center for Microsystems (iMicro), Masdar Institute of Science and Technology, Abu Dhabi, United Arab Emirates
e-mail: mdahlem@masdar.ac.ae

© Springer International Publishing Switzerland 2016
I.M. Elfadel, G. Fettweis (eds.), *3D Stacked Chips*,
DOI 10.1007/978-3-319-20481-9_14

**Fig. 14.1** Illustration of the functionality of an optical add-drop multiplexer (OADM). Wavelength channel $\lambda_n$ is dropped from the main signal stream, while wavelength channel $\lambda_m$ is added

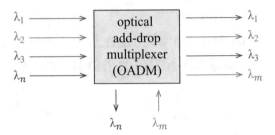

by adding (or dropping) a wavelength channel into (or from) a stream of signals. This concept is illustrated in Fig. 14.1. The added and dropped channels are not necessarily at the same wavelength. This is possible with tunable and reconfigurable optical add-drop multiplexers (ROADMs).

These elements require low-power frequency (wavelength) tuning in order to achieve maximum performance and reconfigurability, as well as maintaining energy efficiency [6, 7]. Silicon-on-insulator (SOI) platforms have been used for building such microring resonator based systems, allowing compact and low-loss propagation at telecom wavelengths. Control of the resonant frequency is essential for channel frequency tuning and post-fabrication frequency trimming (due to fabrication errors), as well as for compensating for potential on-chip temperature variations during operation.

## 14.2 Ring Resonators

The ability to tune the properties of optical add-drop filters is important when designing PICs. Once fabricated, photonic devices can be controlled by introducing a mechanism to modify their intrinsic properties, such as the refractive index. In order to tune the resonant frequencies of a microring resonator, which can be used as an add-drop filter, it is first necessary to formulate its frequency response. One way to analyze this response is by using the $z$-transform. For optical filters, this method was first applied by Moslehi and co-workers to synthesize filter responses of optical fiber systems [8, 9]. Once the $z$-transform is determined, we can evaluate it on the unit circle in order to yield the frequency response of our system. Besides the refractive index, changes in coupling ratios and/or resonator losses can also be used to modify the power balance between different ports in the system.

### 14.2.1 The z-Transform and Transfer Function

The layout and the $z$-transform schematic of a microring resonator are shown in Fig. 14.2 [10]. $X_i(z)$ represents the $z$-transform of the input signal electric field in bus

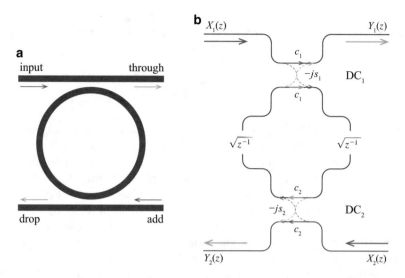

**Fig. 14.2** (**a**) Layout of a ring resonator add-drop filter, and (**b**) its equivalent $z$-transform schematic. The *red arrows* represent the light launched into each bus waveguide input port, and the *blue arrows* the light exiting from each bus waveguide output port. At each directional coupler, the *red dashed arrows* represent the light coupled from the bus waveguides into the microring, and the *green dashed arrows* the light coupled from the microring into the bus waveguides

waveguide $i$, $Y_i(z)$ the $z$-transform of the output signal electric field in bus waveguide $i$, $c_i$ the through-port transmission coefficient in directional coupler $i$ (DC$_i$), and $s_i$ the cross-port transmission coefficient in DC$_i$. $z^{-1}$ represents the unit phase shift delay (due to one full roundtrip) in the microring, defined as

$$z^{-1} = \exp(-j\beta L),  \tag{14.1}$$

where $\beta = (2\pi/\lambda_0)n_{\text{eff}}$ is the propagation constant. $\lambda_0$ is the free-space wavelength, $n_{\text{eff}}$ the effective index of the waveguide, and $L = 2\pi R$ the circumference of the ring with center radius $R$. The phase acquired in each half of the ring resonator corresponds to $\sqrt{z^{-1}}$, as indicated in the figure. The through- and cross-port coefficients in directional coupler $i$ are given by

$$c_i = \sqrt{1 - \kappa_i},  \tag{14.2}$$

$$s_i = \sqrt{\kappa_i},  \tag{14.3}$$

where $\kappa_i$, the power coupling ratio, defines the amount of power coupled from the input port to the cross port of the directional coupler. $\kappa_i$ lies between 0 and 1, 0 being 0 % coupling and 1 being 100 % coupling. The power coupling ratio $\kappa_i$ is calculated from the coupling strength coefficient $\tilde{\kappa}_i$, integrated over the coupling interaction length. Assuming that the wave in each coupler propagates in the $z$ direction, the coupling coefficient $\tilde{\kappa}_i$ can be derived considering two identical parallel rectangular waveguides, and can be expressed for the quasi-TE mode as [11]

$$\tilde{\kappa}_1 = \tilde{\kappa}_2 = \tilde{\kappa} = \frac{\tilde{\omega}\epsilon_0(n_{co}^2 - n_{cl}^2) \iint_A E_{1x}^* E_{2x} \, dx \, dy}{2 \int_{-\infty}^{\infty} \int_{-\infty}^{\infty} E_{1x}^* H_{1y} \, dx \, dy}, \tag{14.4}$$

where $\tilde{\omega}$ is the angular frequency of the wave, $\epsilon_0$ the permittivity of free-space, $n_{co}$ the refractive index of the core, $n_{cl}$ the refractive index of the cladding, and $A$ the cross-sectional area of the waveguide. $E_{1x}, E_{2x}, H_{1y}$ represent the x component of the electric field and the y component of the magnetic field in each waveguide, assuming the structure shown in Fig. 14.2 lies in the $x - z$ plane, where $z$ is the direction of the bus waveguides. We use the same coupling coefficient ($\tilde{\kappa}_1 = \tilde{\kappa}_2 = \tilde{\kappa}$) for DC$_1$ and DC$_2$, since we assume that both couplers are identical (ring and bus waveguides have the same width and height in both couplers) and that the coupling gap between the microring and the bus waveguides is the same.

The output $Y_2(z)$ at the drop port due to input $X_1(z)$ is the (infinite) sum of all the contributions from each roundtrip inside the ring resonator. The first half round trip contributes with

$$Y_2(z)_0 = X_1(z)(-js_1)(\sqrt{\gamma z^{-1}})(-js_2) = -X_1(z)s_1 s_2 \sqrt{\gamma z^{-1}}, \tag{14.5}$$

where $\gamma$ defines the amplitude transmittance for each roundtrip. If we assume $\alpha$ to be the loss in dB per unit length, then $\gamma$ is equal to $10^{-\alpha L/20}$. Each roundtrip contributes with an additional term equal to $c_1 \sqrt{\gamma z^{-1}} c_2 \sqrt{\gamma z^{-1}}$. Multiplying Eq. (14.5) by this term gives the contribution of the $n$th roundtrip:

$$Y_2(z)_n = -X_1(z)s_1 s_2 \sqrt{\gamma z^{-1}} \left(c_1 c_2 \gamma z^{-1}\right)^n. \tag{14.6}$$

Adding up all the contributions leads to

$$Y_2(z) = -X_1(z)s_1 s_2 \sqrt{\gamma z^{-1}} \left(1 + c_1 c_2 \gamma z^{-1} + \left(c_1 c_2 \gamma z^{-1}\right)^2 + \cdots\right), \tag{14.7}$$

which is a geometric series that can be written as

$$Y_2(z) = -X_1(z)s_1 s_2 \sqrt{\gamma z^{-1}} \frac{1}{1 - c_1 c_2 \gamma z^{-1}}. \tag{14.8}$$

Therefore,

$$H_{21}(z) = \frac{Y_2(z)}{X_1(z)} = \frac{-s_1 s_2 \sqrt{\gamma z^{-1}}}{1 - c_1 c_2 \gamma z^{-1}} = \frac{-\sqrt{\kappa_1 \kappa_2 \gamma z^{-1}}}{1 - \sqrt{1 - \kappa_1}\sqrt{1 - \kappa_2}\gamma z^{-1}}. \tag{14.9}$$

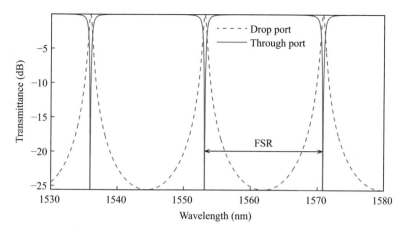

**Fig. 14.3** Plot of the transmittance of the through and drop ports of a lossless ($\gamma = 1$) microring resonator with a radius of $10\,\mu m$ and $10\%$ power coupling ratio between the bus and ring waveguides. The effective index is assumed to be 2.2. The *solid* and *dashed lines* represent the transmittance at the through and drop ports, respectively. The separation between two resonances is known as the free spectral range (FSR)

Repeating the procedure for output $Y_1(z)$ at the through port due to input $X_1(z)$ yields

$$H_{11}(z) = \frac{Y_1(z)}{X_1(z)} = \frac{c_1 - c_2\gamma z^{-1}}{1 - c_1 c_2 \gamma z^{-1}} = \frac{\sqrt{1 - \kappa_1} - \sqrt{1 - \kappa_2}\gamma z^{-1}}{1 - \sqrt{1 - \kappa_1}\sqrt{1 - \kappa_2}\gamma z^{-1}}. \tag{14.10}$$

Assuming equal coupling ratios at both DCs, as mentioned before, the transfer functions in Eqs. (14.9) and (14.10) become

$$H_{21}(z) = \frac{-\kappa\sqrt{\gamma z^{-1}}}{1 - (1 - \kappa)\gamma z^{-1}} \tag{14.11}$$

$$H_{11}(z) = \frac{\sqrt{1 - \kappa}(1 - \gamma z^{-1})}{1 - (1 - \kappa)\gamma z^{-1}}. \tag{14.12}$$

A plot of the transmittance [square of the magnitude of Eqs. (14.11) and (14.12)] at the through and drop ports of a typical microring resonator is shown in Fig. 14.3, where the wavelength response is obtained using Eq. (14.1).

## 14.2.2  Resonator Parameters

### 14.2.2.1  Resonant Wavelength and Free Spectral Range

Equation (14.11) for the drop port has a pole at

$$z = (1 - \kappa)\gamma. \tag{14.13}$$

The pole determines the location of the resonant wavelengths, which can be obtained by replacing $z$ in Eq. (14.13) by $e^{j\beta L}$. This corresponds to evaluating the $z$-transform on the unit circle. Solving for $\beta L$, we get $\beta L = 2\pi q$, with $q \in \mathbb{Z}^+$, which returns the resonant wavelengths:

$$\lambda_q = \frac{2\pi R n_{\text{eff}}(\lambda_q)}{q}. \tag{14.14}$$

$n_{\text{eff}}(\lambda_q)$ is the effective index at wavelength $\lambda_q$, and can be written in terms of the group index $(n_g)$ as

$$n_g(\lambda_q) = n_{\text{eff}}(\lambda_q) - \lambda_q \frac{d n_{\text{eff}}(\lambda)}{d\lambda}\Bigg|_{\lambda=\lambda_q}. \tag{14.15}$$

A change in either the circumference $L$ or the effective index $n_{\text{eff}}$ of the microring, by $\delta L$ or $\delta n_{\text{eff}}$, would modify the resonant wavelength. This change corresponds to an additional phase term $\phi$ that is added to the roundtrip phase, and is determined by

$$e^{-j(2\pi/\lambda_0)(n_{\text{eff}}+\delta n)(L+\delta L)} = z^{-1}e^{-j\phi}, \tag{14.16}$$

where $\phi = (2\pi/\lambda_0)(n_{\text{eff}}\delta L + L\delta n + \delta n\delta L)$. The introduction of any additional phase would further modify the resonant wavelengths. Once fabricated, it is not straightforward to fine tune the physical length of the microring. However, the refractive index can still be modified. Through a first order Taylor expansion of the propagation constant, the change in the resonant wavelength due to a change in the refractive index can be written as

$$\Delta\lambda \approx \lambda \frac{\Delta n_{\text{eff}}}{n_g}. \tag{14.17}$$

The free spectral range (FSR) is defined as the frequency spacing between two adjacent modes inside the resonator (illustrated in Fig. 14.3), and can be written, in terms of wavelength spacing, as

$$\text{FSR}_\lambda \approx \frac{\lambda_0^2}{2\pi R n_g(\lambda_0)}, \tag{14.18}$$

where $\lambda_0$ is the center wavelength between any two resonances. The above equation is valid for large ring radius, where the FSR is small. Modifying the resonant wavelengths will also change the FSR, which is also an important design parameter in filtering applications.

### 14.2.2.2   Coupling and Loss

The power at the drop and through ports at the resonant wavelength condition depends on the power coupling ratio $\kappa$ and the ring losses (quantified through $\gamma$). Evaluating the square of the magnitude of the transfer functions, represented by Eqs. (14.11) and (14.12), at the resonant wavelengths, yields

$$|H_{21}(\lambda_q)|^2 = \frac{\kappa^2 \gamma}{1 - 2(1 - \kappa)\gamma + (1 - \kappa)^2 \gamma^2} \tag{14.19}$$

$$|H_{11}(\lambda_q)|^2 = \frac{(1 - \kappa)(1 - 2\gamma + \gamma^2)}{1 - 2(1 - \kappa)\gamma + (1 - \kappa)^2 \gamma^2} \tag{14.20}$$

for the power at the drop and through ports, respectively. These equations are plotted in Fig. 14.4, as a function of the power coupling ratio $\kappa$, for different $\gamma$ values.

For $\gamma = 1$ (no loss in the microring) and $\kappa \neq 0$, all the power from the input port is routed to the drop port at the resonant wavelength, regardless of the power coupling ratio. This occurs because the wave coupled into the microring and

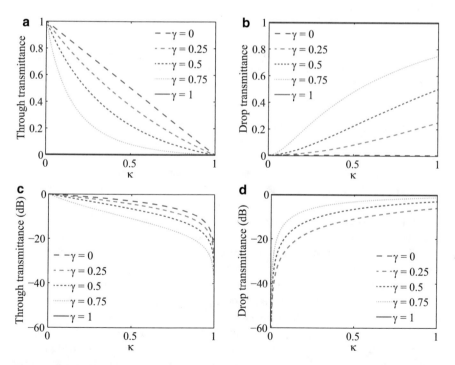

**Fig. 14.4** Transmittance plots at the resonant wavelengths, as a function of the power coupling ratio $\kappa$, for different $\gamma$ values: through-port power in linear (**a**) and dB (**c**) scales; drop-port power in linear (**b**) and dB (**d**) scales

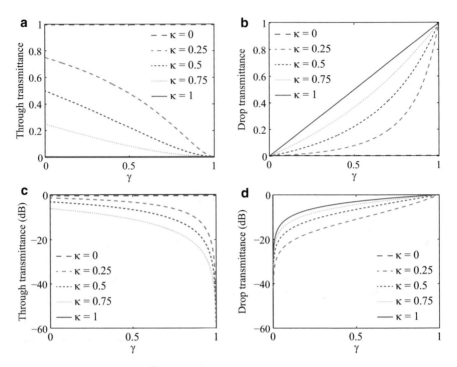

**Fig. 14.5** Transmittance plots at the resonant wavelengths, as a function of the ring loss (through $\gamma$), for different $\kappa$ values: through-port power in linear (**a**) and dB (**c**) scales; drop-port power in linear (**b**) and dB (**d**) scales

back into the upper bus waveguide interferes destructively with the input wave. For lossy rings, as the coupling ratio $\kappa$ increases, the power at the through (drop) port decreases (increases). The losses in the ring can result from several mechanisms such as absorption, bending, scattering, leaky modes, free carriers, or even from perturbations of the evanescent field of the optical mode, as will be seen later.

Equations (14.19) and (14.20) are also plotted as a function of the roundtrip amplitude transmittance $\gamma$, for different $\kappa$ values, and shown in Fig. 14.5. As the ring becomes more lossy (i.e., $\gamma$ decreases), more power is available at the through port. This is due to an incomplete destructive interference between the input wave and the wave coupled back into the upper bus from the ring.

As previously seen in Eq. (14.16), a small change in the effective index or in the ring radius corresponds to an additional phase term that results in tuning of the resonant wavelength. On the other hand, a change in the power coupling ratio $\kappa$ (by changing the coupling gap, coupling length, or even the refractive indices), or in the $\gamma$ coefficient (by inducing losses in the microring) results in a change in the power transferred to the drop or through ports. Hence, tuning the power coupling or controlling the losses in the ring allows for the control of the power balance.

Resonant frequency tuning may find applications in ROADM filters for WDM systems [12], while control of the power balance can be used in switching applications [13]. It is worth mentioning that a change in the ring curvature may also result in a change in the power extinction ratio, due to a change in the effective coupling length of the directional coupler.

Tuning the resonant wavelength through the introduction of free carriers, or even a lossy material to perturb the optical mode, may result in a change in the power balance of the ring, besides a wavelength shift. This is due to a change in the absorption coefficient of the material (related to the imaginary part of the refractive index), which results from a change in the real part of the refractive index. These two quantities are related by the Kramers–Kronig relations, which link the real and imaginary parts of the refractive index [14]. In general, when a wave propagates through a medium, part of it is absorbed. This corresponds to a refractive index described by a complex number. Additional information on absorption can be found in a standard photonics textbook (e.g., [15]).

## 14.3  Tuning Mechanisms

The current section presents a brief overview of some resonant wavelength tuning mechanisms of optical filters based on microresonators, with emphasis on silicon microring resonators.

### 14.3.1  Free Carrier Plasma Dispersion Effect

The injection of free carriers into a material causes a change in its refractive index. This effect is known as the plasma dispersion effect [16]. The free carriers are generated by single- or two-photon absorption from a pump laser, or injected through forward bias of a p-i-n junction. Alternatively, free carriers can be depleted from a previously doped region. The change in the free carrier concentration will therefore affect the optical properties of the material. When carriers are injected into an intrinsic region, the electron–hole pair density increases, which decreases the effective index [16]. When within a resonant cavity, this index decrease will blueshift (move towards lower wavelengths) the resonant wavelength.

Using small footprint devices, free carrier injection has been electrically realized through p-i-n ring resonators in SOI platforms [17], and p-i-n diode structures incorporated into InP microdisk resonators [18], as well as through optical carrier generation in 1D photonic crystal cavities fabricated on SOI platforms [19]. As an example, in [18], a $\Delta n = -2 \times 10^{-3}$ is achieved with 1 mA. However, higher shifts require more current, which may heat up the sample and cause an index change in the opposite direction. In the same paper, a current of 5 mA generated an undesired $\Delta n = +6 \times 10^{-3}$ due to temperature increase. If large $\Delta n$ shifts are needed, this temperature effect is one of the main bottlenecks when using free carrier injection methods. Although carrier injection is slower than carrier depletion, it is still fast compared to other tuning methods presented later in this section (relaxation times of hundreds of picoseconds, limited by the lifetime of the charge carriers [19]).

### 14.3.2 Electro-Optic Effects

Electro-optic effects refer to changes in the optical constants of a material due to an applied electric field, and are broadly divided into changes in absorption or refractive index. They do not involve carrier injection or depletion. Electroabsorption refers to a change in the absorption of a material due to an applied electric field [16]. Other effects, such as the Pockels effect, explain the change in the refractive index of a material with an applied electric field. The Pockels effect is linear with the electric field, but is only observed in materials that lack inversion symmetry (hence, not in silicon). Electro-optic effects have been realized using III–V materials through microdisks [20] and microrings [21]. However, only moderate resonant wavelength shifts are achieved. As an example, [21] demonstrates a 0.105 nm wavelength shift around 1555 nm wavelengths, with an applied voltage of 100 V. This low tuning range (at high voltages) is not ideal for meeting WDM operation requirements.

### 14.3.3 Liquid Crystals

Nematic liquid crystals (NLC) are materials that flow like a liquid but have their molecules oriented as a solid crystal [22]. NLCs are birefringent, i.e., they have a refractive index that is dependent on the polarization and direction of propagation of light. In the presence of an electric field, the NLC molecules align themselves along the field lines. This changes the polarization of the light passing through it, and therefore changes its refractive index. It is therefore possible to achieve electrically tunable optical devices based on NLCs [23]. NLCs can be used as a top and side cladding material for silicon microring resonators. In [23], the NLC molecules align when an electric field is applied between two conductive layers (ITO and silicon substrate), reducing the cladding refractive index and therefore the effective index in the microring. The resulting effect is a blueshift in the resonant wavelength. For an applied 30 V, the resonant wavelengths shift by 0.6 nm. Larger voltages do not result in larger wavelength shifts, due to saturation. This is due to distortion of the NLC director field.

Like the previously mentioned plasma dispersion effect and electro-optic effects, the maximum wavelength tuning range possible with this type of configuration is normally small and not suitable for ROADMs (where several nm may be required to tune between different WDM channels).

### 14.3.4 Thermal Tuning

Thermal tuning mechanisms rely on the thermo-optic effect of materials, where their refractive index changes with temperature. In silicon, the thermo-optic coefficient ($dn/dT$) around 1550 nm and 300 K is about $+1.8 \times 10^{-4} \, K^{-1}$ [24]. Above room temperature, this coefficient increases with a weak quadratic dependence [25]. Similarly, the thermo-optic coefficient of materials such as InP, GaAs, and SiC also increase quadratically with temperature [26].

Several ways to thermally control the resonant frequencies in microrings on silicon platforms have been reported, either by placing microheaters on top of the ring cladding [7, 27, 28], or by directly heating the ring through doping the silicon waveguide as a resistor [29]. Comparing the two approaches, the latter yields higher optical losses, however, lower power consumption and faster tuning speeds (4.4 μW/GHz and 1 μs [29] vs. 17–28 μW/GHz and 7–14 μs [27]).

As a figure of merit, microheaters are credited for their ability to perform large tuning ranges (full FSR for microrings). In [27], 20 nm tuning is demonstrated, and in [29] over 30 nm. Due to the increasing thermo-optic coefficient of silicon with temperature, the resonant wavelength sensitivity increases at higher temperatures, resulting in higher wavelength shifts per temperature change [27]. However, a drawback of microheaters is the large power consumption, as well as the induced thermal crosstalk in adjacent rings.[1] For microrings on SOI platforms, a mechanism to mitigate the adjacent channel crosstalk, as well as to improve the heating efficiency, was reported in [28]. The approach is based on thermal isolation air trenches formed around the microrings. Results show that the tuning power is around 21 mW for one full FSR (19 nm) tuning, a reduction of about 20 % compared to similar structures without trenches. The reported rise and fall times of the microheater are 9 and 6 μs, respectively.

Among the reported tuning methods, thermal tuning is the most widely implemented method for large wavelength shifts, being also very reliable. However, the thermal crosstalk between adjacent channels and the large power consumption for broad tuning ranges are still undesired issues [30].

### 14.3.5 Opto-Mechanical Tuning

Opto-mechanical tuning refers to the integration of microelectromechanical systems (MEMS) with photonic devices. When applied to microresonators, the evanescent field of the resonant mode can be externally perturbed, either from the top [30–32] or from the side [33–35]. Two effects may result as a consequence of such perturbation: change in the effective index of refraction of the mode inside the resonator (since

---

[1] In high order filters that use multiple rings, the microheater used for configuring a specific ring also affects and induces a wavelength shift in the adjacent rings.

the evanescent tail of the mode sees a different cladding), and increase in the optical loss due to coupling/absorption of the propagating mode into/by the perturbing structure. Coupling ratios between bus waveguides and microresonators can also be modified through mechanical movement of suspended bus waveguides [36–38], normally needed for post-fabrication tuning or trimming [39].

Several methods, with different design approaches, have been reported for opto-mechanical wavelength tuning. Top perturbation of a silicon nitride microring was demonstrated using a silica slab connected to an external cantilever [30]. Results show a 25–27 nm wavelength tuning (full FSR of 27 nm) at 1565 nm, which is the widest reported tuning range achieved by evanescent field perturbation. However, the demonstrated mechanism requires an external tuning apparatus, which lacks monolithic integration. An integrated approach using an electrically actuated silicon nitride cantilever on top of a silicon racetrack ring resonator has also been demonstrated [32], with much smaller tuning ranges (122 pm reported, with a maximum tuning range of 5 nm).

Side perturbation wavelength tuning of microrings in SOI platforms was demonstrated in a slot-waveguide ring resonator, by mechanically changing the geometry of the slot through electrostatic actuation [34]. Additional proposed mechanisms include wavelength tuning through side evanescent field perturbation using a silicon cantilever [33, 35], and through the control of the coupling coefficient in a directional coupler connecting two microring resonators [36]. In [34], the height of the inner ring is changed by applying a DC bias across the ring and the silicon substrate. The power needed to achieve a 1 nm wavelength shift is below 100 nW, for TE-polarized light. The mechanisms proposed in [33, 35, 36] allow for larger tuning ranges, estimated to be around 13 nm in the C-band (1530–1565 nm). The mechanism described in [33, 35] forms the basis for the remaining work in this chapter. The authors proposed a side perturbation configuration for interaction between the evanescent field of the mode and a silicon cantilever. The work is done for the TE mode, but can also be done for the TM mode. A high refractive index material (silicon) was chosen for the cantilever, which allows for a large tuning range. Furthermore, in SOI platforms, the waveguides and cantilever can be defined in the same lithography step. Simulation results show about 13 nm resonant wavelength shift around 1550 nm, with larger shifts up to 19 nm for longer wavelengths. Since the evanescent light couples into the cantilever, the quality factor of the microring is minimum at maximum tuning (unbiased cantilever).

Besides resonant wavelength tuning, other approaches have reported to control the power extinction ratio and trim the transmittance at the through and drop ports in microring resonator filters [31, 37, 38]. In [31], a MEMS metal membrane is used to change the quality factor by inducing optical losses, with the device being used as an on/off switch. The work in [37, 38] demonstrates tuning of the coupling ratio between bus and microring waveguides, from the side and from the top, respectively.

Comparing all different wavelength tuning mechanisms described in this section, the main advantages of using MEMS-based approaches are their potentially large tuning range and power efficiency (electrostatic actuation). Challenges still remain in designing low-power, fast, low-loss, and mechanically stable cantilevers for broad wavelength tuning applications. An insight into the mechanical analysis and the trade-off between some of the mentioned parameters are discussed in the following section.

## 14.4   Opto-Mechanical Tuning Using a MEMS Cantilever

Opto-mechanical tuning by electrostatic actuation can potentially provide a broad wavelength tuning range, while requiring low energy for operation. This section covers the general mechanical design analysis that should be taken into account when designing optical MEMS. The approach, proposed earlier by the authors in [33, 35], consists in tuning the resonant frequency of a microring resonator through evanescent field perturbation by a suspended lateral silicon membrane. The presence of the membrane in the near-field of the optical mode changes the effective refractive index, which results in frequency tuning. The optical response of this configuration is evaluated through full 3D finite-difference time-domain (FDTD) numerical computations, for wavelengths between 1440 and 1660 nm. The effects of both lateral gap size and vertical displacement of the silicon cantilever are considered. The mechanics of the structure, in particular the silicon cantilever displacement, is also studied in depth. Both transient and steady-state responses for a step electrostatic actuation are simulated using a finite element method (FEM) software, COMSOL Multiphysics.

The proposed tuning architecture is implemented on a filter formed by a 5 $\mu$m-radius microring resonator with two bus waveguides (input/through and drop ports), with ring-bus gaps of 800 nm [35]. For these dimensions, the microring is under-coupled. For practical applications, the ring-bus gaps need to be carefully calculated, depending on the specific application. The microring resonator is perturbed from the side by vertically displacing a silicon cantilever through electrostatic actu-ation. Figure 14.6 illustrates the top view of the proposed tuning mechanism. The structures are defined in the top silicon layer of a 100 nm-thick SOI wafer, with waveguide width of 600 nm [7, 40]. The lateral gap between the cantilever and the microring is 30 nm, which allows for strong evanescent probing. The beam length $L$ is 50 $\mu$m, measured along the horizontal axis of symmetry, indicated by the dotted line in Fig. 14.6. Considerations on how to select the value for $L$ are presented later in the text. Vertical displacements of the cantilever are performed using a sweeping voltage applied across the silicon cantilever and the silicon substrate, as illustrated in Fig. 14.7.

The cantilever is designed to displace in the normal direction. Without any applied voltage, the cantilever is vertically aligned with the microring resonator. When a voltage is applied, the membrane moves downwards, away from the near-field of the optical mode.

**Fig. 14.6** Illustration of the proposed tuning mechanism (*top view*), using a cantilever defined on the same silicon layer as the microring resonator. The main dimensions used for the optical and mechanical simulations are indicated in the figure

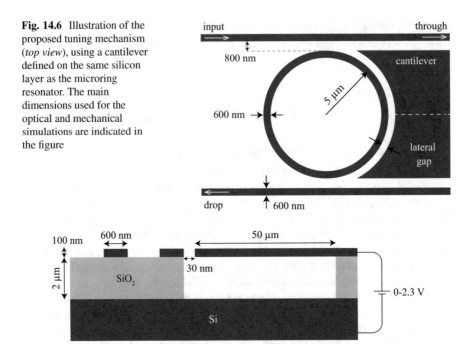

**Fig. 14.7** Cross-section of the proposed tuning mechanism

## 14.4.1 Cantilever Beam Design

The mechanical response of the cantilever can be simplified into a classical driven harmonic oscillator model, as shown in Fig. 14.8. The equation of motion that governs the tip deflection of the beam can be written as

$$m_{\text{eff}}\frac{d^2x}{dt^2} + c\frac{dx}{dt} + k_{\text{eff}}x = F(x), \tag{14.21}$$

where $m_{\text{eff}}$ is the effective mass of the cantilever beam, $k_{\text{eff}}$ the effective stiffness of the beam, and $c$ the damping coefficient. As seen in Fig. 14.7, biasing the silicon beam with a DC voltage corresponds to a system similar to a simple parallel-plate capacitor, where the electric force $F(x)$ applied across the plate is given by

$$F(x) = \frac{\epsilon_0 A V_{\text{DC}}^2}{2(h-x)^2}, \tag{14.22}$$

where $\epsilon_0$ is the electric permittivity of air, $A$ the surface area of the cantilever membrane, $V_{\text{DC}}$ the applied voltage, $h$ the initial distance of the beam from the substrate ($2\,\mu$m), and $x$ the displacement of the beam due to the electrostatic force, measured from the cantilever tip. Equation (14.22) is a simplification of the actual

**Fig. 14.8** Illustration of the cantilever modeled as a classical driven harmonic oscillator (with damping)

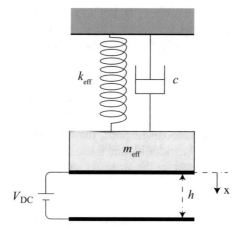

electrostatic force that act on the beam, since it does not include the effects of fringing fields, and assumes a uniform displacement of the cantilever.

Fixing the beam's width to the ring outer diameter ($W = 10.6\,\mu\text{m}$), the thickness to $t = 100\,\text{nm}$ and the initial distance between the top silicon layer and the substrate to $h = 2\,\mu\text{m}$, the length $L$ of the beam remains the only parameter to be determined. Selecting the beam's length determines the cantilever response due to the electrostatic force. Selecting the optimal beam length is important in order to (1) determine the (static) *pull-in* voltage and avoid pulling the beam towards the substrate, and (2) avoid undesirable overshoots by the cantilever due to voltage transients that may also lead to (dynamic) *pull-in* of the beam. The former is considered in the steady-state analysis of the beam, whereas the latter is considered in the respective transient analysis.

### 14.4.2  Steady-State Electromechanical Analysis

The maximum voltage that can be applied to the cantilever, before it pulls in towards the substrate, is known as the *pull-in* voltage. It is necessary to determine this voltage and ensure that the beam can reach the maximum desired displacement before pulling in towards the substrate. The main parameters that define the pull-in voltage of the cantilever are the Young modulus ($E$) and Poisson's ratio ($\nu$) ($E = 170\,\text{GPa}$ and $\nu = 0.28$ for crystalline silicon), the dimensions of the beam, and the initial distance $h$ between the positive and ground terminals. From [41], the pull-in voltage ($V_{\text{pi}}$) is determined by

$$V_{\text{pi}} = \sqrt{\frac{4\gamma_1 B}{\epsilon_0 L^4 \gamma_2^2 \left(1 + \gamma_3 \frac{h}{W}\right)}}, \tag{14.23}$$

with constants $\gamma_1 = 0.07$, $\gamma_2 = 1.00$, and $\gamma_3 = 0.42$ for a cantilever beam. The parameter $B$ is defined as $B = \tilde{E}t^3h^3$, where $\tilde{E}$ is the effective Young modulus, which takes into account effects from the cantilever geometry, defined as

$$\tilde{E} = \frac{E}{1 - \nu^2 \left[ \frac{(W/L)^{1.37}}{0.5 + (W/L)^{1.37}} \right]^{0.98(L/t)^{-0.056}}} . \tag{14.24}$$

In these calculations, the cantilever beam is assumed to be rectangular, with length $L$, width $W$, and thickness $t$. This is a good approximation to the cantilever beam used for the proposed tuning mechanism, illustrated in Fig. 14.6.

### 14.4.3  Transient Electromechanical Analysis

In order to accurately tune the resonant frequencies inside the microring, it is desirable to avoid possible cantilever oscillations that may result from its actuation or release. This is crucial for avoiding unwanted resonant wavelength filtering, maximizing the tuning operation speed, minimizing the power consumption, and avoiding any dynamic pull-in if the beam overshoots. From Eq. (14.21), the cantilever motion can be designed to be critically damped (to yield the fastest tuning speed) if the damping ratio $\zeta = 1$, where $\zeta = c/c_0$. $c$ is the damping coefficient defined earlier, and $c_0$ is the critical damping given by $c_0 = 2m_{\text{eff}}\omega_0$, where $\omega_0$ is the undamped resonant (angular) frequency of the cantilever. In general, the damping coefficient depends on the undamped resonant frequency of the beam, which is determined by the ratio of the effective stiffness to the effective mass:

$$\omega = \sqrt{\frac{k_{\text{eff}}}{m_{\text{eff}}}} . \tag{14.25}$$

The undamped and unbiased natural frequencies of a rectangular cantilever under a load at its tip can be found in [42]. Since the vertical displacement of the beam is nonuniform along its length, a constant applied voltage results in a nonuniform electric field across the beam. The effective mass of the beam at the tip is therefore smaller than the mass $m$ of the beam [42]. This effective mass is equal to $0.25m$ for the fundamental resonant frequency, and $k_{\text{eff}}$ in Eq. (14.25) is replaced by the cantilever stiffness $k = 3EI/L^3$. Since $W > 5t$ in this specific design, the Young modulus should be replaced with the effective Young modulus given by Eq. (14.24). The fundamental unbiased and undamped resonant frequency can then be expressed as

$$f_0 = \frac{\alpha^2}{2\pi} \sqrt{\frac{\tilde{E}I}{mL^3}} , \tag{14.26}$$

where $I = Wt^3/12$ is the area moment of inertia of the cantilever cross-section and $\alpha = 1.875104$ for the fundamental mode. The quantity $\alpha$ has different values for higher-order modes. However, in our analysis, it is sufficient to focus on the fundamental mode, since a DC voltage is applied to the cantilever.

For a cantilever suspended above a substrate, several intrinsic and extrinsic sources of damping can exist, in particular (1) losses into the surrounding (fluid damping), (2) into the material itself (thermoelastic), and (3) into the anchor region (anchor losses). At low pressures, fluid losses are negligible. However, at ambient pressure (1 atm), fluid damping becomes the dominant loss source. Fluid damping exists in two main forms: drag force damping and squeeze film damping [43]. The former becomes evident for extremely thin cantilevers with a high surface to volume ratio [44], and the latter dominates when the gap thickness becomes a few times smaller than the width of the cantilever [43]. Therefore, squeeze film damping becomes the dominant loss effect in most MEMS architectures, and results from squeezing the air film between the cantilever and the substrate. This creates a pressure on the cantilever which opposes its motion [42].

The squeeze film flow can be characterized by a set of quantities, namely the *Knudsen number* (*Kn*), which is the ratio of the mean free path length of the air molecules to the distance between the cantilever and the substrate, the *squeeze number* ($\sigma$), which measures the compressibility of the fluid, and the *Reynolds number* (*Re*), which gives the ratio of the inertial to viscous forces. These can be found in [45]:

$$Kn = \frac{\Lambda}{h}, \tag{14.27}$$

$$\sigma = \frac{12\mu_{\text{eff}}W^2\omega_0}{h^2 P_a}, \tag{14.28}$$

$$Re = \frac{\rho_a h^2 \omega_0}{\mu_{\text{eff}}}. \tag{14.29}$$

$\Lambda$ is the mean free path length of the air molecules (about 65 nm), $P_a$ the ambient pressure ($P_a = 101.325\,\text{kPa}$), $\rho_a$ the air density ($\rho_a = 1.2\,\text{kg/m}^3$), and $\mu_{\text{eff}}$ the effective dynamic viscosity of air, given by:

$$\mu_{\text{eff}} = \frac{\mu}{1 + 9.638 Kn^{1.159}}, \tag{14.30}$$

where $\mu$ is the dynamic viscosity of air under standard temperature and pressure conditions ($\mu = 1.8 \times 10^{-5}\,\text{Ns/m}^2$). When $0.01 < Kn < 0.1$, $\sigma < 1$ and $Re < 1$, the flow is said to lie in the slip flow regime, and both compressibility and inertial effects can be ignored. The (total) unbiased damping coefficient $c$ used in Eq. (14.21) is dominated by the squeeze film damping coefficient, which has been derived in [45]:

$$c_{SQFD} = \frac{768\mu_{\text{eff}}LW^3}{h^3\pi^6} \sum_{m,n=\text{odd}} \frac{\left(m^2\frac{\chi^2}{4} + n^2\right)b_m^2}{(mn)^2\left(\left[m^2\frac{\chi^2}{4} + n^2\right]^2 + \frac{\sigma^2}{\pi^4}\right)}, \quad (14.31)$$

where $\chi = W/L$ is the aspect ratio of the cantilever. $b_m$ is given by

$$
\begin{aligned}
b_m = \frac{1}{\beta_L}&\left[-\frac{m\pi}{2(-1)^{\frac{m-1}{2}}}\frac{2\alpha^3\gamma}{\alpha^4 - \frac{m^4\pi^4}{16}}\right.\\
&+ \frac{m^2\pi^2}{4\left(\alpha^2 - \frac{m^2\pi^2}{4}\right)}[\gamma\sin(\alpha) + \cos(\alpha)]\\
&+ \left.\frac{m^2\pi^2}{4\left(\alpha^2 + \frac{m^2\pi^2}{4}\right)}[\gamma\sinh(\alpha) + \cosh(\alpha)]\right],
\end{aligned}
\quad (14.32)
$$

with

$$\gamma = -[\cosh(\alpha) + \cos(\alpha)]/[\sinh(\alpha) + \sin(\alpha)], \quad (14.33)$$

$$\beta_L = \cosh(\alpha) - \cos(\alpha) + \gamma[\sinh(\alpha) - \sin(\alpha)]. \quad (14.34)$$

The additional stiffness corresponding to the squeeze film can also be computed [45], and is added to the spring constant $k$ to yield $k_{\text{eff}}$. As mentioned, if the squeeze number $\sigma < 1$, the compressibility effect can be ignored, and the stiffness $k$ determined earlier can be used. Equations (14.31)–(14.34) neglect any static bias deflection, and therefore the damping coefficient at a specific applied static voltage cannot be calculated using these equations. However, the damping ratio will increase at higher applied voltages due to (1) an increase in the damping coefficient that results from the reduction of the distance between the cantilever and the substrate (inversely proportional to $h^3$) and (2) a reduction in the critical damping coefficient resulting from the reduction of the cantilever resonant frequency (as a result of the electrostatic spring softening effect [46]). Hence, selecting a beam length $L$ with a critically damped response at the initial cantilever-substrate height ($2\,\mu m$) is sufficient to prevent the beam from oscillating, since the response at higher applied voltages will be over-damped ($\zeta > 1$).

The damping ratio for the unbiased fundamental mode can be obtained from the damping coefficient in Eq. (14.31) and from the critical damping $c_0$, yielding:

$$\zeta = \frac{384\mu_{\text{eff}}W^2}{0.25t\rho_{\text{Si}}\omega_0h^3\pi^6} \sum_{m,n=\text{odd}} \frac{\left(m^2\frac{\chi^2}{4} + n^2\right)b_m^2}{(mn)^2\left(\left[m^2\frac{\chi^2}{4} + n^2\right]^2 + \frac{\sigma^2}{\pi^4}\right)}, \quad (14.35)$$

**Fig. 14.9** Damping ratio (*dashed curve*) and pull-in voltage (*solid curve*) for different cantilever beam lengths. Critical damping occurs at a length around 50 μm

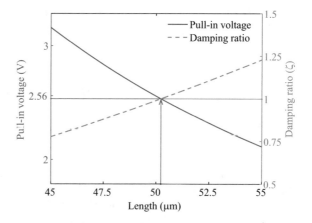

**Table 14.1** Summary of main characteristic parameters computed for $L = 50\,\mu$m: unbiased fundamental frequency, Reynold, Knudsen, and squeeze numbers, pull-in voltage, damping ratio, and minimum settling time (within 1 % of final value)

| $L$ (μm) | $f_0$ (kHz) | $Re$ | $Kn$ | $\sigma$ | $V_{pi}$ (V) | $\zeta$ | Settling time (μs) |
|---|---|---|---|---|---|---|---|
| 50 | $\leq$56 | 0.110 | 0.032 | 0.018 | 2.55 | $\geq$1 | $\geq$20 |

where the cantilever material density is $\rho_{Si} = 2330\,\mathrm{kg/m^3}$. A plot of the damping ratio for cantilever lengths between 45 and 55 μm is shown in Fig. 14.9 (dashed curve). The pull-in voltage is also plotted for the same beam lengths (solid curve). The longer the cantilever beam, the higher the damping ratio, and the smaller the pull-in voltage. The beam length that corresponds to a critically damped system ($\zeta = 1$) is around 50 μm (at the intersection of the two curves), with a pull-in voltage around 2.55 V. The critical damping condition has the shortest settling time (time needed to reach a specific final value within a defined percentage). Higher damping ratios correspond to longer settling times. Table 14.1 summarizes some of the characteristic parameters obtained for $L = 50\,\mu$m: unbiased fundamental resonant frequency, Reynold number, Knudsen number, squeeze number, pull-in voltage, expected damping ratio, and minimum expected settling time (within 1 % of final value). The settling time is estimated by solving Eq. (14.21), using a unit step function for the external force. The $\leq$ and $\geq$ signs refer to a decrease or increase in the corresponding parameters, occurring when the cantilever is biased.

## 14.4.4 Numerical Simulations

The electromechanical response (steady-state and transient) of the cantilever for different applied voltages is computed through the finite element method.

Finite-difference time-domain simulations are also performed in order to determine the resonant wavelength shifts in the microring resonator, for different cantilever heights.

### 14.4.4.1   Finite Element Method Electromechanical Simulations

For the steady-state response, the vertical displacement of the cantilever as a function of the applied voltage is computed using the Maxwell stress tensor, which represents a more accurate description of the distributed load than the simplified force given by Eq. (14.22). For the transient computations, the full Navier–Stokes equations are solved, coupled with the force calculated from the Maxwell stress tensor for the electrostatic solution, yielding the beam response as a function of time. In the transient analysis, wall boundaries far away from the sides are used.

Numerical results reveal that the cantilever tip at the center (dotted line in Fig. 14.6) is displaced by approximately 500 nm when the applied voltage reaches 2.3 V. The beam pulls in towards the substrate, and no longer operates, when the applied voltage reaches and exceeds 2.55 V, which is consistent with the estimated pull-in voltage calculated before. Due to the circular end shape of the beam, the tip experiences slightly different vertical displacements at the edge and at the center of the cantilever (around 10 % difference at 2.3 V). The steady-state beam deformation is illustrated in Fig. 14.10, for a few different applied voltages (0, 1.2, 1.7, and 2.3 V). The steady-state tip displacement as a function of the applied voltage is displayed in Fig. 14.11 (solid curve). The biased cantilever fundamental resonant frequency is also simulated, and shown in the figure (dashed curve). The reduction in the fundamental frequency is evident, and is due to the electrostatic spring softening effect that occurs when the cantilever is biased.

The damped response of the cantilever (solid curves) for a voltage step function (dashed curves) is shown in Fig. 14.12, for 0–1 V (at $t = 0 \mu s$) and 1–0 V (at $t = 50 \mu s$) in (a), and 0–2.3 V (at $t = 0 \mu s$) and 2.3–0 V (at $t = 100 \mu s$) in

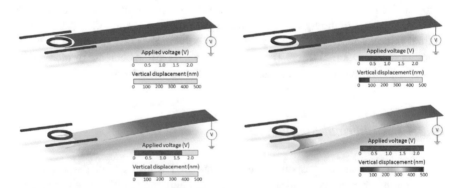

**Fig. 14.10** Illustration of the steady-state cantilever deformation, for applied voltages of 0, 1.2, 1.7, and 2.3 V

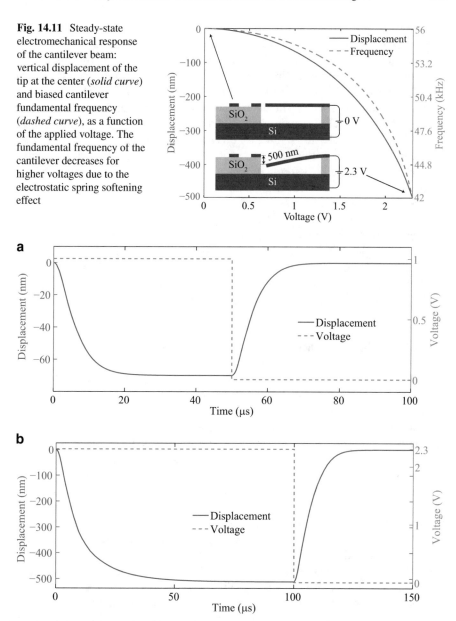

**Fig. 14.11** Steady-state electromechanical response of the cantilever beam: vertical displacement of the tip at the center (*solid curve*) and biased cantilever fundamental frequency (*dashed curve*), as a function of the applied voltage. The fundamental frequency of the cantilever decreases for higher voltages due to the electrostatic spring softening effect

**Fig. 14.12** Mechanical transient response of the cantilever beam for an applied step voltage of (**a**) 0–1 and 1–0 V, and (**b**) 0–2.3 and 2.3–0 V

(b). The settling time values (within 1 % of final value) obtained from the figure are approximately 22 μs (for 0–1 V), 21 μs (for 1–0 V), 54 μs (for 0–2.3 V), and 22 μs (for 2.3–0 V). The over-damped response of the cantilever is evident in the transitions to large voltages (e.g., 0–2.3 V), corresponding to larger settling times that result from the smaller air gap between the cantilever and the substrate. It is important to note that the (unbiased) settling times obtained from the FEM simulations are higher than the ones predicted in the previous section, which only accounted for the squeeze film damping effect. In the full FEM simulations, additional damping sources should be incorporated (such as thermoelastic and anchor). Although not dominant, these additional loss mechanisms could account for 5–10 % of the total settling time, for this particular configuration.

The computed Von Mises stress, which value indicates whether or not a specific material will plastically deform, is around 4.8 MPa at the anchor point of the cantilever with the silica, when 2.3 V is applied. This value is below the yield stress of crystalline silicon, at about 7 GPa [47]. Hence, fracture is not likely to occur when the voltage is swept between 0 and 2.3 V.

Ideally, the cantilever does not consume power at a static deflection. However, power is dissipated during actuation, for changing its position from one state to another. The stored energy in the cantilever at a static deflection is in the form of electrical and strain energy. This total stored energy is around 1.75 fJ when 1 V is applied, and around 10.5 fJ when 2.3 V is applied.

### 14.4.4.2   Finite-Difference Time-Domain Optical Simulations

The effect of both the lateral gap and vertical displacement of the cantilever on the optical resonant frequency of the microring resonator is studied through 3D FDTD simulations [48], for the transverse-electric (TE) polarized mode. Figure 14.13a shows the shift in the resonant wavelength (of different modes) with the membrane lateral gap, when compared to the unperturbed values (gap far away from the microring). A lateral gap of 30 nm is seen to enable wavelength shifts of about 13 nm for resonances around 1550 nm, with larger shifts up to 19 nm for longer wavelengths. For this fixed lateral gap, the microring can now be tuned by displacing the cantilever in the vertical direction. Figure 14.13b shows the corresponding wavelength shifts due to the vertical membrane displacement, for a fixed lateral gap of 30 nm. Resonances at longer wavelengths experience larger shifts, since the corresponding mode is less confined in the waveguide core and extends further into the cladding. The FDTD results from Fig. 14.13b can be combined with the FEM steady-state cantilever displacement results from Fig. 14.11 (solid curve), in order to plot the change in the resonant wavelength of the microring resonator due to the applied voltage, in Fig. 14.13c. When the cantilever is close to the microring, a portion of the optical mode leaks into the beam. This leakage results in a lower quality factor, which can be a limiting aspect for this tuning configuration in some applications. In order to visualize how both the wavelength and quality factor of the resonator are affected by the cantilever, the intensity drop-port spectra for different vertical displacements (with a lateral gap of 30 nm) are computed for the resonance around 1550 nm. The respective surface color map (dB scale) is shown in

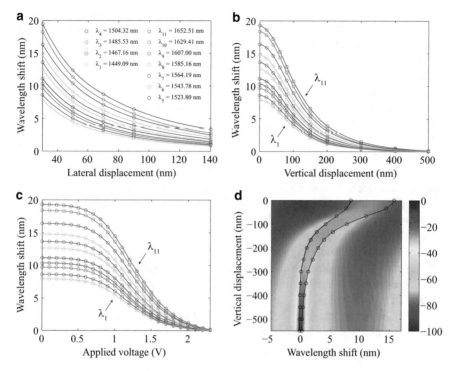

**Fig. 14.13** Tuning effect for different cantilever positions: (**a**) wavelength shift as a function of the lateral gap, (**b**) wavelength shift as a function of the cantilever vertical displacement, for a lateral gap of 30 nm, (**c**) wavelength shift as a function of the applied voltage, and (**d**) surface color map of the drop-port spectra for different cantilever vertical displacements (lateral gap of 30 nm), for the resonance around 1550 nm. The intensity map is in dB scale, and the *shaded region* represents the full-width at half-maximum

Fig. 14.13d. The shaded region represents the full-width at half-maximum (FWHM) of the resonance for the different displacements, which decreases (increase in quality factor) as the cantilever moves towards the substrate.

## 14.4.5   System Limitations

For implementing a system based on the current tuning architecture, it is necessary to understand and mitigate the effect of noise in the system, as well as other limitations. Some of such limitations and their undesirable effects are mentioned in this subsection. These may rise from operation under harsh environment conditions, such as high temperature, humidity, or pressure.

First, as mentioned before, if the DC voltage applied to the cantilever exceeds the pull-in voltage, the mechanical restoring force of the cantilever can no longer overcome its opposing electric force, resulting in static pull-in.

For a mass-spring oscillator, the root-mean-square (rms) displacement resulting from thermal agitation is written as $\langle x \rangle = \sqrt{k_B T / k}$ [49], where $k_B$ is the Boltzmann constant, $T$ the absolute temperature, and $k$ the spring stiffness. For the current cantilever design, the estimated rms displacement due to thermal noise at room temperature is around 1 nm, which is not significant. However, a different combination of temperature/stiffness may become a limiting factor if the resulting rms displacement is larger.

Due to humidity, condensation of water molecules between the silicon cantilever and the substrate may occur, giving rise to capillary forces. If these capillary forces exceed the restoring mechanical force of the cantilever, they will cause stiction and the beam may eventually stick to the substrate [42]. Hence, humid conditions should be avoided for optimal operation.

When operating under higher pressures, the effect of the air layer between the cantilever and the substrate increases, translated into higher squeeze film damping. This increase affects the settling time, and must be considered for determining the maximum tuning speeds.

Finally, it is important to keep in mind minimum feature sizes, and adjust the design to match possible fabrication constraints.

## 14.5   Conclusion

This chapter has studied the design and numerical validation (both mechanical and optical) of a frequency tuning mechanism based on a MEMS implementation, using a cantilever that probes the evanescent field of the optical mode of a microring resonator. The effects of both lateral gap size and vertical displacement of the silicon cantilever are computed for wavelengths around 1550 nm (for the TE mode). The results have shown a maximum tuning range of the center resonant wavelength of about 13 nm, for a lateral gap of 30 nm. The vertical displacement of the cantilever membrane can be controlled by electrical actuation for low-power tuning, with a sweeping voltage between 0 and 2.3 V for vertical displacements of 0–500 nm. For DC operation, settling times (within 1 % of final value) are between 21 and 54 μs. No power is required to keep the cantilever at a specific position (under ideal conditions), and stored energies can be as low as several fJ. This is a promising approach that could play a significant role in building power-efficient ROADMs for WDM applications, enabling high-bandwidth connectivity for on-chip and chip-to-chip architectures in 3D integrated chip stacks.

# References

1. G.T. Reed, Device physics: the optical age of silicon. Nature **427**(6975), 595–596 (2004)
2. R. Soref, The past, present, and future of silicon photonics. IEEE J. Sel. Top. Quantum Electron. **12**(6), 1678–1687 (2006)
3. M. Lipson, Guiding, modulating, and emitting light on silicon-challenges and opportunities. J. Lightwave Technol. **23**(12), 4222–4238 (2005)
4. G. Keiser, *Optical Communications Essentials* (McGraw-Hill, New York, 2003)
5. A. Khilo, S.J. Spector, M.E. Grein, A.H. Nejadmalayeri, C.W. Holzwarth, M.Y. Sander, M.S. Dahlem, M.Y. Peng, M.W. Geis, N.A. DiLello et al., Photonic ADC: overcoming the bottleneck of electronic jitter. Opt. Express **20**(4), 4454–4469 (2012)
6. T. Barwicz, M.A. Popović, F. Gan, M.S. Dahlem, C.W. Holzwarth, P.T. Rakich, E.P. Ippen, F.X. Kärtner, H.I. Smith, Reconfigurable silicon photonic circuits for telecommunication applications, in *Lasers and Applications in Science and Engineering* (International Society for Optics and Photonics, Bellingham, 2008), p. 68 720Z
7. M.S. Dahlem, C.W. Holzwarth, A. Khilo, F.X. Kärtner, H.I. Smith, E.P. Ippen, Reconfigurable multi-channel second-order silicon microring-resonator filterbanks for on-chip WDM systems. Opt. Express **19**(1), 306–316 (2011)
8. B. Moslehi, J.W. Goodman, M. Tur, H.J. Shaw, Fiber-optic lattice signal processing. Proc. IEEE **72**(7), 909–930 (1984)
9. K.P. Jackson, S.A. Newton, B. Moslehi, M. Tur, C.C. Cutler, J.W. Goodman, H. Shaw, Optical fiber delay-line signal processing. IEEE Trans. Microwave Theory Tech. **33**(3), 193–210 (1985)
10. C.K. Madsen, J.H. Zhao, *Optical Filter Design and Analysis* (Wiley-Interscience, New York, 1999)
11. K. Okamoto, *Fundamentals of Optical Waveguides* (Academic, New York, 2010)
12. M.A. Popovic, T. Barwicz, M.S. Dahlem, F. Gan, C.W. Holzwarth, P.T. Rakich, H.I. Smith, E.P. Ippen, F.X. Krtner, Tunable, fourth-order silicon microring-resonator add-drop filters, in *ECOC*, 2007
13. B. Little, H. Haus, J. Foresi, L. Kimerling, E. Ippen, D. Ripin, Wavelength switching and routing using absorption and resonance. IEEE Photon. Technol. Lett. **10**(6), 816–818 (1998)
14. K. Shore, D. Chan, Kramers-Kronig relations for nonlinear optics. Electron. Lett. **26**(15), 1206–1207 (1990)
15. B.E. Saleh, M.C. Teich, *Fundamentals of Photonics*. Wiley Series in Pure and Applied Optics (Wiley, New York, 2007)
16. R.A. Soref, B.R. Bennett, Electrooptical effects in silicon. IEEE J. Quantum Electron. **23**(1), 123–129 (1987)
17. Q. Xu, B. Schmidt, S. Pradhan, M. Lipson, Micrometre-scale silicon electro-optic modulator. Nature **435**(7040), 325–327 (2005)
18. K. Djordjev, S.-J. Choi, S.-J. Choi, P. Dapkus, Microdisk tunable resonant filters and switches. IEEE Photon. Technol. Lett. **14**(6), 828–830 (2002)
19. S. Schönenberger, T. Stöferle, N. Moll, R.F. Mahrt, M.S. Dahlem, T. Wahlbrink, J. Bolten, T. Mollenhauer, H. Kurz, B. Offrein, Ultrafast all-optical modulator with femtojoule absorbed switching energy in silicon-on-insulator. Opt. Express **18**(21), 22485–22496 (2010)
20. K. Djordjev, S.-J. Choi, S.-J. Choi, P. Dapkus, Vertically coupled InP microdisk switching devices with electroabsorptive active regions. IEEE Photon. Technol. Lett. **14**(8), 1115–1117 (2002)
21. A. Guarino, G. Poberaj, D. Rezzonico, R. Degl'Innocenti, P. Günter, Electro-optically tunable microring resonators in lithium niobate. Nat. Photon. **1**(7), 407–410 (2007)
22. J. Prost, *The Physics of Liquid Crystals*, vol. 83 (Oxford University Press, Oxford, 1995)
23. W. De Cort, J. Beeckman, R. James, F.A. Fernández, R. Baets, K. Neyts, Tuning of silicon-on-insulator ring resonators with liquid crystal cladding using the longitudinal field component. Opt. Lett. **34**(13), 2054–2056 (2009)

24. J. Komma, C. Schwarz, G. Hofmann, D. Heinert, R. Nawrodt, Thermo-optic coefficient of silicon at 1550 nm and cryogenic temperatures. Appl. Phys. Lett. **101**(4), 041905 (2012)
25. G. Cocorullo, F. Della Corte, I. Rendina, Temperature dependence of the thermo-optic coefficient in crystalline silicon between room temperature and 550 K at the wavelength of 1523 nm. Appl. Phys. Lett. **74**(22), 3338–3340 (1999)
26. F.G. Della Corte, G. Cocorullo, M. Iodice, I. Rendina, Temperature dependence of the thermo-optic coefficient of InP, GaAs, and SiC from room temperature to 600 K at the wavelength of 1.5 μm. Appl. Phys. Lett. **77**(11), 1614–1616 (2000)
27. F. Gan, T. Barwicz, M. Popovic, M. Dahlem, C. Holzwarth, P. Rakich, H. Smith, E. Ippen, F. Kärtner, Maximizing the thermo-optic tuning range of silicon photonic structures, in *Photonics in Switching* (IEEE, San Francisco, CA, 2007), pp. 67–68
28. P. Dong, W. Qian, H. Liang, R. Shafiiha, N.-N. Feng, D. Feng, X. Zheng, A.V. Krishnamoorthy, M. Asghari, Low power and compact reconfigurable multiplexing devices based on silicon microring resonators. Opt. Express **18**(10), 9852–9858 (2010)
29. M.R. Watts, W.A. Zortman, D.C. Trotter, G.N. Nielson, D.L. Luck, R.W. Young, Adiabatic resonant microrings (ARMs) with directly integrated thermal microphotonics, in *Conference on Lasers and Electro-Optics* (Optical Society of America, Washington, DC, 2009), p. CPDB10
30. P.T. Rakich, M.A. Popovic, M.R. Watts, T. Barwicz, H.I. Smith, E.P. Ippen, Ultrawide tuning of photonic microcavities via evanescent field perturbation. Opt. Lett. **31**(9), 1241–1243 (2006)
31. G.N. Nielson, D. Seneviratne, F. Lopez-Royo, P.T. Rakich, Y. Avrahami, M.R. Watts, H. Haus, H.L. Tuller, G. Barbastathis, Integrated wavelength-selective optical MEMS switching using ring resonator filters. IEEE Photon. Technol. Lett. **17**(6), 1190–1192 (2005)
32. S. Abdulla, L. Kauppinen, M. Dijkstra, M. De Boer, E. Berenschot, H. Jansen, R. De Ridder, G. Krijnen, Tuning a racetrack ring resonator by an integrated dielectric MEMS cantilever. Opt. Express **19**(17), 15864–15878 (2011)
33. H. Shoman, M.S. Dahlem, Electrically-actuated cantilever for planar evanescent tuning of microring resonators in SOI platforms, in *International Conference on Optical MEMS and Nanophotonics (OMN), 2014* (IEEE, Glasgow, Scotland, 2014), pp. 141–142
34. C. Errando-Herranz, F. Niklaus, G. Stemme, K.B. Gylfason, A low-power MEMS tunable photonic ring resonator for reconfigurable optical networks, in *28th IEEE International Conference on Micro Electro Mechanical Systems (MEMS)* (IEEE, Estoril, Portugal, 2015), pp. 53–56
35. H. Shoman, M.S. Dahlem, Architectures for evanescent frequency tuning of microring resonators in micro-opto-electro-mechanical SOI platforms, in *SPIE OPTO*, International Society for Optics and Photonics, 2015, p. 936706
36. T. Mamdouh, D. Khalil, A MEMS tunable optical ring resonator filter. Opt. Quant. Electron. **37**(9), 835–853 (2005)
37. M.-C.M. Lee, M.C. Wu, MEMS-actuated microdisk resonators with variable power coupling ratios. IEEE Photon. Technol. Lett. **17**(5), 1034–1036 (2005)
38. J. Yao, D. Leuenberger, M.-C.M. Lee, M.C. Wu, Silicon microtoroidal resonators with integrated MEMS tunable coupler. IEEE J. Sel. Top. Quantum Electron. **13**(2), 202–208 (2007)
39. B. Little, S.T. Chu, Theory of loss and gain trimming of resonator-type filters. IEEE Photon. Technol. Lett. **12**(6), 636–638 (2000)
40. M.A. Popovic, T. Barwicz, E.P. Ippen, F.X. Kärtner, Global design rules for silicon microphotonic waveguides: sensitivity, polarization and resonance tunability, in *Conference on Lasers and Electro-Optics* (Optical Society of America, Washington, DC, 2006), p. CTuCC1
41. R.K. Gupta, Electrostatic pull-in test structure design for in-situ mechanical property measurements of microelectromechanical systems (MEMS). Ph.D. dissertation, Massachusetts Institute of Technology, 1998
42. M.I. Younis, *MEMS Linear and Nonlinear Statics and Dynamics*, vol. 20 (Springer, Berlin, 2011)
43. M. Bao, H. Yang, Squeeze film air damping in MEMS. Sens. Actuators A Phys. **136**(1), 3–27 (2007)

44. J. Yang, T. Ono, M. Esashi, Energy dissipation in submicrometer thick single-crystal silicon cantilevers. J. Microelectromech. Syst. **11**(6), 775–783 (2002)
45. A.K. Pandey, R. Pratap, Effect of flexural modes on squeeze film damping in MEMS cantilever resonators. J. Micromech. Microeng. **17**(12), 2475–2484 (2007)
46. C. Liu, *Foundations of MEMS* (Pearson Education Limited, Essex, England, 2012)
47. K.E. Petersen, Silicon as a mechanical material. Proc. IEEE **70**(5), 420–457 (1982)
48. A.F. Oskooi, D. Roundy, M. Ibanescu, P. Bermel, J.D. Joannopoulos, S.G. Johnson, MEEP: a flexible free-software package for electromagnetic simulations by the FDTD method. Comput. Phys. Commun. **181**(3), 687–702 (2010)
49. T.B. Gabrielson, Mechanical-thermal noise in micromachined acoustic and vibration sensors. IEEE Trans. Electron Devices **40**(5), 903–909 (1993)

# Chapter 15
# Athermal Photonic Circuits for Optical On-Chip Interconnects

**Peng Xing and Jaime Viegas**

## 15.1 Background

During the last four decades, integrated electronic circuits have achieved great success, changing our lives significantly. The integration level of electronic circuitry is steadily increasing, with billions of transistors integrated in a single CPU chip. However, restricted by quantum effects, transistor sizes cannot be made infinitely small. Also due to power dissipation issues, alternative materials, processes, and architectures have been sought to solve the near-future computational scaling bottleneck.

Several methods are proposed to overcome this bottleneck. One promising method is to find a substitute material based on which the boundary could be pushed and the devices could be made smaller. Graphene [1], a single atomic layer of carbon is a very strong candidate, as transistors based on graphene could be made atomically thin, and still retain an electronic mobility higher than can be achieved in silicon, pushing the IC operating frequency higher. Until now, a transistor with the size of 10 atoms [2] has been reported. However graphene electronics is only a reality within the laboratory and major developments in the growth of high quality single crystal monoatomic graphene sheets are required for mass production of graphene electronics with high-yield.

A distinct approach is quantum computing [3], where the paradigm is totally changed. The theoretical potential is huge but the challenge of building a compact quantum computer with the level of complexity of current ULSI devices operating at room temperature seems hard to tackle within the next decades.

P. Xing • J. Viegas (✉)
Department of Electrical Engineering and Computer Science, Institute Center for Microsystems (iMicro), Masdar Institute of Science and Technology, Abu Dhabi, United Arab Emirates
e-mail: pxing@masdar.ac.ae; jviegas@masdar.ac.ae

© Springer International Publishing Switzerland 2016
I.M. Elfadel, G. Fettweis (eds.), *3D Stacked Chips*,
DOI 10.1007/978-3-319-20481-9_15

Another approach to continue the progress of electronic systems is to improve the data transfer between and within the microchips [4].

Photonics has many advantages over electronics in data communication. First, optical communication allows higher data transfer rate due to its high carrier frequency, in the order of hundreds of terahertz. Second, the optical signal is more stable as it is not influenced by electromagnetic interference. Third, optical communication may consume less energy, depending on the bitrate-channel length product. The scalability of the optical channel is more than ten thousand times larger than copper line.

In summary, the best choice for continuing to improve the performance of electronic integrated circuits is to integrate photonics and electronics on the same chip, with the photonic circuit taking the role of transmitting data between different electronic circuits.

Some photonic devices such as modulators, filters, and lasers based on III–V materials have already entered the market in the last several decades, being the backbone of such commodities as fiber-to-the-home (FTTH) data communication and the hardware support for the World Wide Web. For classic telecommunication applications in which different components are separated and linked by fiber, these devices are the best candidates. However, for photonic integrated circuits fabricated in the same platform as the electronic devices, III–V based components are usually not an option due to process incompatibility with silicon manufacturing lines and high-cost.

Using silicon as the optical medium makes it possible to integrate photonic and electronic devices in the same substrate. As the most widely used material in current semiconductor industry, silicon is intrinsically CMOS compatible. Furthermore, silicon is transparent in the wavelength range used in telecommunications (wavelengths longer than $1.3 \mu m$), enabling direct coupling of optical fiber technology (used in medium and long distance telecommunications) with the computational hardware backbone of data centers, servers, and supercomputing clusters.

Additionally, in silicon-on-insulator (SOI) substrates, silicon has a high refractive index than the underlying silica. This allows more compact devices built with sub-micrometer waveguides and sharp bends, therefore leading to greater photonic integration [5, 6].

In summary, silicon photonics could provide a platform for integration of photonic and electronic devices in the same chip for optical interconnect applications which enable high data streams within and between microchips with small footprint.

### 15.1.1 Challenges in Silicon Photonics

Although silicon photonics has so many advantages, there are still many challenges for such technology.

Due to its indirect band gap, silicon is a very inefficient light emitter. Generally, photon emission in semiconductors is the result of the recombination of hole-electron pairs. This process is called radiative recombination; but for silicon, due to the indirect band gap, the radiative process probability is reduced by the fact that this process requires the mediation of a phonon, a quantized lattice vibration of the semiconductor crystal, to meet the requirement of momentum conservation. Meanwhile, non-radiative processes such as Auger recombination have a higher probability of occurrence, with faster recombination times. Therefore, the light emitting efficiency of silicon is extremely low and not enough for serving as the light source. This is in stark contrast with semiconductor crystals of elements belonging to group III and V of the periodic table of the elements, and their varying alloys. Current technology requires the use of an external light source, based on these III–V materials, something that can be accomplished with packaging level integration or even at wafer/die level, with bonding techniques.

The high guiding propagation losses caused by scattering on the sidewalls of the waveguides are also a serious challenge for silicon photonics. A large number of devices are designed based on submicron waveguides. Devices with higher losses consume more energy and have higher power requirements for the light source. This will also block the applications of some small size waveguides.

The refractive index of the silicon is highly sensitive to temperature. The thermo-optic coefficient for silicon is three times larger than other semiconductor materials. This is good for the applications in temperature sensors. But when the photonic devices are integrated with electronic devices, the performance of the photonic devices will be affected by the fluctuation of the substrate temperature. Some methods have been proposed to solve this problem. But they are neither energy efficient nor CMOS compatible.

The high refractive index of silicon allows high confinement of the light in the silicon waveguide but the effective index of the waveguide is strongly dependent on the geometry of the waveguide cross-section. Therefore, silicon photonic devices are very sensitive to fabrication variability. Taking the Mach–Zehnder interferometer (MZI) as an example, the change of the width or the height of the waveguide in any arm will lead to the change of the optical length of that arm and the transmission spectra.

The high loss of the optical connections between the fiber and integrated circuits is another problem. In general, there are two ways to couple the light into or out of integrated circuits. One is to couple the light directly into the device via a cleaved facet. Another one is vertical coupling with the help of a grating coupler. However these two methods do not have very high energy efficiency and the manual alignment process is not efficient for mass production.

## 15.2  Towards Thermally Stable Photonic Circuits

A few techniques have been developed for stabilizing the temperature dependent behavior of optical devices and circuits, based on design with different materials, usage of active heating and geometrical design of light paths that passively compensate for thermal varying loads. In all cases we assume a temperature variation in time, on the timescale of milliseconds to tens of seconds, but uniform in the spatial extend of the optical device.

### 15.2.1 Tuning Thermal Optical Coefficients

Optical devices based on silicon photonics have great sensitivity to thermal variations [7, 8]. Thermal fluctuations lead to a change in refractive index in semiconductors and dielectrics. The high sensitivity to thermal changes has been used in the design of high-resolution temperature sensors [9]. However thermal fluctuations are a major concern in optical systems, as they lead to phase-noise that unbalances the optimal output of an optical system, inducing intensity or spectral fluctuations that impair the system performance. Many approaches have been proposed and successfully implemented to address the high temperature sensitivity of silicon photonics devices. One successful approach is to stabilize the temperature of the device by heating it using a Ti heater which is fabricated on top of the device [10]. This kind of device is not very complicated to fabricate. However, the Ti heater consumes a significant amount of energy and raises the temperature of neighboring areas on the chip. Since so many efforts have been made to decrease the operating temperature of integrated circuits, this approach nullifies those efforts.

Another method is to cover the silicon photonic devices with some polymers which has negative thermo-optic coefficient and could counteract the positive thermo-optic coefficient of silicon [11, 12]. For example, if a SOI waveguide is covered with this kind of polymer, its effective index will not change with the fluctuation of temperature as the refractive indexes of the silicon and polymer change in different direction. Unfortunately, polymers are not compatible with standard CMOS technology and cannot be used in massive production. More research is under way on searching for appropriate materials.

### 15.2.2 Athermal Photonic Circuits by System Design

The most promising solution was proposed by Uenuma and Motooka [13] where a temperature-independent silicon MZI was realized utilizing a combination of wide and narrow waveguides. Although the two arms of the interferometer have different optical path, their optical path will have the same response even in the presence of temperature fluctuations, so the optical path difference will be stable as well as the transmitting wavelength. Although this idea was novel, the authors did not fully take into account the dependence of the waveguide effective index on the temperature and wavelength. Therefore their fabricated device still presented a significant dependency on temperature. Based on the idea from Uenuma and Motooka, Guha, and co-workers [14] proposed an improved design with better performance. They calculated the effective index for the modes of different waveguide widths using a full vector finite element solver. Their fabricated device had a very small spectral shift with temperature (5 pm/K). But their design also has some shortcomings. The loss is relatively high due to utilization of very narrow waveguide with a width less than 200 nm. Also, the footprint is rather large which is not conducive to compact integration.

### 15.2.2.1 Proposed Athermal Interferometer for Filter Implementation

In a MZI, assuming that the effective index and length of two arms are $n_1$, $L_1$ and $n_2$, $L_2$, respectively, at temperature $T$, then the phase condition for destructive interference at wavelength $\lambda_0$ can be expressed as:

$$m\lambda_0 = n_1 L_1 - n_2 L_2 \qquad (15.1)$$

Here $m$ is a half integer for the destructive interference. If the temperature changes by $\Delta T$ and the consequential spectral shift is $\Delta\lambda$, the effective index at temperature $T + \Delta T$ and wavelength $\lambda_0 + \Delta\lambda$ is $n + \partial n/\partial T \cdot \Delta T + \partial n/\partial\lambda \cdot \Delta\lambda$. Then Eq. (15.1) can be expressed as:

$$m(\lambda_0 + \Delta\lambda) = \left(n_1 + \frac{\partial n_1}{\partial T}\Delta T + \frac{\partial n_1}{\partial\lambda}\Delta\lambda\right)L_1 - \left(n_2 + \frac{\partial n_2}{\partial T}\Delta T + \frac{\partial n_2}{\partial\lambda}\Delta\lambda\right)L_2 \qquad (15.2)$$

The MZI's spectral shift with temperature at wavelength $\lambda_0$ can be derived by combining Eqs. (15.1) and (15.2) as shown in Eq. (15.3) in which $n_{g,1}$ and $n_{g,2}$ correspond to the group index of two arms

$$\frac{\Delta\lambda}{\Delta T} = \lambda_0 \frac{\frac{\partial n_1}{\partial T} \cdot L_1 - \frac{\partial n_2}{\partial T} \cdot L_2}{n_{g,1} \cdot L_1 - n_{g,2} \cdot L_2} \Rightarrow 0 \qquad (15.3)$$

To make the MZI temperature insensitive, the spectral shift with temperature should be zero. Then, a first requirement for athermal operation as proposed by Wang et al. [11] and Teng et al. [12] can be derived as Eq. (15.4). However, the behavior of the MZI designed following the first requirement has a great dependency on the wavelength which means that the spectral shift with temperature at $\lambda_0$ is zero and increases very fast when the wavelength diverges from $\lambda_0$.

$$\frac{\partial n_1}{\partial T} \cdot L_1 - \frac{\partial n_2}{\partial T} \cdot L_2 = 0 \qquad (15.4)$$

Two approaches are deployed to derive the second requirement for designing a broadband athermal MZI. The first one is described by Eq. (15.5) where the free spectral range (FSR) of the MZI is defined by Eq. (15.6). If the spectral shift with temperature at spectral minima $\lambda_0$ is zero by construction, Eq. (15.5), and the FSR change with temperature is set to zero, the spectral shift at the spectral minima $\lambda_0 \pm \text{FSR}$ should also be zero. By this way, zero spectral shift with temperature can be achieved over a wide spectral range.

$$\frac{\partial \text{FSR}}{\partial T} = 0 \tag{15.5}$$

$$\text{FSR} = \frac{\lambda^2}{n_{g,1} \cdot L_1 - n_{g,2} \cdot L_2} \tag{15.6}$$

Another approach to derive the second requirement for athermal operation is to solve Eq. (15.7) in which the wavelength dependency of the device temperature sensitivity is brought to zero. Both these approaches described in Eqs. (15.5) and (15.7) give the same result shown in Eq. (15.8) which is called the second requirement for athermal operation.

$$\frac{\partial (\Delta\lambda / \Delta T)}{\partial \lambda} = 0 \tag{15.7}$$

$$\frac{\partial^2 n_1}{\partial T \partial \lambda} \cdot L_1 - \frac{\partial^2 n_2}{\partial T \partial \lambda} \cdot L_2 = 0 \tag{15.8}$$

In the design of an integrated MZI, the first step is to set its waveguide profiles which include the material and the geometry. In silicon photonics, silicon is used as the material of the waveguide core, surrounded by silica cladding. The present discussion is about minimizing the effect of the high thermo-optical coefficient of silicon with a judicious choice of the geometrical layout of the waveguides composing the photonic device.

The most commonly used SOI platforms for silicon photonics use a silicon device layer of 220 nm on a buried silicon oxide 2 μm deep, all on a silicon handle ranging in thickness 500–800 μm. The waveguide cross-section on the aforementioned SOI platform is shown in Fig. 15.1a as the constituent elements of the MZI. In SOI based waveguides, it is usually hard to change the height of the waveguide due to process constraints. So the only easily changeable dimension of the waveguides is the waveguide width. For our design, the first step is simplified as setting the waveguide width. We start with a simple design that the waveguide in the first arm of MZI has width $w_1$, length and the second arm has $w_2$, $L_2$, assuming light mode is TE$_0$. To make it temperature insensitive and broadband, the lengths $L_1$ and $L_2$ must fulfill the design constraints including FSR requirement in Eq. (15.6) and two requirements for athermal operation in Eqs. (15.4) and (15.8) in which $\partial n_{\text{eff}}/\partial T$, $\partial^2 n_{\text{eff}}/\partial \lambda \partial T$ and group index can be computed with a numerical method.

If two arms in the MZI have the same waveguide $w_1 = w_2$, Eqs. (15.4) and 15.8 will give the solution $L_1 = L_2$ meaning that the two arms are identical and this cannot meet the FSR requirement. Thus the waveguide width in the two arms should be different $w_1 \neq w_2$. When the waveguides are different, the two requirements for

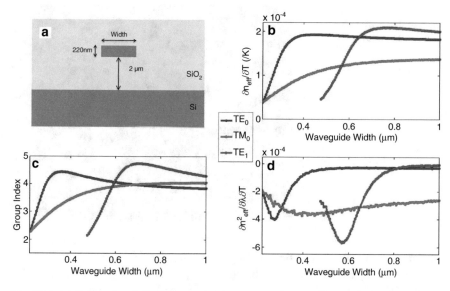

**Fig. 15.1** (**a**) Schematic of the silicon-on-insulator waveguide. (**b**) Thermo-optic coefficient of $TE_0$, $TM_0$, and $TE_1$ mode of the waveguide as a function of waveguide width. (**c**) Group index of $TE_0$, $TM_0$, and $TE_1$ mode of the waveguide as a function of the waveguide width. (**d**) Dependency of thermo-optic coefficient on wavelength of $TE_0$, $TM_0$, and $TE_1$ mode of the waveguide as a function of waveguide width

athermal operation, Eqs. (15.4) and (15.8), will give one set of solutions ($L_1$, $L_2$) resulting in that the MZI will have a fixed FSR. However, in some applications, the control over the FSR of the MZI in the design is necessary. To enable independent FSR tuning while maintaining athermal operation, another waveguide (width $w_3$ and length $L_3$) is introduced in the second arm. Then the two requirements for athermal operation become the first two equations in Eq. (15.9). There are infinite number of solutions ( $L_1$, $L_2$, $L_3$) for these two equations. Therefore, the FSR in Eq. (15.9) could be set to any number by changing $L_1$, $L_2$, and $L_3$.

$$\begin{cases} \frac{\partial n_1}{\partial T} \cdot L_1 - \frac{\partial n_2}{\partial T} \cdot L_2 - \frac{\partial n_3}{\partial T} \cdot L_3 = 0 \\ \frac{\partial^2 n_1}{\partial T \partial \lambda} \cdot L_1 - \frac{\partial^2 n_2}{\partial T \partial \lambda} \cdot L_2 - \frac{\partial^2 n_3}{\partial T \partial \lambda} \cdot L_3 = 0 \\ \text{FSR} = \frac{\lambda^2}{n_{g,1} \cdot L_1 - n_{g,2} \cdot L_2 - n_{g,3} \cdot L_3} \end{cases} \quad (15.9)$$

In summary, the design process could be described as follows:

1. Choose the waveguide widths ($w_1$, $w_2$, $w_3$) which are different from each other;
2. Set the FSR;
3. Calculate $\partial n_{\text{eff}}/\partial T$, $\partial^2 n_{\text{eff}}/\partial \lambda \partial T$ and group index for three waveguides with a numerical method;
4. Find the solution ( $L_1$, $L_2$, $L_3$) for Eq. (15.9).

**Fig. 15.2** Visible light microscope image of the fabricated devices

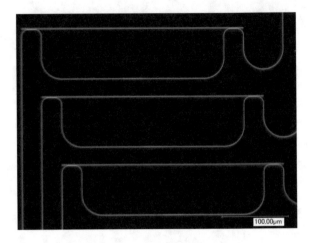

**Simulation Results** Theoretically, for any combination of $(w_1, w_2, w_3)$, there is a solution $(L_1, L_2, L_3)$. Different combination of $w_1, w_2, w_3$ will give different $L_1, L_2, L_3$ and result in a different device footprint. To find the design with the smallest device footprint, we need to find the solutions for various combinations of $w_1, w_2, w_3$. Also, different propagating modes in the same waveguide have different waveguide properties ($\partial n_{\text{eff}}/\partial T$, $\partial^2 n_{\text{eff}}/\partial\lambda\partial T$ and group indexes). Changing the mode will also lead to different waveguide lengths ($L_1, L_2, L_3$) and thus a different device size. So waveguide properties needed in Eq. (15.9) for quasi $TE_0$, $TM_0$, and $TE_1$ mode of 0.2–1.0 μm wide waveguides are computed.

The schematic of the SOI waveguides used in our simulation is shown in Fig. 15.1a. The waveguide height is fixed to be 220 nm with 2 μm buried silicon oxide. The waveguides are covered by 2 μm silicon oxide. The effective index of the $TE_0$, $TM_0$, and $TE_1$ mode for 0.2–1.0 μm wide waveguide is calculated using a full vector finite element method solver at temperature from 20 to 40 °C and wavelength from 1.50 to 1.60 μm. The refractive index of silicon and silica at different wavelengths and temperatures are from [15, 16]. The effective index of the $TE_0$, $TM_0$, and $TE_1$ mode are assumed to be linearly dependent on both temperature and wavelength. Since the waveguide effective index has a quadratic dependency on the wavelength, the thermal sensitivity will have a small quadratic dependency on wavelength. By fitting the computed effective index to temperature and wavelength the following parameters can be derived: $\partial n_{\text{eff}}/\partial T$, $\partial^2 n_{\text{eff}}/\partial\lambda\partial T$, and group index. The results are plotted in Fig. 15.1b–d.

**Design and Fabrication** The designed MZI was fabricated at the Masdar Institute microfabrication facility and also on a commercial silicon photonics foundry (IME-Singapore) using a SOI process to validate the feasibility of the design on a commercial platform. Figure 15.2 displays an optical micrograph of a section of a fabricated device.

The schematic of the designed device is shown in Fig. 15.3, with the corresponding light polarization states in each interferometer arm. The input and output

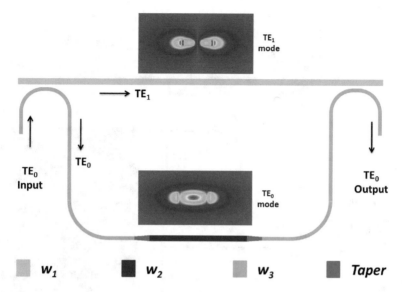

**Fig. 15.3** Schematic of the designed MZI with insets of the TE mode profiles at each interferometer arm

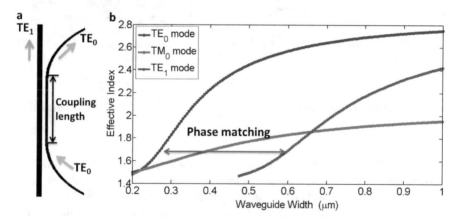

**Fig. 15.4** (**a**) Schematic of the directional coupler. (**b**) Phase matching condition on dispersion diagram

couplers are designed with different waveguide widths for different mode coupling from one waveguide to another, based on a phase matching criteria between the fundamental $TE_0$ mode in one waveguide and the $TE_1$ mode in the other waveguide, as represented in Fig. 15.4. The design parameters are summarized in Table 15.1.

**Measurements** Figure 15.5 depicts the setup used to characterize the fabricated devices. Light from a tunable laser (Agilent 81600B) is coupled into the waveguide through a lensed fiber. The TE polarized input is achieved using a polarization

**Table 15.1** Waveguide length of two arms of the designed MZI

| Arm | Mode | Waveguide width (nm) | Waveguide length (μm) |
|---|---|---|---|
| 1 | TE$_1$ | 600 | 294.6 |
| 2 | TE$_0$ | 400 | 25.8 |
| 2 | TE$_0$ | 270 | 374.6 |

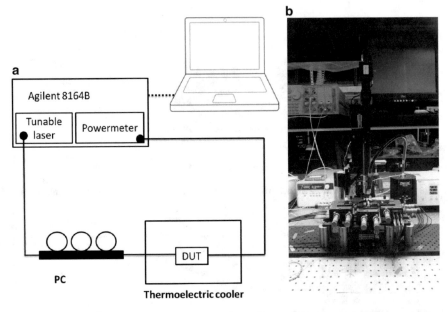

**Fig. 15.5** (**a**) Sketch of the characterization setup. (**b**) Photograph of the optical characterization setup

controller. The devices under test are placed on a thermoelectric cooler which is used to control the temperature of the chip with the devices under test. The light coupled out of the device is received by a power meter (Agilent 81636B). A laptop is used to communicate with the Agilent 8164B mainframe in which the tunable laser and power meter are built.

The measured spectra at different temperatures ranging from 20 to 60 °C over the spectral range from 1.5 to 1.6 μm are shown in Fig. 15.6a. Due to fabrication variability, the gap between two waveguides in the directional coupler changes from 400 nm (as designed) to 365 nm (as fabricated). So the light is not split equally into two arms, which leads to the high transmission at the spectral minima. Nevertheless, this will not affect the thermal sensitivity of the device. Due to the very small temperature sensitivity (less than 5 pm/K), it is very hard to measure the spectral shift from the original spectra. So we get the fitted spectra which is Fig. 15.6b. By comparing Fig. 15.6a, b, we find that the fitted spectra represent the original spectra very well. The spectral shift at different wavelengths in Fig. 15.6c is measured from the fitted spectra.

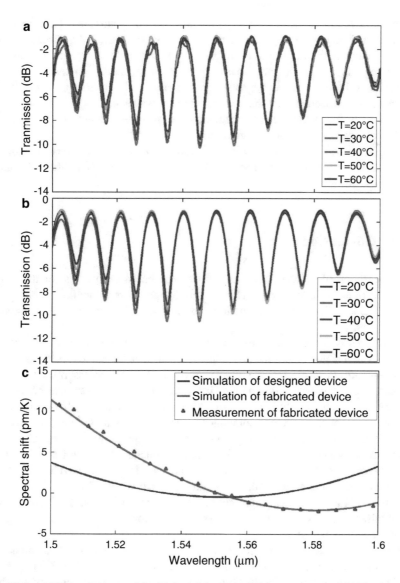

**Fig. 15.6** (**a**) Measured spectra of the fabricated device at different temperature. (**b**) Fitted spectra of the fabricated device at different temperatures. (**c**) Simulated spectral shift with temperature of the designed device, and simulated and measured spectral shift with temperature of the fabricated device

As shown in Fig. 15.6c, the measurement results of the fabricated device match very well with the simulated results. Over the spectral range from 1.54 to 1.60 μm, the spectral shift with temperature is less than 2.5 pm/K. In either the designed device or the fabricated device, the dependency of the thermal sensitivity on wavelength is not reduced to zero while the second requirement [Eq. (15.8)] is satisfied. This is due to the inaccuracy of the approximation in Eq. (15.2) that the effective index of the waveguides is linearly dependent on the wavelength. In reality, the waveguide's effective index has a quadratic dependency on the wavelength as discussed before. This is why the spectral shift with wavelength is parabolic. However, the dependency of the spectral shift on wavelength is brought to a very small number using the second requirement. So near zero temperature sensitivity can be achieved over a wide spectral range.

Also, even with the presence of a fabrication error, the thermal sensitivity of the device does not largely differ from the designed device. This is another advantage of this design.

## 15.3   Conclusion

The integration of silicon photonics with silicon-based electronics for ultra-high bandwidth data exchange between highly parallel multicore computing architectures will enable significant power reduction and bandwidth increase in next generation data servers.

Nevertheless, some challenges remain in order to fully integrate electronics and photonics in a single chip. One possible approach to tackle some of the challenges is 3D integration of electronic chips (memory and microprocessors), III–V laser sources, and the photonic communication channel. The later can also be used as an ultra-high bandwidth link to the external world. In such approach, the combined thermal loads of the different components, such as laser module and its drivers, and microelectronics core, will lead to unavoidable thermal fluctuations. Given the large thermal sensitivity of most photonics components, this is an issue that must be addressed by proper design. In this chapter we have described an improved approach to design a broadband, temperature insensitive, all-silicon MZI. We have used the MZI as the typical photonic device, for which it is easy to demonstrate the requirements and conditions for athermal operation in a broad spectral range in the infrared telecom window.

Following the design methodology presented, the measured results on an actual device show that the interferometer has a near zero (2.5 pm/K) thermal dependency over more than 60 nm of spectral range near 1550 nm which is much wider than what has been reported in prior work. The discussed approach is also applicable to MZIs working on other wavelengths and made of other materials other than silicon and waveguide geometry.

## References

1. H.P. Boehm, R. Setton, E. Stumpp, Nomenclature and terminology of graphite intercalation compounds (IUPAC recommendations 1994). Pure Appl. Chem. **66**(9), 1893–1901 (1994)

2. L.A. Ponomarenko, F. Schedin, M.I. Katsnelson, R. Yang, E.W. Hill, K.S. Novoselov, A.K. Geim, Chaotic Dirac Billiard in graphene quantum dots. Science **320**(5874), 356–358 (2008)
3. M.N. Leuenberger, D. Loss, Quantum computing in molecular magnets. Nature **410**(6830), 789–793 (2001)
4. J.D. Meindl, Beyond Moore's law: the interconnect era. Comput. Sci. Eng. **5**(1), 20–24 (2003)
5. R. Soref, The past, present, and future of silicon photonics. IEEE J. Sel. Top. Quantum Electron. **12**(6), 1678–1687 (2006)
6. T. Baehr-Jones, T. Pinguet, P. Lo Guo-Qiang, S. Danziger, D. Prather, M. Hochberg, Myths and rumors of silicon photonics. Nat. Photon. **6**(4), 206–208 (2012)
7. D.A.B. Miller, Rationale and challenges for optical interconnects to electronic chips. Proc. IEEE **88**(6), 728–749 (2000)
8. F.G. Della Corte, M. Bellucci, G. Cocorullo, M. Iodice, I. Rendina, Measurement and exploitation of the thermo-optic effect in silicon for light switching in optoelectronic integrated circuits, in *Proceedings of SPIE 3953, Silicon-based Optoelectronics II*, 2000, p. 127. doi: 10.1117/12.379604
9. G.-D. Kim, H.-S. Lee, C.-H. Park, S.-S. Lee, B.T. Lim, H.K. Bae, W.-G. Lee, Silicon photonic temperature sensor employing a ring resonator manufactured using a standard CMOS process. Opt. Express **18**(21), 22215–22221 (2010)
10. P. Dong, R. Shafiiha, S. Liao, H. Liang, N.-N. Feng, D. Feng, G. Li, X. Zheng, A.V. Krishnamoorthy, M. Asghari, Wavelength-tunable silicon microring modulator. Opt. Express **18**(11), 10941–10946 (2010)
11. X. Wang, S. Xiao, W. Zheng, F. Wang, Y. Hao, X. Jiang, M. Wang, J. Yang, Athermal silicon arrayed waveguide grating with polymer-filled slot structure, in *2008 5th IEEE International Conference on Group IV Photonics*, September 2008, pp. 253–255
12. J. Teng, P. Dumon, W. Bogaerts, H. Zhang, X. Jian, X. Han, M. Zhao, G. Morthier, R. Baets, Athermal silicon-on-insulator ring resonators by overlaying a polymer cladding on narrowed waveguides. Opt. Express **17**(17), 14627–14633 (2009)
13. M. Uenuma, T. Motooka, Temperature-independent silicon waveguide optical filter. Opt. Lett. **34**(5), 599–601 (2009)
14. B. Guha, A. Gondarenko, M. Lipson, Minimizing temperature sensitivity of silicon Mach-Zehnder interferometers. Opt. Express **18**(3), 1879–1887 (2010)
15. B.J. Frey, D.B. Leviton, T.J. Madison, Temperature dependent refractive index of silicon and germanium (2006). arXiv:physics/0606168, pp. 62732J–62732J–10
16. D.B. Leviton, B.J. Frey, Temperature-dependent absolute refractive index measurements of synthetic fused silica (2008). arXiv:0805.0091 [physics]

# Chapter 16
# Integrated Circuits for 3D Photonic Transceivers

**Ronny Henker, Guido Belfiore, Laszlo Szilagyi, and Frank Ellinger**

## 16.1 Introduction to Laser Driver ICs

The use of optical inter and intraconnects for short distance data communication is coming increasingly into research focus as they provide superior properties at high date rates in comparison to their electrical counterparts. Optical links can handle bandwidths up to few THz and therefore, transmit data with rates up to Tbit/s-range while signal latencies as well as losses are rather low and primarily defined by the physical connection and alignment of the components. Furthermore, optical connections show less electromagnetic interferences and can provide a high link parallelization either by dense physical links or by multi-carrier transmission over one optical medium. However, to complement or even replace electrical interconnects with distances shorter than a few centimeters, optical inter and intraconnects have to be improved with regard to their higher energy efficiency.

In general, an optical transmission system, as described in detail in Sect. 11.2, consists of three functional blocks:

1. the optical transmission medium, which can be a free space transmission or material-based wave-guided structure like optical fibers or onboard/integrated optical waveguides,
2. the components for the electro-optical conversion, namely a laser or optical modulator at the transmitter side and a photodetector/-diode at the receiver side, and
3. the electrical circuitry that includes the amplification and regeneration of the signal at the transmitter and receiver side.

R. Henker (✉) • G. Belfiore • L. Szilagyi • F. Ellinger
Technische Universität Dresden, Chair for Circuit Design and Network Theory,
01069 Dresden, Germany
e-mail: ronny.henker@tu-dresden.de; guido.belfiore@tu-dresden.de;
laszlo.szilagyi@tu-dresden.de; frank.ellinger@tu-dresden.de

© Springer International Publishing Switzerland 2016
I.M. Elfadel, G. Fettweis (eds.), *3D Stacked Chips*,
DOI 10.1007/978-3-319-20481-9_16

Even though such a system is for optical signal transmission, the major part of the transceiver consists of electrical integrated circuits (ICs). At the transmitter part, several parallel electrical channels with small bit rates are multiplexed (MUX) to a high bit rate signal. This signal will be provided to the laser by a laser diode driver (LDD) which enables best possible electrical matching and sufficient electro-optical modulation. At the receive side, the photocurrent of the photodiodes (PD) will be transformed into an amplified voltage by a transimpedance (TIA) and limiting amplifier (LA). Finally, a clock-data-recovery (CDR) extracts the signal clock and refreshes the data signal. Afterwards the electrical signal is demultiplexed (DMUX) to the electrical sub data rates again.

In order to accommodate electrical links, the ICs in optical links have to fulfill several requirements: that include high bandwidth and low power consumption resulting in high energy efficiency. Furthermore, as higher integration comes into play with 3D packaging, compact design, and new integration methodologies as well as circuit structures are required. To optimize the IC designs the advantages of various semiconductor technologies can be exploited for a given application. It has been shown that SiGe BiCMOS IC technology can achieve a well-balanced performance to meet most of the requirements described above [1]. Furthermore, novel vertical inductor structures, which can be used for inductive peaking for instance, have been proposed to enable more compact chips [2, 3].

Most of the published optical links are designed for off-chip and off-board communications. VCSEL diodes are a good candidate to be used in high-speed, power-efficient optical intraconnects. As explained in Sect. 13.1.1.2, the light is emitted vertically allowing an easy connection between different ICs in the 3D chip stack. The light can also be coupled into on-chip optical transmission lines using 45° mirrors or grating couplers. Although the structure of the currently available VCSELs need to be optimized for the use in optical intraconnects, according to the state of the art (Table 16.1), in ultra-short range optical transmission, direct modulated VCSELs are more power-efficient than modulator-based optical links. However, direct modulation of VCSELs is currently possible up to bandwidths of

**Table 16.1** Comparison between the proposed VCSEL driver and the state of the art

| Ref. | Approach | Technology | DR (Gbit/s) | FOM (mW/Gbit/s) | Comment |
|------|----------|-----------|-------------|-----------------|---------|
| [9] | M, QAM | InP bipolar | 224 | 14.3 | DP-16QAM modulation |
| [10] | M, OOK | 130 nm SiGe BiCMOS | 56 | 10.82 | Driver + CW laser, 50 Gbit/s error free |
| [5] | V, OOK | 90 nm CMOS | 25 | 2.4 | DR measured optically |
| [11] | V, OOK | 32 nm SOI | 35 | 0.856 @ 25 Gbit/s | Common anode VCSEL |
| [4] | V, FFE | 130 nm SiGe BiCMOS | 71 | 13.4 | 2-tap FFE |
| [8] | V, 4-PAM | 130 nm SiGe BiCMOS | 56 | 2.05 | 4-PAM DR measured electrically |

M = modulator, V = VCSEL

only 25 GHz. Thus, the use of several techniques in combination with the ICs to enhance the link bandwidth is required. This will be explained later in Sects. 16.3 and 16.4. By now, the achieved world record data rate for a VCSEL-based optical link is at 71 Gbit/s by applying a two-tap feed forward electrical equalization [4]. However, the total power consumption of the transmitter was 950 mW and of the receiver 860 mW which results in an overall energy per bit of approximately 25 pJ/bit. Such a power consumption is of course too high to use this transceiver for optical on-chip or chip-to-chip communication in a 3D chip stack. Therefore, more energy-efficient solutions have to be found when data rates of more than 1.4× of the available VCSEL bandwidth must be reached.

Basically, there are three options which can be used and incorporated by the link ICs to achieve high data rates in optical intraconnects:

1. *Link parallelization*: the signal is transmitted using a parallel optical link.
2. *Pre-emphasis and equalization techniques*: the signal that drives the VCSEL is adjusted in order to compensate the low pass behavior of the VCSEL.
3. *Advanced modulation schemes*: e.g., multilevel modulation (*n*-PAM) where the signal can be modulated and transmitted by more bits per symbol.

This chapter gives an overview about recent advances on (LDD) IC design for optical inter-/intraconnects and discusses the potential to achieve compact high-speed and very energy-efficient circuits. The three different ways for the increase of the transmitted data rates are first reviewed. Successful IC implementations are shown and trade-offs are explained. Last, novel vertical integrated inductor structures are introduced which enable highly compact chip design while using inductors for additional bandwidth peaking.

## 16.2  Parallelization of Laser Driver ICs

One possibility to reach high data rates is to parallelize the optical link. In this fashion the data rate of each low speed link is given by the target data rate divided by the number of links. As suggested by Table 16.1, the power consumption of a parallel link can be lower than the power consumption of a single high-speed optical link when the target data rate is at the edge of the IC and VCSEL technologies possibility. A slower link can be realized using CMOS instead of SiGe BiCMOS technology. Doing so, the driver and receiver circuit, running at lower data rate, can be integrated with the digital processor. On the other hand, the high transient frequency of SiGe technologies is preferable in very high speed driver and receiver circuits.

The proposed transmitter circuit for this task is a 25 Gbit/s common cathode laser driver realized in 90 nm CMOS technology. The driver published in [5] provides the bias voltage and current to the VCSEL. Since the bias voltage of the VCSEL exceeds the breakdown voltage of the thin-oxide transistors, the driver is realized using isolated wells. The chip micrograph is shown in Fig. 16.1. The active area

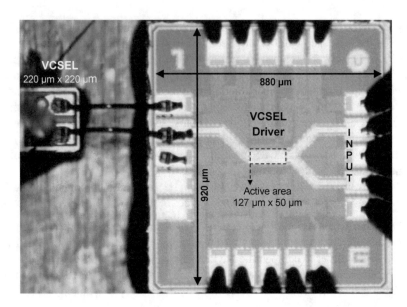

**Fig. 16.1** Chip micrograph of a 25 Gbit/s VCSEL driver bonded to a commercial 14 Gbit/s VCSEL and connected to on-chip measurement probes

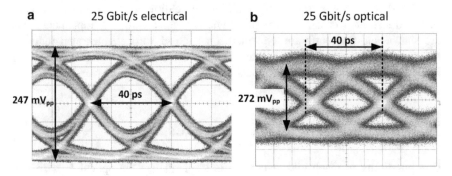

**Fig. 16.2** (**a**) Electrical and (**b**) optical eye diagrams at 25 Gbit/s. A 55 GHz measurement amplifier and a 30 Gbit/s photodiode are used at the receiver side

is only $127\,\mu\text{m} \times 50\,\mu\text{m}$ and the power consumption is $60\,\text{mW}$ which makes this design suitable for high-speed parallel connections. The laser driver is bonded to a commercial 14 Gbit/s VCSEL. The optical connection can run error free up to 25 Gbit/s as shown in the eye diagram of Fig. 16.2 and the BER bathtub test reported in Fig. 16.3.

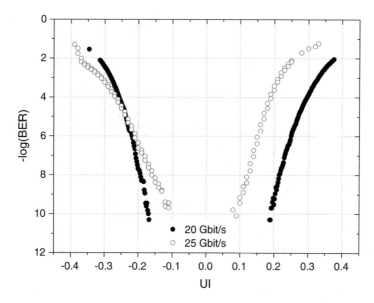

**Fig. 16.3** Bathtub curves at 20 and 25 Gbit/s of the optical transmitter showing the bit-error rate (BER) vs the unit interval (UI). In the receiver side a standard purpose 50 Ω measurement amplifier and a 40 Gbit/s photodiode are used

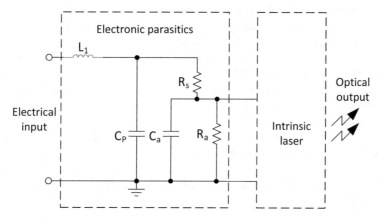

**Fig. 16.4** VCSEL block diagram. The intrinsic laser block is described by the VCSEL rate equations

## 16.3   Pre-emphasis and Equalization Laser Drivers

Pre-emphasis is a technique used to compensate the low pass behavior of the optical components. The VCSEL model is represented in Fig. 16.4. The model is composed of two blocks: the electrical interfaces and the intrinsic model. The electrical interface can often be approximated with the circuit depicted in Fig. 16.4.

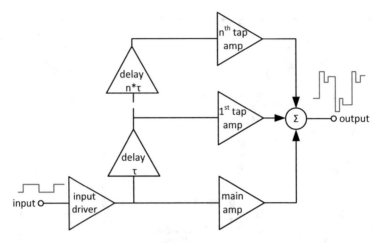

**Fig. 16.5** Concept of a $n$-tap feed forward equalization (FFE) technique applied to a laser driver

On the other hand, the intrinsic laser model can be described using the VCSEL rate equations and has a second order low pass filter behavior. As a first step, the laser driver should feature a peaking that aims to compensate the low pass behavior of the VCSEL electrical interface and chip-to-VCSEL connection. The peaking is usually realized using inductive peaking or emitter degeneration in the laser driver. For an even more accurate overall bandwidth extension the intrinsic laser model should be considered as well since it directly affects the optical eye diagram. In order to have a better control over the optical eye diagram decreasing jitter and over-peaking, a $n$-tap feed forward equalizer (FFE) can be used in the laser driver circuit [6]. Figure 16.5 shows an example of a $n$-tap equalizer. FFE consists in combining different replica of the input signal properly amplified, that are shifted and summed up creating an equalized signal.

This technique is used in both the modulation of a 26 GHz bandwidth VCSEL and in the receiver circuit thus allowing a 71 Gbit/s optical connection [4]. The drawback is that the power consumption is high compared to parallelization and multilevel modulation. To reach this state-of-the-art data rate, SiGe BiCMOS technology is used.

## 16.4 Advanced Modulation Laser Drivers

Using a pre-modulated signal to drive the VCSEL, the link bandwidth can be extended. One example of modulation that can be applied to the VCSEL is amplitude modulation. Amplitude modulation consists in driving the VCSEL with $n$ amplitude levels. In this case the bit rate (BR) can be higher than the link bandwidth (BW):

$$BR > BW * \log(n) \tag{16.1}$$

**Fig. 16.6** Chip micrograph
of the 4-PAM laser driver IC.
The pad limited chip size is
0.59 mm$^2$

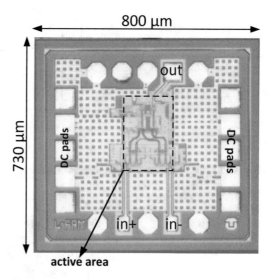

It has already been proven that the VCSEL diode can be modulated with an 8-PAM signal. In this case, using a 25 GHz VCSEL, data rates higher than 112 Gbit/s can be reached. The drawback of using $n$-PAM over OOK is the power penalty. The power penalty expressed in decibel for and $n$-PAM modulation is

$$P_{p\text{DR,dB}} = 10 \log_{10} \left[ \frac{n-1}{\sqrt{\log_2(n)}} \right] \text{ dB} \tag{16.2}$$

This implies that in order to have the same SNR, the optical transmitted power using 4-PAM has to be 3.8 dB higher than on–off key (OOK) [7].

The proposed IC for this task is a 56 Gbit/s 4-PAM laser driver [8]. The chip micrograph is shown in Fig. 16.6. The pad limited chip size is 0.59 mm$^2$. The power consumption of the driver is 115 mW including the single to differential 50 Ω converters (SDCs) and the common cathode VCSEL. Figure 16.7 depicts the driver block diagram. The circuit converts two single-ended input signals to a 4-level single-ended output. The internal signals are differential, and emitter degeneration is the only peaking technique applied in the design of this driver. It is possible to change the amplification of the single- to differential-ended converter in order to adjust the four output levels. The emitter degeneration is used in the last stages to compensate the parasitics given by the VCSEL which is modelled as a 310 fF capacitor in parallel to a 90 Ω resistor.

The electrical output eye diagram at 56 Gbit/s is reported in Fig. 16.8. The eye diagram is measured on chip using a wafer probe station. At the maximum data rate of 56 Gbit/s, the eye input signal starts to degrade and the parasitic effect of the cables, attenuators, and probes used in the setup reduces the eye amplitude and creates jitter. At 40 Gbit/s, the output eye amplitude with 50 Ω load is 320 mV$_{pp}$

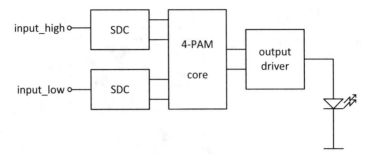

**Fig. 16.7** Block diagram of the proposed 4-PAM laser driver

**Fig. 16.8** Eye diagram at 56 Gbit/s. The eye scale is: $dx = 6\,ps/div$, $dy = 42\,mV/div$

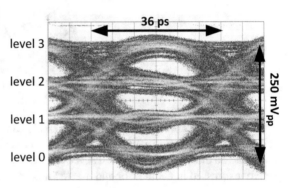

considering an input eye amplitude of $150\,mV_{pp}$. As can be seen in Table 16.1, this is a power-efficient solution for high-speed short-range optical connections.

## 16.5 Vertical Inductors

With the increasing need of high data rates, peaking methods such as inductive peaking are required. Conventional on-chip spiral inductors are realized by creating conductive coils using one or more metal layers. In conventional structures the spiral is oriented in parallel to the substrate. The main drawback of inductive peaking using conventional on-chip inductors is the large chip area occupied by inductors that are therefore very cost-inefficient. Unlike conventional inductors where the spiral plane is oriented in parallel to the substrate, in vertical inductors the spiral plane is perpendicular to the substrate. The goal of vertical inductors is to save chip area. One example of vertical inductor layout can be seen in Fig. 16.9. In [2] the first study of vertical inductors is reported where the parameters are swept and the simulations with a EM simulator are compared with analytical calculations. The inductance at low frequencies of the vertical inductor can be calculated like the sum of the self inductance of the metal ($L_{self}$) plus the mutual inductance between the metal layers [12].

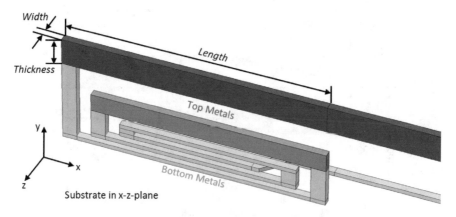

**Fig. 16.9** Vertical inductor cross section. Unlike conventional spiral inductors, in vertical inductor the spiral is oriented perpendicularly (*XY*-plane) to the chip substrate (*XZ*-plane)

$$L_{\text{tot}} = L_{\text{self}} + M_+ + M_- \qquad (16.3)$$

The mutual inductance is positive $(M_+)$ in metals with the same current vector direction and negative $(M_-)$ when the current vector direction is opposite. In the vertical inductor the metal layer stack is defined by the technology. Since the vertical distance between metals is limited, the metal with opposite current direction is close to each other. This increases the negative mutual inductance and reduces the total inductance. To compensate the high negative mutual inductance longer metal lines need to be used. The drawback of using longer metal lines is that the series resistance increase causing a decrease in the quality factor. More vias and larger metals can be used to reduce the series resistance of the inductor. Moreover, inductive peaking is often realized using an inductor in series with a resistor. In this case the low Q-factor is not a problem and can be easily compensated by changing the resistor in series to the inductor [3].

In [3] vertical inductors are manufactured in a 130 nm SiGe BiCMOS technology. The measurement results are compared with simulations shown in Fig. 16.10. The S-parameters are measured using a 67 GHz vector network analyzer. The 67 GHz RF probes and cables are also calibrated with short-open-load-through de-embedding standards. The pads are de-embedded using an on-chip short-open structure. Table 16.2 reports the parameters of the manufactured vertical inductor.

In [3] the vertical inductor is also compared with conventional inductors resulting in a higher inductance per unit of area. Vertical inductors can also be extended to a 3D chip stack. One example of this application can be seen in Fig. 16.11 where two chips are connected via flip-chip on an interposer. Chip number one uses the metal layers and vias of the second chip in order to have high inductance value occupying a minimum amount of area. Furthermore, with the incorporation of through-silicon vias (TSVs) a larger inductor loop can be created stacking also more than two chips.

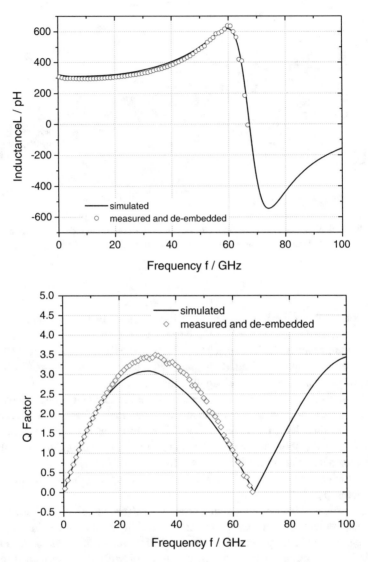

**Fig. 16.10** Comparison between simulations and de-embedded measurements of vertical inductor's inductance (*top*) and Q factor (*bottom*)

## 16.6 Conclusions

In this chapter different integrated circuit implementations usable for optical intra-connects in 3D chip stack integration have been described. The major requirements are high bandwidths and low power consumption to enable high energy efficiency. Three methods have been discussed for improving the bandwidth, data rate, and

**Table 16.2** Values of the manufactured and measured vertical inductor reported in [3]

| Size (μm²) | Inductance value (pH) | Resonance frequency (GHz) | Quality factor | Ind. per unit area (nH/mm²) |
|---|---|---|---|---|
| 150 × 5 | 318 | 66.5 | 3.1 @ 30 GHz | 424 |

**Fig. 16.11** Possible application of vertical inductors in a 3D chip stack by flip-chipping of two active chips on an interposer using solder balls

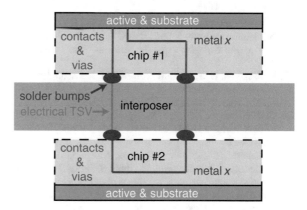

power consumption in optical intraconnects to enabled the integrated laser drivers: link parallelization, pre-emphasis, and advanced or multilevel modulation. While parallelization increases the complexity of the systems significantly, pre-emphasis provides high data rates at a single link. However, this comes at the expenses of high power consumption and therefore low energy efficiency. One solution which effectively trades off bandwidth, power consumption, and compactness, might be advanced modulation formats such as PAM. Furthermore, novel vertical inductive structures were described which can be used for inductive peaking of the circuit's bandwidth while the required chips area increases only marginal.

Optimizing both, optical components and electrical driving ICs, can pave the way toward compact high-speed energy-efficient optical intraconnect for the data communication within a 3D integrated chip stack in future. Hence, optical intraconnects could be a promising alternative to copper-based links even for ultra-short-range data transmission.

# References

1. D. Schoeniger, R. Henker, S. Schumann, F. Ellinger, A low-noise energy-efficient inductor-less 50 Gbit/s transimpedance amplifier with high gain-bandwidth product in 0.13 μm SiGe BiCMOS, in *Proceedings of the International Semiconductor Conference Dresden—Grenoble*, 2013
2. G. Belfiore, R. Henker, F. Ellinger, New design approach of vertical inductors for high-frequency integrated circuits, in *IEEE IMOC*, 2013

3. G. Belfiore, R. Henker, F. Ellinger, Measurement and application of vertical inductors in high-speed broadband circuit. Electron. Lett. **50**(25), 1915–1917 (2014)
4. D. Kuchta, A. Rylyakov, F. Doany, C. Schow, J. Proesel, C. Baks, P. Westbergh, J. Gustavsson, A. Larsson, A 71 Gb/s NRZ modulated 850 nm VCSEL-based optical link. IEEE Photon. Technol. Lett. **27**(6), 577–580 (2015)
5. G. Belfiore, L. Szilagyi, R. Henker, F. Ellinger, Common cathode VCSEL driver in 90 nm CMOS enabling 25 Gbit/s optical connection using 14 Gbit/s 850 nm VCSEL. Electron. Lett. **51**(4), 349–351 (2015)
6. A.V. Rylyakov, C.L. Schow, J.E. Proesel, D.M. Kuchta, C. Baks, N.Y. Li, C. Xie, K.P. Jackson, A 40-Gb/s, 850-nm, VCSEL-based full optical link, in *Optical Fiber Communication Conference and Exposition and the National Fiber Optic Engineers Conference (OFC/NFOEC)*, 2012
7. K. Szczerba, P. Westbergh, E. Agrell, M. Karlsson, P.A. Anderekson, A. Larsson, Comparison of intersymbol interference power penalties for OOK and 4-PAM in short-range optical links. IEEE J. Lightwave Technol. **31**(22), 3525–3534 (2013)
8. G. Belfiore, L. Szilagyi, R. Henker, U. Joerges, F. Ellinger, Design of a 56 Gbit/s 4-PAM inductor-less VCSEL driver IC in 130 nm BiCMOS technology, in *IET CD&S*, 2015, pp. 213–220
9. N. Itabashi, T. Tatsumi, T. Ikagawa, N. Kono, M. Seki, K. Tanaka, K. Yamaji, Y. Fujimura, K. Uesaka, T. Nakabayashi, H. Shoji, S. Ogita, A compact low-power 224-Gb/s DP-16QAM modulator module with InP-based modulator and linear driver ICs, in *IEEE Compound Semiconductor Integrated Circuit Symposium (CSICs)*, 2014
10. G. Denoyer, C. Cole, A. Santipo, R. Russo, C. Robinson, L. Li, Y. Zhou, A. Chen, B. Park, F. Boeuf, S. Cremer, N. Vulliet, Hybrid silicon photonic circuits and transceiver for 50 Gb/s NRZ transmission over single mode fiber. J. Lightwave Technol. **33**(6), 1247–1254 (2015)
11. J.E. Proesel, B.J. Lee, C.W. Baks, C.L. Schow, 35-Gb/s VCSEL-based optical link using 32-nm SOI CMOS circuits, in *OFC*, 2013
12. H.M. Greenhouse, Design of planar rectangular microelectronic inductors. IEEE Trans. Parts Hybrids Packag. **10**(2), 101–109 (1974)

# Chapter 17
# Review of Interdigitated Back Contacted Full Heterojunction Solar Cell (IBC-SHJ): A Simulation Approach

**Ayesha A. Al-Shouq and Adel B. Gougam**

## 17.1 Introduction

The power of the sun has been harvested for the last few decades [1] with an ascendant curve in terms of technology development that by virtue of enhancing the energy efficiency of the system allowed for a substantial decrease in the overall system cost. The latter makes the solar-based technologies competitive with the other existing energy sources. In this review article we focus on silicon-based (denoted thereafter by: c-Si) photovoltaic system which constitutes about 90 % [2] of the market share in terms of PV technologies deployed. Few advantages of the silicon material that is the main ingredient in the solar cells discussed in this review article are listed below.

The availability of the raw material silicon is one of the reasons for the widespread use of this type of solar cells [2] (silicon is obtained after chemical process of silicon dioxide in the form of sand). It is the second most abundant element in the earth's crust, its non-toxicity is an advantage over other materials. The c-Si cells have exhibited excellent stability (light, temperature) [3], the technology deployed for their fabrication is mature and has benefited greatly from various processes developed for the microelectronic industry in the last 50 years.

In the race for Si-based highly efficient (above 20 %) solar cells, the interdigitated back contacted (IBC) solar cell and the heterojunction Si cell with an intrinsic thin amorphous layer (SHJ for silicon heterojunction) stand out [4]. By virtue of having all contacts on one side of the solar cell, one can integrate this power source (solar cell) in a 3D way to power various other devices, the approach can be mechanical integration in the first phase then potentially move to a monolithic

A.A. Al-Shouq • A.B. Gougam (✉)

Department of Mechanical and Materials Engineering, Masdar Institute of Science and Technology, Abu Dhabi, United Arab Emirates

e-mail: aalshouq@masdar.ac.ae; agougam@masdar.ac.ae

© Springer International Publishing Switzerland 2016

I.M. Elfadel, G. Fettweis (eds.), *3D Stacked Chips*,

DOI 10.1007/978-3-319-20481-9_17

integration of the solar cell and the devices to be powered. The IBC structure concept was introduced by Lammert and Schwartz in 1975 [5]. In this design, both metal contacts and emitter are located on the non-sun facing surface (back of the solar cell). The contacts are deposited in an interdigitated configuration. This design is also called back junction solar cell. The other highly efficient solar cell design is the heterojunction solar cell developed by Tanaka et al. back in the early eighties of the last century [6]. Our aim in this review article is to study via simulation the combination of these two approaches in a single device denoted thereafter as IBC-SHJ. It is anticipated that it will allow for an enhancement of the cell performance as it combines the advantages of moving all metal contacts to the rear side, on the one hand, and hence reducing shadow losses and using the exceptional passivating properties of a-Si thin films to the crystalline Si-based material [7] which has a direct impact on the recombination processes of the charge carriers. Introducing heterojunction with thin a-Si layers will improve the open circuit voltage while the introduction of the IBC configuration will enhance the short circuit current density and simplifies integration of the cells into modules. This concept has been investigated by the Institute of Energy Conversion (IEC) at the University of Delaware [8]. A variation of the previous design using in addition point contacts was developed by Helmholtz Zentrum Berlin (HZB) and reported thus far an efficiency of 13.9 % [9]. The highest conversion efficiency reported for an IBC-SHJ structure is 25.7 % by Panasonic Co. [10] (champion cell with area 143.7 cm$^2$) (Fig. 17.1).

In recent years, many publications have demonstrated that the IBC-SHJ cells have the capability of reaching efficiencies up to 25 % as reported in Table 17.1.

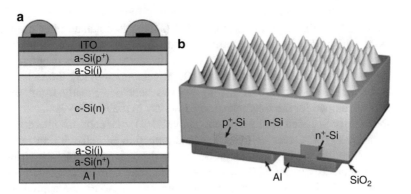

**Fig. 17.1** (a) HIT silicon cell. (b) IBC silicon cell

**Table 17.1** Overview of reported IBC-HIT silicon cell efficiencies in the last 5 years (all silicon substrates used are FZ type except the Panasonic Co. reported to be CZ substrate)

| Organization | Year | Substrate | Area cm$^2$ | Efficiency (%) |
|---|---|---|---|---|
| Helmholtz Zentrum Berlin Institute [9] | 2009 | Fz, p-type c-Si | 1 | 13.9 |
| Ecole Polytechnique Federale de Lausanne [11] | 2014 | Fz, n-type c-Si | 9 | 21.5 |
| LG Electronics Advanced Research Institute, Korea [12] | 2012 | Fz, n-type c-Si | 2 | 23.4 |
| INES-CEA [13–16] | 2013 | Fz, n-type c-Si | 150 | 12.2 |
| | 2010 | Fz, n-type c-Si | 25 | 12.7 |
| | 2012 | Fz, n-type c-Si | 5 | 19 |
| | 2011 | Fz, n-type c-Si | 25 | 15.7 |
| University of Delaware [17] | 2010 | Fz, n-type c-Si | NM | 15 |
| Panasonic, Japan [10] | 2014 | Cz, n-type c-Si | 143.7 | 25.6 |

## 17.2 Cell Modelling

The experimental development of the various layers forming an IBC-SHJ solar cell requires an optimization of a large number of parameters, a task that can be rendered even more complex if the said parameters are interdependent. Computer modelling can constitute an effective preliminary tool to identify the effect of the various parameters, allowing for resource usage minimization and for shortening the experimental development cycle of the solar cell. The availability of powerful software packages on the market is another incentive to follow this path. The modelling of the IBC-SHJ solar cell as opposed to a non-back contacted cell prevents us from neglecting the second physical dimension (along the $x$-axis in Fig. 17.2 below), and as such the simulation tool used has to at least account for the traditional growth direction ($z$-axis) and the interdigitation direction ($x$-axis). For the present work, we have chosen to use the package TCAD from Synopsys (www.synopsys.com/tools/TCAD) in its 2D format. The 3D format was used for some of our simulations and the final results were found to be very similar to the 2D approach with a much larger computation time. Hence $x$–$z$-based simulation is quoted in all the results presented in this review. Adapting some of the existing codes in the software to simulate properly the IBC-SHJ cell was quite daunting. In particular, incorporating adequately the a-Si layer properties required some major changes to existing codes from the supplier. The model used in these software packages tend to simplify the density of states magnitude and their distribution in the amorphous

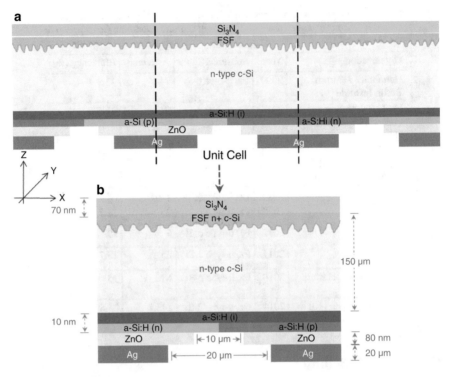

**Fig. 17.2** IBC-HIT silicon cell

layers. An adjustment to the code in this regard is necessary to simulate accurately the heterojunction-based solar cell [18]. The behavior of IBC-SHJ silicon solar cells depends on several parameters as mentioned above. We categorize these parameters into four groups: c-Si substrate parameters, front surface passivation parameters, amorphous silicon (a-Si) parameters, and cell geometry parameters.


## 17.2.1  Impact of c-Si Substrate

The impact of c-Si substrate on the solar cell characteristics can be attributed to various parameters such as the doping level (wafer electrical resistivity), dopant type, substrate's thickness, and the minority carrier lifetime. Several groups have studied the effect of one or more of these properties on the cell's performance. Figure 17.3 [19] shows the relationship between silicon wafer resistivity and doping concentration. The substrate resistivity plays a major role in the performance of the device, especially for values less than 2.5 $\Omega$-cm corresponding to a silicon wafer doping level in the range $10^{15}$–$10^{16}$ cm$^{-3}$. Varying resistivity has shown its influence on cell's characteristics in our simulations. All cell's characteristics are

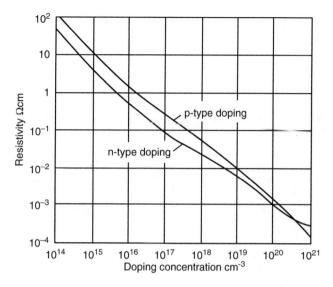

**Fig. 17.3** The relationship between wafer resistivity and doping concentration [19]

affected within a narrow range of doping level (see Fig. 17.4). The fill factor (FF) degradation can be attributed to an increase of the series' resistance of the cell as substrate's resistivity is increased. Within the same range of resistivity variation both the open circuit voltage ($V_{oc}$) and the short circuit current density ($J_{sc}$) are enhanced. This can be explained by carrier lifetime increase as highlighted in our simulation in Fig. 17.5 for $V_{oc}$ and by a decrease of Auger recombination mechanisms for $J_{sc}$ improvement. Overall the open circuit voltage and the short circuit density are found to be improving with resistivity up to a saturation point at resistivity $\rho = 1.6\,\Omega$-cm. We note that our simulations do find similar trends to those reported by other groups [4, 8, 18, 20]. We do find however a larger improvement of $V_{oc}$ of up to 22 mV in our case vs. 7 mV for some other studies. The optimal substrate resistivity is found to be at 0.85 $\Omega$-cm corresponding to a doping level of $\approx$5 $\times$ 10$^{15}$ cm$^{-3}$.

The absorber layer thickness is anticipated to have an effect on the overall efficiency of the cell. Many solar manufacturers are looking into reducing the amount of silicon used as a one way of overall reduction of the $/W ratio. The standard thickness for c-Si-based solar cells is between 210 and 270 μm, studying the influence of substrate thickness on the cell's behavior becomes essential.

It has been reported that the increase of substrate thickness causes the $V_{oc}$ to decrease regardless of the quality of the substrate as identified by the bulk lifetime. Indeed, for thick substrates, the bulk recombination of minority carriers dominates [7]. This fact is observed in our simulations as well as seen in Fig. 17.6. The effect of the substrate thickness on $J_{sc}$ is dependent on the quality of the wafer, Diouf et al. study shows that $J_{sc}$ is enhanced significantly with the substrate thickness decrease if the bulk lifetime is around 0.1 ms. On the other hand, for the high minority carrier

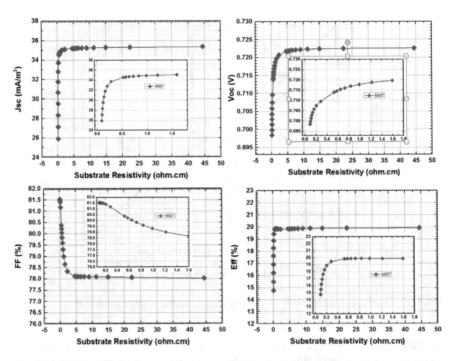

**Fig. 17.4** The impact of substrate resistivity on the *JV* parameters

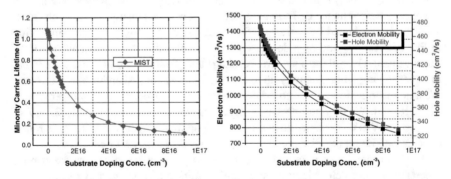

**Fig. 17.5** Relationships between minority carrier lifetime, mobilities, and substrate doping level

lifetime case ($\tau_B > 1$ ms), $J_{sc}$ values decrease as the substrate thickness is reduced or at best have an optimum value for a thickness of $\approx 150\,\mu$m. Our simulations confirm similar findings, the optimum thickness for absorber substrates with lifetime $\geq 1$ ms was found to be in the range 150–200 $\mu$m displaying an efficiency above 20 % as shown in Fig. 17.6.

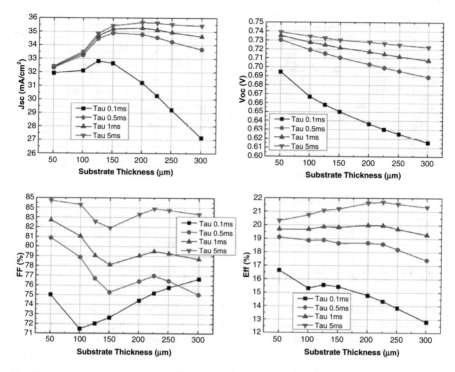

**Fig. 17.6** Impact of the substrate quality on *JV* parameters as a function of substrate thickness

## 17.2.2   *Impact of Front Surface Passivation*

Back contacted solar cells are known to be highly sensitive to the front surface recombination. The generation of the carriers occurs near the front surface. These photo-generated carriers will diffuse through the bulk to reach the contacts in the rear side of the cell. So it is crucial to prevent these carriers from recombining at the surface. That can be achieved by having a good passivating layer in the front surface that will chemically passivate silicon dangling bonds (recombination centers), and if the front layer is doped, an added field effect will enhance the passivation effect via coulomb repulsion of the majority carriers. It has been reported that adding a front surface field (FSF) layer (n++ for n type wafer) will minimize the impact of the front surface recombination velocity (Front SRV) on the cell's performance [4, 21] which is in agreement with our simulation (see Fig. 17.7).

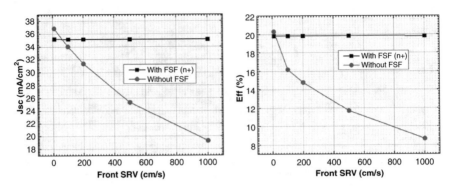

**Fig. 17.7** Impact of FSF as a function of the front SRV

## 17.2.3 Impact of Amorphous Silicon Layer

### 17.2.3.1 Role of Intrinsic Silicon Layer as Passivation

Amorphous silicon (a-Si) has been a subject of several studies [8, 18, 21]. The impact of a-Si layer has been investigated as a buffer if undoped and as an emitter layer when doped. The following properties have been studied for a-Si layers: doping level, thickness, bandgap, and dangling bonds density, in addition to defect states density at the c-Si/a-Si hetero-interfaces.

In order to reduce the interface recombination between the substrate and the doped layers, a good passivation at the back should be implemented. The introduction of a thin i-aSi layer (intrinsic) has been studied by several groups [18, 21]. Studies showed that adding such a layer improves both $J_{sc}$ and $V_{oc}$ by reducing interface defects, but results in a negative impact on the fill factor FF [18, 21] that can be explained by the enhancement of the valence band offset at the hetero-interface in addition to an inversion layer formation at this same interface. Our choice of simulation parameters of i-aSi as a buffer layer between the emitter and the substrate confirms the $V_{oc}$ and $J_{sc}$ behavior as reported by others, however, we have been able to avoid the S-shaped $JV$ curve formation, and a decent FF and efficiency are obtained (Table 17.2).

Insertion of i-aSi layer as a buffer layer between back surface field layer (BSF) and the substrate does not affect the $J-V$ curve shape [18, 21]. Improving FF can be achieved by tuning the i-aSi buffer properties: thickness, bandgap, and conductivity.

The thickness of the rear intrinsic amorphous silicon layer has a significant influence on FF. Reducing the buffer thickness shows a significant improvement of the FF with very little effect on $J_{sc}$ [8, 18, 21]. On the other hand, the impact of thickness on $V_{oc}$ differs from one study to other. Both Mejin Lu et al. and Allen et al. groups showed that the thicker i-layer, the higher the $V_{oc}$ achieved [8, 21] while no change on $V_{oc}$ is reported in Callozzo et al. study [18]. Our simulation showed that the effect on the open circuit voltage of the thickness is rather small. FF is the predominant factor affecting the efficiency as confirmed in our simulation (Fig. 17.8).

**Table 17.2** The influence of rear i-aSi layer in the output solar characteristics

| Group | Presence of rear i-aSi layer | FF (%) | $J_{sc}$ (mA/cm$^2$) | $V_{oc}$ (mV) | Efficiency (%) |
|---|---|---|---|---|---|
| Masdar (MIST) | No i-layer | 80.96 | 35.12 | 702.5 | 20.48 |
| | i-layer (thickness 5 nm and Bulk lifetime 1 ms) | 78.09 | 35.19 | 721.7 | 19.83 |
| Callozzo et al. [18] | No i-layer | 80.85 | 32.89 | 668 | 17.77 |
| | i-layer (thickness 5 nm and Bulk lifetime 0.6 ms) | 61.15 | 33.50 | 696 | 14.26 |
| Lu et al. [21] | No i-layer | 73 | ~27.5 | ~600 | – |
| | i-layer (thickness 10 nm and Bulk lifetime > 5 ms) | ~37 | ~36 | ~683 | – |

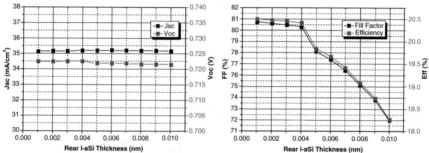

**Fig. 17.8** The influence of rear i-aSi thickness on the output of the solar cell characteristics

This behavior of FF can be explained in light of the energy band diagram of the structure as shown in Fig. 17.9 below. From Fig. 17.9a, b, when the buffer layer thickness is reduced, the S-shape for the $J$–$V$ curve is gradually minimized to completely disappear. As the thickness reduces, the energy band rearranges as seen in Fig. 17.9c. The band bending at the interface c-Si/aSi is reduced with the i-layer thickness such that the potential barrier for minority carriers is lowered which allows for optimum extraction of carriers across the junction [21].

Enhancing the buffer layer electrical conductivity is another way to reduce the impact of the buffer layer on the emitter strip and hence the bands re-alignment. This increases the collection of minority carriers, and as such improves the FF value. Doping lightly the buffer layer with the same dopant type as the emitter has been reported to significantly improve the FF [21]. No change in $J_{sc}$ and $V_{oc}$ were reported in this case. Our simulations show a significant improvement of FF for a doping level above 8e17 cm$^{-3}$ as shown in Fig. 17.10.

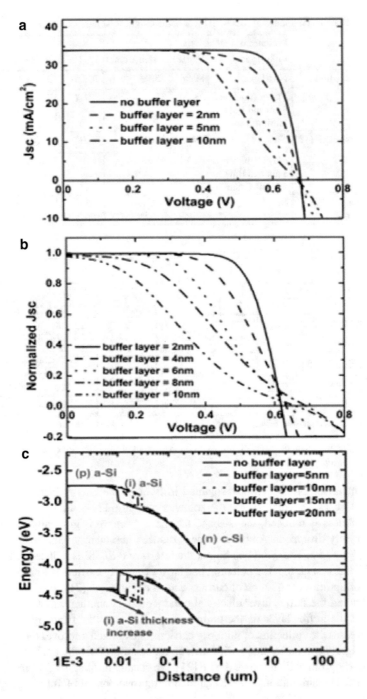

**Fig. 17.9** (**a**) Experimental J-V curve. (**b**) Simulated J-V for IBC-SHJ cells. (**c**) Equilibrium band diagram for different buffer thickness [21]

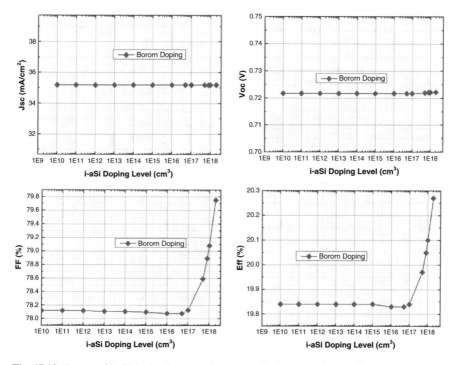

**Fig. 17.10**   Impact of i-aSi doping level on all output solar characteristics

**Fig. 17.11**   Equilibrium band diagram for different doping level of the buffer

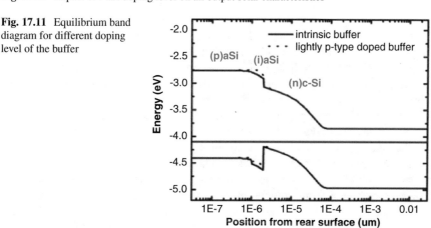

The fill factor (FF) behavior can be explained again using the band diagram (Fig. 17.11). A lightly doped layer has a lower potential barrier for the minority carriers compared to an intrinsic one. This will ease the minority carriers transport and will eventually improve the FF.

Tuning the buffer bandgap improves FF. Experimentally, the bandgap can be modified by adjusting the deposition temperature of the intrinsic layer, the $H_2/SiH_4$

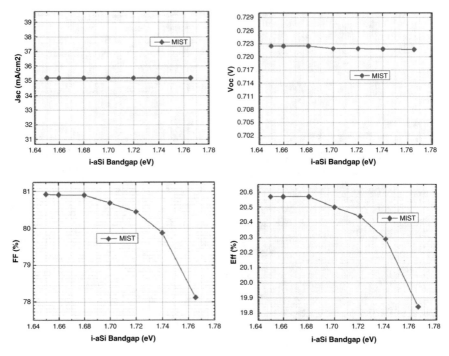

**Fig. 17.12** Influence of intrinsic a-Si bandgap on cell's performance

ratio and/or RF plasma power [8]. It was reported that reducing the bandgap of the buffer increases FF and the S-shape behavior is then addressed. There is no significant change on both $J_{sc}$ and $V_{oc}$. We note that the bandgap is lower for the doped a-Si than for the intrinsic a-Si. S-shaped $JV$ has been reported for a bandgap above 1.72 eV [8, 21]. Our simulations results are in agreement with the reported data with the FF showing a sharp decrease for bandgaps above 1.68 eV (Fig. 17.12).

As can be seen in Fig. 17.13, the increase of the bandgap makes it more difficult for the minority carriers to cross the junction.

The geometrical design of the back side of the cell can also be used to one's advantage to allow for a better passivation and better current collection at a potential cost of a more complex process to implement. For instance, adding a small opening in the i-aSi buffer, so that the emitter and the BSF layers can contact directly the absorber wafer has been shown to enhance FF and alleviate the formation of the S-shaped $JV$ curve [22]. Figure 17.14 shows clearly the effect of modifying the IBC-SHJ cell with openings.

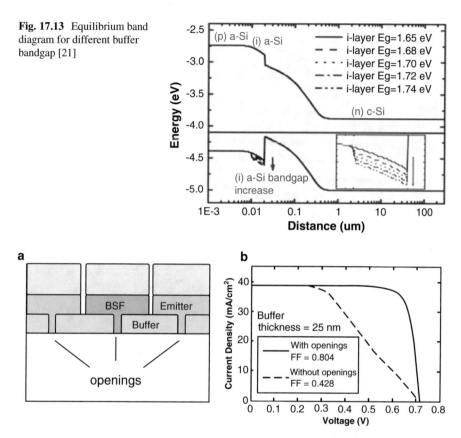

**Fig. 17.13** Equilibrium band diagram for different buffer bandgap [21]

**Fig. 17.14** (**a**) SHJ-IBC cell with openings. (**b**) *JV* curve

### 17.2.3.2    Doped a-Si Layers Optimization

The other layers requiring optimization are the doped a-Si layers (emitter and BSF). As anticipated, the recombination losses would increase with large density of defects in doped layers. This occurs even if these layers are ultrathin (few nm). The study of defects density on the cell's performance is usually summed up in varying the mid-gap defect density and in some cases by varying the capture cross sections of these recombination centers. The reported study [8] simulated a structure without an intrinsic a-Si layer. A significant decrease of FF was observed as the mid-gap defect density in the emitter layer is increased while fixing the mid-gap defect density of the BSF layer. On the other hand, there is no change in FF if the mid-gap defect density in BSF is varied [8]. In our study we did use a structure with an intrinsic amorphous silicon as a passivating layer. The doping levels for emitter and BSF layers were set to 1e19 cm$^{-3}$. As can be seen from Fig. 17.15, both $J_{sc}$ and $V_{oc}$ are independent of the defect density in the both doped amorphous layers as long as

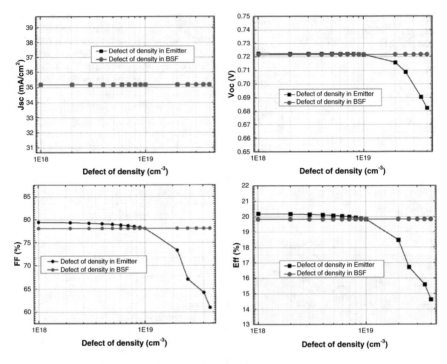

**Fig. 17.15** Impact of defect density on cell's performance

the defect density is less than the doping level of the layer for the emitter or BSF. FF is insensitive to the defect concentrations in BSF while there is a clear drop with defect concentration in emitter layer.

The interface recombination rate depends on the defect states density and their capture cross section of carriers (probability of recombination). The defects at the interface c-Si/a-Si:H have a direct impact on the device's performance. The influence of interface states density at the c-Si/a-Si:H hetero-interface has been investigated in Diouf et al. study [4]. Both $V_{oc}$ and FF are sensitive to concentration of defect states. The higher density at the c-Si/doped aSi interface, the lower $V_{oc}$ and FF. $J_{sc}$ is reportedly sensitive to the density of defect states at the c-Si/BSF interface. It degrades as the density of defects is increased.

The study [4] investigates the interface recombination by varying the capture cross section of one carrier while fixing both the defect density and the capture cross section of the other carrier. It was observed that as one increases the capture cross section of the majority carriers (electrons at the BSF interface) or that of minority carriers (holes at the emitter interface), $V_{oc}$ is decreased due to a formation of an accumulation (or inversion) layer at the BSF interface (or emitter interface). The interface here refers to that with c-Si (n). The emitter parameters have been investigated by various groups. As mentioned above, the bandgap of the emitter layer will have an impact on the overall performance. Jeyakumar et al.

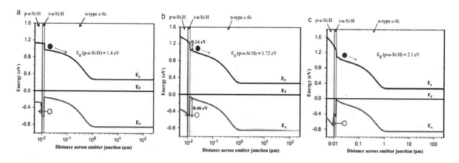

**Fig. 17.16** (**a**) Band diagram for bandgap = 1.4 eV. (**b**) Band diagram for bandgap = 1.2 eV. (**c**) Band diagram for bandgap = 2.1 eV

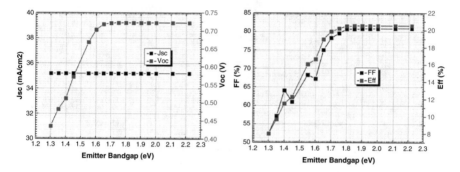

**Fig. 17.17** Influence of emitter bandgap on cell performance

group reported that when increasing the bandgap, $J_{sc}$ remains constant owing to a constant potential barrier height between the buffer and the absorber layers as shown in Fig. 17.16 [23]. Our simulation showed that both $V_{oc}$ and FF increase as the bandgap is increased and the saturation occurs around 1.7 eV. No change in $J_{sc}$ is observed when the emitter bandgap increases as shown in Fig. 17.17.

## 17.2.4   Back Side Geometry Optimization

### 17.2.4.1   Impact of Back Side Geometry

Optimizing the geometry of the rear side plays a critical role in the cell's behavior. Several studies have investigated the influence of emitter width, BSF width, gap (gap between contacts) width, and number of fingers, in addition to the surface recombination velocity (SRV) of the gap and emitter contact fraction.

$J_{sc}$ and FF are sensitive to the dimensions of emitter, BSF and the gap between the two. Widening the BSF and gap strips results in the degradation of both $J_{sc}$ and FF if they are not well passivated. However, larger emitter strips show higher $J_{sc}$

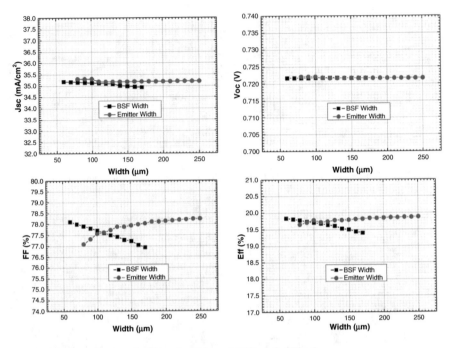

**Fig. 17.18** The influence of widths of emitter and BSF on the FF values

values, but, on the other hand, it causes FF to drop. Wider emitter region means a larger junction area. More minority carriers reach the emitter and thus more photo-generated current to be collected. For majority carriers, additional distance is added to carriers to reach the BSF. The additional distance translates into an additional series resistance and thus reduces FF value. Choosing the optimum width for the emitter is hence important to get higher efficiencies [4, 8, 20, 21]. $V_{oc}$ is nearly independent of doped layers width, a result confirmed in our simulations. The short circuit current density $J_{sc}$ decreases with BSF width and increases slightly with emitter width as shown in Fig. 17.18. That shows an agreement with the other studies. However FF increases as the emitter width increases which is not the case in study [4].

Herasimenka et al. work mentions that adding small opening to the buffer as seen in Fig. 17.14 will cancel the effect of wide BSF strips [22]. Figure 17.19 shows that the FF remains almost unchanged with the increase of the width of n-strip width.

The passivation of the gap between the doped layers has an impact on the electrical cell characteristics. Higher gap width could reduce both $J_{sc}$ and FF as mentioned before. However, if this gap is well passivated, it will have a little impact on the cell's performance [8, 20]. The influence of the gap surface passivation is simulated by varying the SRV. Higher SRV would result in degrading both $J_{sc}$ and $V_{oc}$.

**Fig. 17.19** The influence of widths of emitter and BSF with and without openings on the FF values

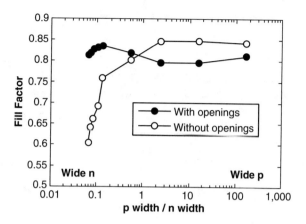

It is necessary to determine an optimum number of fingers per cell that would provide the best performance. Reducing the number of fingers has a slight negative effect on $V_{oc}$ and $J_{sc}$. FF is more sensitive to the number of fingers. No S-shaped $J$–$V$ curve has been observed while varying the number of fingers per cell. Increasing the number of strips enhances the device performance, until we reach a plateau regime where the number of fingers has no longer any positive effect on the performance. It has shown that 4p-3n (for cell width of 6500 μm) is an optimum number for a good device performance [8].

### 17.2.4.2   Impact of Metal Coverage

As one moves to see the effect of the metallic layers, emitter coverage by the metal contact has been explored in various studies as well as in our present study. Collection of minority carriers is a key parameter to control the cell performance. The full metal coverage is not an option in an IBC cell as one would short the emitter and BSF contacts. The emitter contact fraction has been studied by Desrues et al. and [14], Diouf et al. [20]. Emitter contact fraction affects the $J_{sc}$ and FF values. Neither $J_{sc}$ nor $V_{oc}$ is reported to change in Diouf's et al. work, while there is a change in $J_{sc}$ with the contact fraction in Desrues' et al. work. FF drops significantly for higher emitter contact coverage. Our simulation confirmed the influence of emitter contact coverage in both FF and $J_{sc}$ as seen in Fig. 17.20.

## 17.3   Conclusion

We have simulated a full IBC-SHJ solar cell sweeping the various parameters and layers involved in the cell design and have benchmarked our work against different groups working in the field. The substrate, aSi layer properties have been

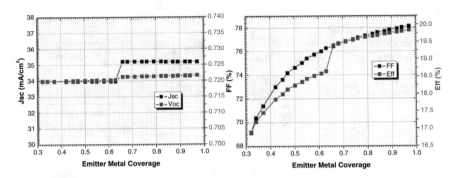

**Fig. 17.20** Impact of emitter contact fraction in *JV* parameters

incorporated in our modelling. We have found that 2D modelling leads to very similar results of a full 3D approach for the IBC-SHJ cell. With our goal being the reduction of wafer thickness, our simulations show that for a bulk lifetime above 1 ms one obtains efficiencies above 20 % with thicknesses as low as 150 μm. We conclude that i-layer presence is indeed needed for proper passivation at the back side of the solar cell and if its thickness is kept small, the S-shaped curves would be alleviated. A good control of the doping level and/or bandgap of the said layer as well as that of the doped layers will have an important impact on the final performance. The FSF layer is an important component as well, because the e–h pair generation occurs near the front end of the cell, one has to ensure that a good front passivation is in place to minimize recombination at that interface. Last but not least we have highlighted optimum designs including direct contact between emitter/BSF and the absorbing layer as well as proper metal coverage.

**Acknowledgements** This project was funded by Mubadala and ATIC (Abu Dhabi Technology Investment company) with the twinlab TU Dresden-Masdar Institute initiative.

# References

1. E.Van Kerschaver, G. Beaucarne, Back-contact solar cells: a review. Prog. Photovolt. Res. Appl. **14**(2), 107–123 (2006)
2. International technology roadmap for photovoltaic. http://www.itrpv.net/ (2015)
3. M. Lu, Silicon heterojunction solar cell and crystallization of amorphous silicon. Ph.D. thesis, 2008
4. D. Diouf, J.P. Kleider, T. Desrues, P.J. Ribeyron, Study of interdigitated back contact silicon heterojunctions solar cells by two-dimensional numerical simulations. Mater. Sci. Eng. B **159–160**, 291–294 (2009)
5. R.J. Schwartz, M.D. Lammert, Silicon solar cells for high concentration applications. Electron Devices Meet. **21**, 350–352 (1975)

6. T. Matsuyama, T. Sawada, S. Tsuda, S. Nakano, H. Hanafusa1, M. Tanaka, M. Taguchi, Y. Kuwano, Development of new a-si/c-si heterojunction solar cells: Acj-hit (artificially constructed junction-heterojunction with intrinsic thin-layer). Jpn. J. Appl. Phys. **31**(11), 3518–3522 (1992)

7. D. Diouf, J.-P. Kleider, C. Longeaud, *Two-Dimensional Simulations of Interdigitated Back Contact Silicon Heterojunctions Solar Cells* (Springer, Berlin, 2011), pp. 483–519

8. J. Allen, Interdigitated back contact silicon heterojunction solar cells: analysis with 2D simulations. Ph.D. thesis, 2011

9. R. Stangl, J. Haschke, M. Bivour, L. Korte, M. Schmidt, K. Lips, B. Rech, Planar rear emitter back contact silicon heterojunction solar cells. Sol. Energy Mater. Sol. Cells **93**(10), 1900–1903 (2009)

10. Panasonic hit solar cell achieves world's highest energy conversion of 25.6 at research level. http://news.panasonic.com/press/news/official.data/all-e.html#date (April 2014)

11. A. Tomasi, B. Paviet-Salomon, D. Lachenal, S. Martin de Nicolas, A. Descoeudres, J. Geiss-buhler, S. De Wolf, C. Ballif, Back-contacted silicon heterojunction solar cells with efficiency. IEEE J. Photovoltaics **4**(4), 1046–1054 (2014)

12. K.-s. Ji, H. Syn, J. Choi, H.-M. Lee, D. Kim, The emitter having microcrystalline surface in silicon heterojunction interdigitated back contact solar cells. Jpn. J. Appl. Phys. **51**(10S), 10NA05 (2012)

13. S. De Vecchi, T. Biavin, T. Desrues, F. Souche, D. Muoz, M. Lemiti, P.-J. Ribeyron, New metallization scheme for interdigitated back contact silicon heterojunction solar cells. Energy Procedia **38**, 701–706 (2013)

14. T. Desrues, P.J. Ribeyron, A. Vandeneynde, A.S. Ozanne, F. Souche, D. Muoz, C. Denis, D. Diouf, J.P. Kleider, B-doped a-si:h contact improvement on silicon heterojunction solar cells and interdigitated back contact structure. Phys. Status Solidi C **7**(3–4), 1011–1015 (2010)

15. T. Desrues, S. De Vecchi, F. Souche, D. Munoz, P.J. Ribeyron, Slash concept: a novel approach for simplified interdigitated back contact solar cells fabrication, in *2012 38th IEEE Photovoltaic Specialists Conference (PVSC)*, 2012, pp. 001602–001605

16. T. Desrues, S. De Vecchi, F. Souche, D. Diouf, D. Munoz, M. Gueunier-Farret, J.-P. Kleider, P.J. Ribeyron, Development of interdigitated back contact silicon heterojunction (ibc si-hj) solar cells. Energy Procedia **8**, 294–300 (2011)

17. B. Shu, U. Das, J. Appel, B. McCandless, S. Hegedus, R. Birkmire, Alternative approaches for low temperature front surface passivation of interdigitated back contact silicon heterojunction solar cell, in *2010 35th IEEE Photovoltaic Specialists Conference (PVSC)*, pp. 003223–003228, 2010

18. A. Callozzo, Numerical simulation of interdigitated back contact hetero-junction solar cells. Ph.D. thesis, 2011

19. C. Bulucea, Recalculation of Irvin's resistivity curves for diffused layers in silicon using updated bulk resistivity data. Solid-State Electron. **36**, 489–493 (1993)

20. T. Desrues P.-J. Ribeyron D. Diouf, J.P. Kleider, Interdigitated back contact a-si:h/c-si heterojunction solar cells modelling: limiting parameters influence on device efficiency, in *23rd European Photovoltaic Solar Energy Conference and Exhibition*, Valencia, 2008, pp. 1949–1952

21. M. Lu, U. Das, S. Bowden, S. Hegedus, R. Birkmire, Optimization of interdigitated back contact silicon heterojunction solar cells: tailoring hetero-interface band structures while maintaining surface passivation. Prog. Photovolt. Res. Appl. **19**(3), 326–338 (2011)

22. S. Herasimenka, K. Ghosh, S. Bowden, C. Honsberg. 2d modeling of silicon heterojunction interdigitated back contact solar cells, in *2010 35th IEEE Photovoltaic Specialists Conference (PVSC)*, 2010, pp. 001390–001394

23. R. Jeyakumar, T.K. Maiti, A. Verma, Influence of emitter bandgap on interdigitated point contact back heterojunction (a-si:h/c-si) solar cell performance. Sol. Energy Mater. Sol. Cells **109**, 199–203 (2013)

# Index

© Springer International Publishing Switzerland 2016
I.M. Elfadel, G. Fettweis (eds.), *3D Stacked Chips*,
DOI 10.1007/978-3-319-20481-9

Printed in the United States
By Bookmasters